Biomolecules and Materials Based Approaches in Biomedical Field: 2nd Edition

Biomolecules and Materials Based Approaches in Biomedical Field: 2nd Edition

Guest Editors

Sabrina Morelli
Antonella Piscioneri
Chien-Chung Chen
Loredana De Bartolo

MDPI

Basel • Beijing • Wuhan • Barcelona • Belgrade • Novi Sad • Cluj • Manchester

Guest Editors
Sabrina Morelli
Institute on Membrane Technology
National Research Council of Italy
Rende
Italy

Antonella Piscioneri
Institute on Membrane Technology
National Research Council of Italy
Rende
Italy

Chien-Chung Chen
Graduate Institute of Biomedical Materials and Tissue Engineering
Taipei Medical University
Taipei
Taiwan

Loredana De Bartolo
Institute on Membrane Technology
National Research Council of Italy
Rende
Italy

Editorial Office
MDPI AG
Grosspeteranlage 5
4052 Basel, Switzerland

This is a reprint of the Special Issue, published open access by the journal *Biomolecules* (ISSN 2218-273X), freely accessible at: https://www.mdpi.com/journal/biomolecules/special_issues/558QV8DI89.

For citation purposes, cite each article independently as indicated on the article page online and as indicated below:

Lastname, A.A.; Lastname, B.B. Article Title. *Journal Name* **Year**, *Volume Number*, Page Range.

ISBN 978-3-7258-6780-6 (Hbk)
ISBN 978-3-7258-6781-3 (PDF)
https://doi.org/10.3390/books978-3-7258-6781-3

Contents

Preface

The Special Issue provides a comprehensive overview of the state of the art in biomolecule- and material-based approaches in biomedical applications, with a particular focus on recent advancements and innovative strategies. It aims to consolidate current knowledge by bringing together high-quality original research articles and systematic reviews that highlight the use of biomolecules and advanced materials for a wide range of biomedical purposes. The motivation behind this collection is to showcase the latest developments in biomolecule- and material-based biomedical research and to foster the integration and advancement of the existing knowledge across disciplines. The Special Issue is intended for researchers and scientists working in biomolecular science, materials science, and related biomedical research fields.

Sabrina Morelli, Antonella Piscioneri, Chien-Chung Chen, and Loredana De Bartolo
Guest Editors

biomolecules

MDPI

Article

Serum PTH ≥ 40 pg/mL as a Marker of Bone Fragility and Vitamin D Deficiency in Periodontitis Patients: Biochemical, Densitometric and Genetic Evidence

Giada Marroncini [1,2,*], Serena Martinelli [2,*], Francesco Petrelli [3,4], Francesco Bombardiere [1], Antonio Sarnataro [1] and Francesco Saverio Martelli [1]

1 Biomolecular Diagnostic Laboratories, Via N. Porpora, 50144 Florence, Italy; francesco.bombardiere@biomoleculardiagnostic.com (B.F.); antonio.sarnataro@bdmail.it (S.A.); francesco.martelli@ednmail.it (M.F.S.)

2 Department of Clinical and Experimental Medicine, University of Florence, 50139 Florence, Italy

3 Department of Biology, School of Pharmacy, University of Rome "Tor Vergata", 00133 Rome, Italy; francesco.petrelli.1@unil.ch

4 Department of Biomedical Sciences, University of Lausanne, 1005 Lausanne, Switzerland

* Correspondence: giada.marroncini@bdmail.it (M.G.); serena.martinelli@unifi.it (M.S.)

Abstract

(1) Background: this study aimed to determine whether a serum parathyroid hormone (PTH) threshold of 40 pg/mL represents a clinically relevant risk factor for vitamin D (VitD) deficiency and reduced bone mineral density (BMD). It also investigated potential genetic interactions influencing PTH regulation and skeletal health in patients with periodontitis. (2) Methods: a cross-sectional analysis was conducted on 1038 periodontitis patients (35–75 years). Serum PTH, VitD, calcium (Ca), phosphate (P), and urinary parameters were assessed. Dual-energy X-ray absorptiometry (DXA) was used to evaluate BMD in 261 subjects. Vitamin D Receptor (VDR) and estrogen receptor alpha (ERα) polymorphisms were genotyped, and composite genetic risk scores were calculated. Statistical analyses included correlation tests, subgroup comparisons, and regression models. (3) Results: sixty-two percent of individuals had PTH > 40 pg/mL, which was associated with significantly lower 25(OH)D and Ca levels and reduced T-scores ($p < 0.05$). PTH levels negatively correlated with BMD (Pearson's $r = -0.159$, $p = 0.0105$). Patients with higher ERα polymorphism scores showed increased PTH values ($p < 0.05$), while VDR variants demonstrated a positive but no significant trend. (4) Conclusions: a PTH threshold of 40 pg/mL identifies individuals at higher risk of VitD deficiency and skeletal fragility, even without overt hypercalcemia. Genetic factors, particularly ERα variants, may contribute to elevated PTH levels, suggesting value in integrating biochemical, densitometric, and genetic screening for early bone health risk stratification.

Keywords: parathyroid hormone (PTH); vitamin D (VitD); bone mineral density (BMD); genetic polymorphisms (SNPs)

1. Introduction

Parathyroid hormone (PTH) is an 84-amino acid polypeptide secreted exclusively by the parathyroid glands. Its release is tightly regulated by circulating levels of ionized calcium (Ca), with hypocalcemia serving as the primary stimulus for increased PTH secretion [1].

https://doi.org/10.3390/biom15111600

PTH is essential for maintaining Ca and phosphate homeostasis and plays a pivotal role in regulating bone mineral density [2]. In the kidneys, PTH enhances tubular reabsorption of Ca, reduces urinary Ca excretion, and stimulates renal 1-alpha-hydroxylase activity, promoting the conversion of 25-hydroxyvitamin D (25(OH)D)to its active form, 1,25-dihydroxyvitamin D (1,25$(OH)_2$D) [3]. This active form increases intestinal Ca absorption, contributing to the restoration of serum Ca levels [4].

Physiologically, PTH levels fluctuate in response to serum Ca concentrations [5]. When Ca levels drop, PTH secretion increases to mobilize Ca from the bone, a readily accessible reservoir, into the bloodstream. Once normocalcemia is restored, PTH levels decrease, allowing bone-forming osteoblast activity to resume [4].

VitD also plays a crucial role in maintaining Ca and phosphate balance, partly by regulating the synthesis and secretion of PTH from the parathyroid glands through several coordinated mechanisms. It directly suppresses transcription of the PTH gene by binding to VitD receptors in the parathyroid glands, thereby reducing hormone synthesis [6]. Additionally, VitD increases intestinal Ca absorption, helping maintain normal serum Ca levels, and indirectly inhibits further PTH release through the Ca-sensing receptor [7]. When 25(OH)D levels fall below approximately 50–60 nmol/L, synthesis of 1,25$(OH)_2$D decreases, leading to reduced Ca absorption and a compensatory increase in PTH secretion to preserve Ca homeostasis [8].

In the setting of primary hyperparathyroidism (PHPT), PTH secretion becomes dysregulated, leading to elevated hormone levels despite normal or high serum Ca concentrations. This inappropriate secretion is most commonly due to parathyroid adenomas but can also result from familial hypocalciuric hypercalcemia [9]. In contrast, secondary hyperparathyroidism (SHPT) represents an adaptive response to hypocalcemia, often caused by vitamin D (VitD) deficiency, chronic kidney disease, or malabsorption syndromes. In SHPT, elevated PTH levels reflect a compensatory mechanism to increase Ca availability through bone resorption, enhanced renal Ca retention, and stimulation of intestinal absorption [10,11].

Chronic elevation of PTH, whether primary or secondary, exerts significant effects on bone remodeling [12]. PTH directly stimulates osteoclast differentiation and activation via the Receptor Activator of NF-κB Ligand (RANKL) signaling pathway, leading to increased bone resorption [13]. While transient PTH elevations may promote bone formation (an effect leveraged in intermittent PTH analog therapy), sustained elevations favor catabolism over anabolism, resulting in reduced bone mass and compromised bone microarchitecture [11,14]. Consequently, this imbalance contributes to the pathophysiology of osteopenia and osteoporosis, marked by diminished bone strength and increased fracture susceptibility.

Osteoporosis is a chronic, degenerative skeletal disorder defined by decreased bone mass and disruption of bone microarchitecture, increasing the risk of fragility fractures and functional decline [15]. Bone integrity is maintained by continuous remodeling, mediated by osteoblasts, responsible for osteoid synthesis and mineralization, and osteoclasts, which resorb bone matrix. This balance is delicately modulated by systemic hormones such as PTH, VitD, and local cytokines [16].

In individuals with hyperparathyroidism (HPT), the persistent stimulation of osteoclast activity leads to excessive Ca release from the skeleton, often accompanied by hypercalcemia [17,18]. Notably, elevated serum Ca in this context reflects bone demineralization rather than adequate intake or absorption. The resultant structural deterioration underlies the well-established association between HPT and osteoporosis [19]. VitD deficiency frequently coexists with PHPT and SHPT, impairing intestinal Ca absorption and further exacerbating PTH hypersecretion. Additionally, VitD deficiency and elevated PTH levels have been independently associated with increased cardiovascular risk. Mechanistically, high PTH may contribute to hypertension, vascular calcification, smooth muscle dysfunction, and myocardial hypertrophy [20]. PTH has also been implicated in systemic inflammation through suppression of anti-inflammatory cytokines such as interleukin-10 (IL-10), whose levels increase following VitD supplementation [21].

Moreover, the coexistence of low 25(OH)D and elevated PTH has been identified as an independent risk factor for sudden cardiac death in older adults without pre-existing cardiovascular disease [22]. In cases where PTH levels remain elevated without hypercalcemia or an identifiable secondary cause, the condition is termed normocalcemic primary hyperparathyroidism (NPHPT), an increasingly recognized variant that may still confer skeletal and cardiovascular risk [22].

In this study, we demonstrate that a PTH threshold of 40 pg/mL is a significant risk factor for the development of VitD deficiency and the onset of osteoporosis. Elevated PTH levels were inversely correlated with serum 25(OH)D and Ca concentrations, as well as bone mineral density.

2. Materials and Methods

2.1. Study Population

This cross-sectional study examined 1038 patients who visited a dental clinic between January 2022 and January 2024. The included participants were affected by periodontitis and had available data on serum PTH, 25(OH)D, serum and urinary Ca, or phosphate as well as sex, age, bone mineralization data and instrumental examinations. The main exclusion criteria were (i) the VitD supplementation, (ii) the intake of medication known to affect the concentration of PTH or VitD, including antiepileptic medicine, immune modulating medicine or intake of medically prescribed supplementary PTH, 25(OH)D, Ca, or phosphate. Therefore, the cohort included 1038 participants eligible for analysis, consisting of 437 men and 601 women, aged 35 to 75 years.

The inclusion criteria were: age > 35 and <75 years, male and female participants, ability and willingness to provide written informed consent, ability to comply with study procedures and visits. Participants were required to have stable clinical conditions, no acute illness or hospitalization within the previous four weeks, and preserved renal function (estimated glomerular filtration rate [eGFR] $\geq$ 60 mL/min/1.73 m^2). Individuals were also required to have resided in the same geographical area for at least six months to ensure comparable sunlight exposure. Subjects receiving vitamin D or calcium supplementation were included only if on a stable dose for at least three months or if not receiving supplementation during the same period. Patients with hypercalcemic primary hyperparathyroidism, hypoparathyroidism, or chronic kidney disease were excluded. Furthermore, individuals with certain medical conditions or medication histories—such as diagnoses of hypophosphatasia, idiopathic hypocalciuric hypercalcemia, or osteomalacia, and those using bisphosphonates, denosumab, romosozumab, or teriparatide—were also excluded during data retrieval.

The informed consent form for participation to the study was distributed to all participants and signed [23].

2.2. Biochemical Tests

For each participant, fasting blood samples were obtained by venipuncture and serum sample were separated the day of clinical examination. PTH was analyzed use CLIA methods (Snibe Diagnostic, Maglumi 800, Shenzhen, China) inter- and intra-assay coefficients of variation 9.28% and 5.74% for low concentrations, and 9.34% and 5.16% for high concentrations. The same method was used for 25(OH)D (Snibe Diagnostic, Maglumi 800, China) with inter- and intra-assay coefficients of variation 6.25% and 3.45% for low concentrations, and 6.13% and 3.17% for high concentrations. Serum and urinary Ca (Ca/S) and phosphate (P/S) were measured by standard automated methods in an BA-220 Automatic Chemistry System autoanalyzer (BioSystem, Barcelona, Spain). All analysis followed standard optimization and quality control as per laboratory policy.

2.3. Bone Densitometry

For the Bone mineral density (BMD) analysis, we included 261 subjects and measurements were performed according to the WHO criteria [24]. BMD was assessed using (Dual-energy X-ray absorptiometry) (DXA) measurement. BMD at the lumbar spine (L2–L4) and proximal femur (neck, Ward's triangle, greater trochanter) was measured by dual-energy X-ray absorptiometry (Dexa).

2.4. DNA Genotyping

Patient genomic DNA was isolated from EDTA blood samples or swab using a silica-based, phenol- and chloroform-free process in spin-column and 96-well-plate formats (QIAcube, Qiagen, Hilden, Germany). Each DNA sample was genotyped for Vitamin D Receptor (VDR) ApaI (rs64978 G>T), VDR BsmI (rs63980 G>A), VDR TaqI (rs1056 T>C), VDR FokI (rs30920 T>C), estrogen receptor alpha (ER-alpha) PvII (IVS1-397 T/C) and ER-alpha XbaI (IVS1-351 A/G) using Real Time qPCR (Quant Studio5, ThermoFisher Scientific, Waltham, Massachusetts, USA). Single-Nucleotide Polimorphisms (SNPs) were analyzed using Genotyping Software (Genotyping, qPCR, version 2020.1.4-Q1-20-build1) (Thermo Fisher Scientific, Waltham, Massachusetts, USA) and archived in the Cloud.

2.5. Statistical Analysis

Data analysis was performed by the computer program GraphPad Prism Version 8.4.3 for Windows (GraphPad Software). The statistical significance of value differences was evaluated by Studen's *t*-test using GraphPad Prism Version 8.4.3 for Windows. A P value of less than 0.05 was considered significant.

Data were expressed as mean ± SE (standard error) or ± SD (standard deviation) as detailed in figure legends. Normally distributed variables, skewed variables, and categorical variables were described using mean ± standard deviation, median (upper quartile, lower quartile), and frequency (percentage), respectively. Serum VitD levels were divided into three groups, deficiency (<20 ng/mL), insufficiency (20–30 ng/mL), and sufficiency (≥30 ng/mL). Comparisons between groups at different VitD levels were made using ANOVA or Kruskal–Wallis tests for continuous variables and chi-square tests for categorical variables. For post hoc tests, we used Bonferroni for multiplicity adjustment. Between-group comparisons of VitD levels between patients with osteopenia and osteoporosis were made using independent samples t-tests. Between-group comparisons of categorical variables between patients with osteopenia and osteoporosis were made using chi-square tests and Fisher exact method. Spearman and Pearson's correlation analysis were used to test the correlations between serum VitD and PTH levels with lumbar and femoral T-score.

3. Results

In our study, 1038 individuals of both sexes and different age groups were analyzed. The mean age of participants was 57.08 ± 10.27 years, with ages ranging from 35 to 75 (57.89 ± 10.07 for females and 57.76 ± 10.5 for males). Vitamin D levels were lower during summer and winter (summer, 38.15 ng/mL (47,3%) and winter, 32.02 ng/mL (52,5%)). The average 25(OH)D levels were 35.1 ng/mL ± 17.9, and the mean PTH level was 51.08 pg/mL ± 27.4, both within the normal reference ranges, as were the Ca and P levels in serum and urine.

Notably, 60% of the individuals had hypovitaminosis (25(OH)D < 30 ng/mL), and 62% had elevated PTH levels (>40 pg/mL). Indeed, other parameters such as Free Triiodothyronine (FT3) and Free Thyroxine (FT4) thyroid hormones, as well as bone alkaline phosphatase (B-ALP), were below the minimum reference range, with values of 2.9 ± 0.5 ng/dL, 5.05 ± 4.8 ng/dL, and 19.2 ± 10.6 U/L, respectively. Lumbar, femoral and feet BMD were assessed and 62.8% of the individuals had T-scores > −1.0, significant for osteopenia ($p \leq 0.002$ compared with reference ranges) (Table 1).

Table 1. Population setting.

Variables	Mean Values	Standard Deviations (SD)	Reference Values
Age/years	61.5	±12.1	
PTH (pg/mL)	51.08	±27.4	10–65 pg/mL
25(OH)D (ng/mL)	35.1	±17.9	>30 ng/mL
Serum Calcium	9.2	±1.7	9–11 mg/dL
Serum Phosphate	3.4	±0.6	2.5–4.5 mg/dL
Urinary Ca/24 h (mg/dL)	222.8	168.5	100–250 mg/24 h
Urinary P/24 h (mg/dL)	492.5	267.5	400–1300 mg/24 h
Homocysteine (μmol/L)	13.4	±6.7	<15 μmol/L
Uric acid (mg/dL)	4.6	±0.6	2.6–7.2 mg/dL
Glycemia (mg/dL)	89.2	±17.6	70–100 mg/dL
Total cholesterol (mg/dL)	188.7	±79.6	<200 mg/dL
LDL (mg/dL)	139.7 *	±53.3	<100 mg/dL
FT3 (pmol/L)	2.9 *	±0.5	3–8 pmol/L
FT4 (pmol/L)	5.05 *	±4.8	9–23 pmol/L
TSH (μIU/mL)	5.2	±9.1	0.4–5.5 μIU/mL
Bone alkaline phosphatase (B-ALP (U/L)	19.2 *	±10.6	50–220 U/L
Lumbar spine (L1-L4) T-score	−1.38 *	±1.56	>−1.0
Femoral neck T-score	−1.65 *	±0.93	>−1.0
Femoral trochanter T-score	−1.36 *	±1.02	>−1.0
Feet T-score	−2.02 *	±0.67	>−1.0

Summary of the characteristics of the patients enrolled in this study (n = 1038). The mean ± SD was reported and compared to the specific reference range; (* = values that deviate from the reference range).

3.1. Baseline Characteristics

3.1.1. Population Analysis for PTH Subgroups

The population was divided into two groups according to PTH levels: subgroup1 (<40 pg/mL) and subgroup2 (>40 pg/mL). 62% of individuals had PTH > 40 pg/mL and the distribution based on age showed a statistically significant increase in PTH among 56- and 75-year-olds (Figure 1).

3.1.2. Characterization of Subgroups

When individuals were divided by PTH level of 40 pg/mL, the difference between VitD and serum Ca were statistically significant. Therefore, in subgroup1 (PTH < 40 pg/mL) the 25(OH)D mean was 34.22 mg/dL ± 1.026 and in subgroup2 (PTH > 40 pg/mL) was 26.55 mg/dL ± 0.63. Analogously, the serum Ca mean was 9.34 mg/dL ± 0.11 and 8.8 mg/dL ± 0.13, respectively. Based on our results, the PTH value of 40 pg/mL is able to distinguish individuals with SHPT characterized by hypersecretion of PTH, hypovitaminosis D and hypocalcemia (Figure 2).

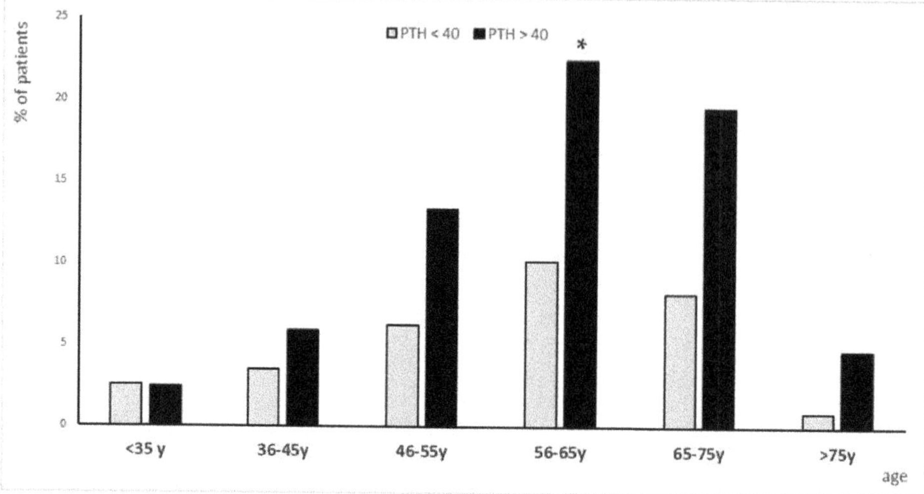

Figure 1. Population subgroups distributed by age. The population (n = 1038) was divided into two groups based on PTH levels < 40 (gray bars) or >40 (black bars). Bar graphs represent the % of individuals clustered by age (<35 years, from 36 to 45 years, from 46 to 55 years, from 56 to 65 years, from 65 to 75 years and >75 years old). * = $p \leq 0.05$ compared with subgroup with PTH< 40 pg/mL.

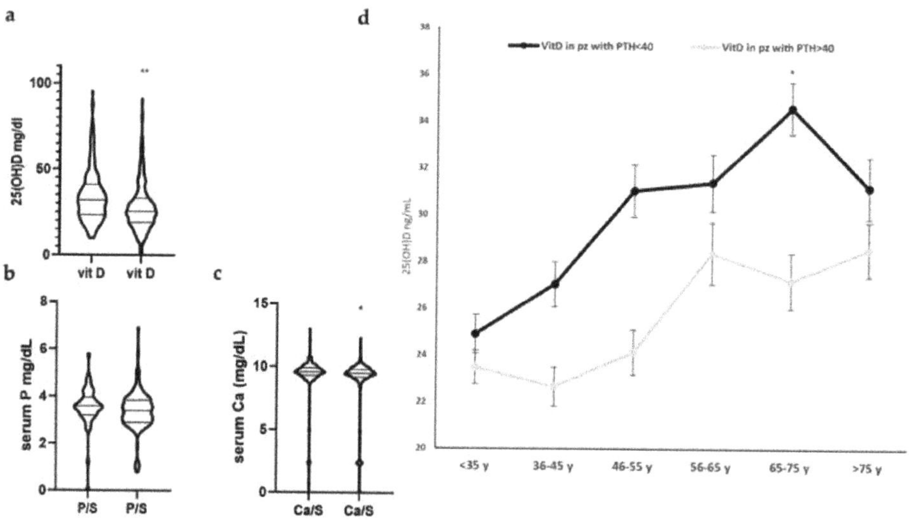

Figure 2. Patients divided according on PTH level of 40 pg/mL showed statistically significant differences on 25(OH)D, serum Ca and P. (**a–c**) Scatter dot plots in the subgroup of patients of 25(OH)D, serum P and Ca, respectively. Results are expressed as mean ± SD. (**d**) Graphical representation of 25(OH)D levels in the two subgroups of patients (PTH levels <40 gray line, or ≥40 black line) based on age ranges. (* = $p < 0.05$ and ** = $p < 0.02$ were considered statistically significant).

3.2. Bone Mineral Density (BMD)

A total of 261 patients with complete biochemical and densitometric data were included in the BMD analysis. Based on serum PTH concentrations, the cohort was divided into two groups: 146 patients (56%) with high-PTH group (PTH levels > 40 pg/mL) and 115 patients (44%) with normal or low PTH group (PTH levels ≤ 40 pg/mL). The high-PTH group exhibited a significantly higher mean PTH value (62.88 ± 23.56 pg/mL) compared to the low-PTH group (27.75 ± 7.85 pg/mL). Interestingly, the mean 25(OH)D level was also higher in the PTH > 40 group (39.4 ± 24.8 ng/mL) than in those with PTH ≤ 40 pg/mL (32.9 ± 15.5 ng/mL), possibly reflecting or altered hormonal regulation. Of these patients, 33.1% had a normal T-score (T-score > −1), 36.8% had osteopenia (T-score between −1 and −2.5), and 29.8% had osteoporosis (T-score < −2.5) according to WHO guidelines [24,25] (Table 2).

Table 2. Clinical and biochemical characteristics of BMD patients stratified by PTH levels.

	PTH > 40	PTH ≤ 40
N (patients)	146	115
PTH (Mean)	62.8 ± 23.5	27.7 ± 7.8
25(OH)D (Mean)	39.3 ± 24.8	32.8 ± 15.4
T-score Mean (Mean)	−1.6 ± 0.8	−1.3 ± 0.8

Summary of the main variables in patients with serum parathyroid hormone (PTH) levels higher than 40 pg/mL (n = 146) versus those with PTH ≤ 40 pg/mL (n = 115). Data are presented as mean ± SD. The high PTH group showed significantly elevated PTH concentrations compared to the low PTH group, while their serum VitD levels were slightly, but not significantly, higher. The mean T-score (calculated as the average of lumbar and femoral measurements) was lower in the PTH > 40 group (–1.6 ± 0.8), demonstrating a reduced BMD in this population.

In terms of bone status, the mean T-score (average of lumbar spine and femur) was lower in patients with PTH > 40 pg/mL (–1.6 ± 0.90) than in those with PTH ≤ 40 pg/mL (–1.3 ± 0.90), suggesting greater skeletal involvement in the former group (Table 2).

A correlation analysis was conducted to explore the relationship between serum PTH levels and the mean T-score. A statistically significant inverse correlation was observed (Pearson's r = –0.159, p = 0.0105; Spearman's ρ = –0.142, p = 0.0228), indicating that higher PTH values are associated with lower BMD. This finding is consistent with the hypothesis that elevated PTH, especially in the context of SHPT, may contribute to bone demineralization, even in patients with normal serum Ca levels (Figure 3a).

A scatterplot with linear regression line shows the relationship between serum PTH concentrations and the mean BMD T-score, calculated as the average of lumbar spine (L1–L4) and total femur values. A statistically significant inverse correlation was observed (Pearson's r = –0.159, p = 0.0105; Spearman's ρ = –0.142, p = 0.0228), indicating that higher PTH levels are associated with lower BMD. The black regression line and confidence interval reflect the linear fit and variability of the data. To further evaluate the diagnostic performance of serum PTH as a continuous variable in predicting osteopenia/osteoporosis, a receiver operating characteristic (ROC) curve analysis was performed. The optimal threshold was determined using Youden's index, and sensitivity, specificity, and AUC with 95% confidence intervals were reported (Figure 3b).

Moreover, Figure 4 shows that patients with PTH ≤ 40 pg/mL consistently have higher T-scores across all measured skeletal regions (lumbar spine (L1–L4), femoral region, and feet) compared to those with PTH > 40 pg/mL. The difference was most pronounced at the lumbar spine (difference statistically significant) and foot levels, where the gap between groups approached or exceeded 0.5 SD.

While femoral T-scores showed the least divergence between the two groups, the overall trend suggests a more pronounced reduction in BMD among individuals with high PTH levels.

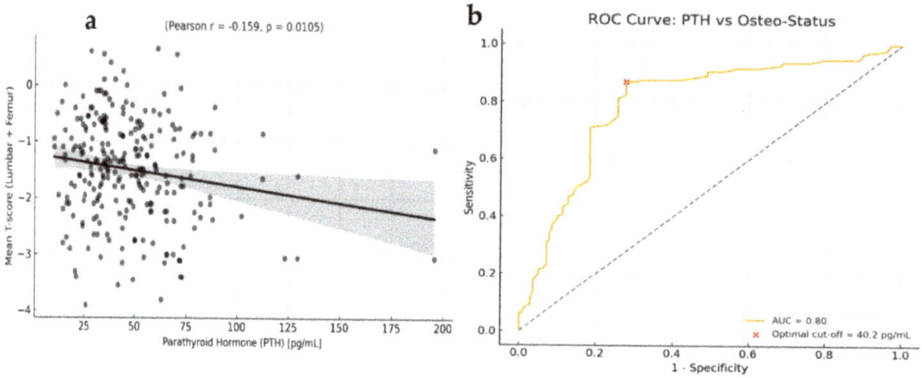

Figure 3. Statistical analysis of serum PTH levels for predicting osteopenia/osteoporosis. (**a**) Correlation between circulating PTH concentrations and the mean T-score (lumbar spine and femur), revealing a significant inverse relationship. (**b**) Receiver operating characteristic (ROC) curve analysis demonstrated good discriminative performance, with an area under the curve (AUC) of 0.797. The optimal threshold, determined using Youden's index, was 40.2 pg/mL. This cut-off yielded a sensitivity of 87% (95% CI: 81–91%) and a specificity of 72% (95% CI: 64–80%). The red dot indicates the optimal decision point.

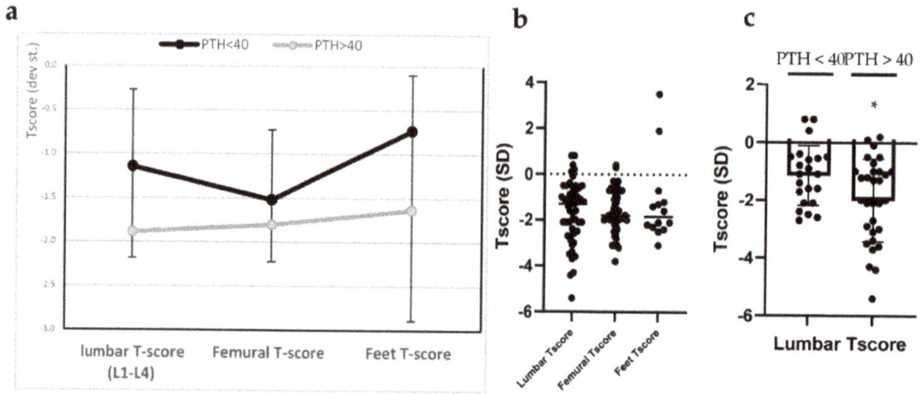

Figure 4. Comparison of T-scores across anatomical sites between patients with PTH $\leq$ 40 pg/mL and PTH > 40 pg/mL by *t*-test. (**a**) Line graph depicting mean T-scores ($\pm$SD) at the lumbar spine (L1–L4), femoral region, and feet in two groups stratified by serum PTH levels. Patients with PTH $\leq$ 40 pg/mL (black line) exhibited consistently higher T-scores at all measured skeletal sites compared to those with PTH > 40 pg/mL (gray line). The difference was most pronounced in the lumbar spine and foot measurements. (**b**) Bar graph showing the distribution of patients with osteopenia and osteoporosis at each anatomical site (lumbar spine, femur, feet), stratified by PTH levels; (**c**) Box plots illustrating the variation in T-scores across the three anatomical sites in the two PTH groups. Median T-scores were significantly lower in the subgroup of patients with PTH > 40 pg/mL (* = $p \leq 0.05$ vs. PTH < 40 pg/mL subgroup).

3.3. Interaction Between Genetic Polymorphisms and PTH Levels

To evaluate the potential contribution of genetic predisposition to PTH regulation, we assessed the cumulative burden of polymorphisms within genes involved in VitD signaling and estrogen metabolism. Specifically, we genotyped each individual for four VDR polymorphisms (ApaI (rs64978 G>T), BsmI (rs63980 G>A), TaqI (rs1056 T>C), and FokI (rs30920 T>C)) as well as two ER-alpha polymorphisms (PvII (IVS1-397 T/C) and

XbaI (IVS1-351A/G)). To quantify the overall genetic load, we adopted a simple additive scoring model: a score of 2 was assigned for altered homozygous polymorphic genotypes, 1 for heterozygous genotypes, and 0 for wild type. The individual scores across all six single-nucleotide polymorphisms (SNPs) were summed, yielding a composite polymorphism score ranging from 0 to 8 for VDR SNPs and 0 to 4 for ER SNPs. This cumulative score was then correlated with circulating PTH concentrations in the study population. As shown in Figure 5, a positive trend was observed between the genetic polymorphism score and PTH levels, suggesting a potential interaction whereby a higher burden of polymorphic alleles may predispose individuals to elevated PTH levels. The PTH variation among groups reached statistical significance in patients with higher ER SNPs score ($p < 0.05$). Additionally, VDR SNPs scores showed an increase with increasing PTH concentrations, although this trend was not statistically significant. Indeed, the graphical representation demonstrates a consistent increase in PTH values mean with increasing genetic load.

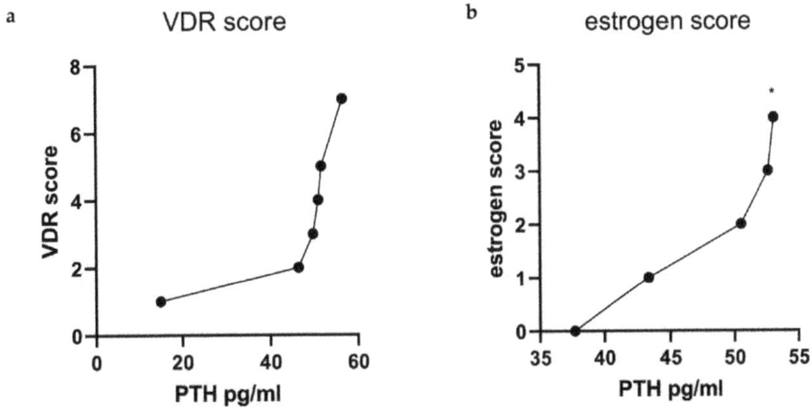

Figure 5. Relationship between cumulative genetic polymorphism burden and PTH levels. Patients were genotyped for six SNPs associated with VitD and estrogenic signaling: (**a**) VDR ApaI (rs64978), BsmI (rs63980), TaqI (rs1056), FokI (rs30920) and (**b**) ERα PvII (IVS1-397 T/C) and XbaI (IVS1-351 A/G). A cumulative polymorphism score was computed by assigning a value of 2 for altered homozygous polymorphic genotypes, 1 for heterozygous, and 0 for wild-type at each locus. The sum of all scores was then plotted against serum PTH levels. The graph shows a positive trend between the number of genetic variants and PTH concentration, with a stepwise increase in mean PTH observed across score categories, both when PTH was plotted against VDR score (**a**) and against ER score (**b**). (* = $p < 0.05$ was considered statistically significant).

4. Discussion

In this study, we demonstrate for the first time that a PTH threshold of 40 pg/mL represents a clinically significant risk factor for both 25(OH)D deficiency and the early onset of osteoporosis. Elevated PTH levels were inversely correlated with serum VitD and Ca concentrations, as well as BMD, highlighting the central role of PTH in skeletal homeostasis. These findings support the concept that sustained elevations in PTH, even in the absence of overt hypercalcemia, may promote bone loss through mechanisms compatible with SHPT, commonly triggered by chronic VitD insufficiency (Figure 6).

Our results are consistent with previous reports identifying elevated PTH as one of the earliest biochemical indicators of VitD depletion, often preceding detectable changes in bone density [26–29]. By identifying a specific threshold beyond which PTH becomes pathologically relevant, our findings contribute to a more nuanced understanding of the VitD–PTH–bone axis and emphasize the importance of early detection and intervention to mitigate SHPT-related skeletal deterioration.

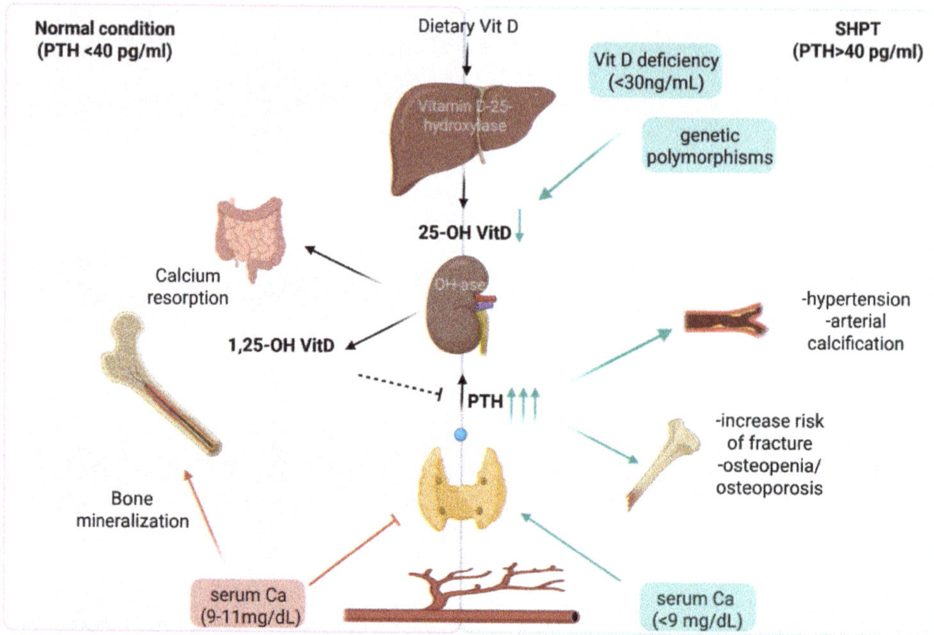

Figure 6. Schematic representation of the physiological differences between normal parathyroid hormone (PTH) levels and secondary hyperparathyroidism (SHPT). On the left, individuals with PTH < 40 pg/mL show normal calcium absorption from the gut, sufficient conversion of vitamin D into its active form (1,25-OH Vit D), proper bone mineralization, and stable serum calcium levels (9–11 mg/dL). On the right, SHPT (PTH > 40 pg/mL) is driven by vitamin D deficiency (<30 ng/mL) and genetic polymorphisms that enhance PTH secretion. This biochemical imbalance leads to reduced serum calcium (<9 mg/dL), increased bone resorption, and diminished bone mineral density, resulting in osteopenia, osteoporosis, and a higher risk of fractures. Moreover, elevated PTH is associated with hypertension and arterial calcification, underscoring its systemic impact. Overall, the illustration highlights the 40 pg/mL PTH threshold as a critical marker of bone fragility and cardiovascular risk, emphasizing the importance of monitoring and managing vitamin D status and PTH levels to maintain skeletal and systemic health. Created with Biorender.

Although femoral T-scores showed the least divergence between groups, the overall trend indicated a more pronounced reduction in BMD among individuals with elevated PTH levels, reaching statistical significance at the lumbar spine. This is consistent with the known catabolic effects of PTH on trabecular bone, particularly at weight-bearing skeletal sites [30]. Moreover, inclusion of the foot T-score (a site rarely reported in bone densitometry studies) offered additional insight into peripheral skeletal involvement. The lower T-scores observed at this site suggest early skeletal fragility in patients with chronic SHPT and highlight the potential value of evaluating nontraditional sites in assessing systemic bone loss [31]. These data reinforce the systemic impact of elevated PTH on both cortical and trabecular bone, particularly at weight-bearing regions. Importantly, this pattern supports the utility of PTH stratification in clinical bone health assessments and suggests that even borderline elevations in PTH may justify closer monitoring and early therapeutic intervention, particularly when accompanied by decreasing T-scores [32,33].

Importantly, our findings were corroborated by ROC curve analysis, which confirmed the relevance of a PTH threshold around 40 pg/mL in discerning patients with low bone mass. The optimal cut-off of 40.2 pg/mL showed good sensitivity and specificity, supporting the clinical utility of this value for identifying individuals at increased skeletal risk.

Our findings also suggest a genetic component to PTH regulation and its downstream effects on bone metabolism. Specifically, variations in the VDR and ERα genes may modulate individual responses within the Ca–VitD–PTH axis [34]. Patients with higher ERα polymorphism scores exhibited significantly elevated PTH levels, suggesting that genetic variants may influence either PTH secretion or tissue sensitivity. While the trend observed for VDR polymorphisms did not reach statistical significance, the overall direction of the data supports the hypothesis of a gene–environment interaction affecting skeletal outcomes [35,36].

These results are particularly relevant given the composition of our study population, which consisted of individuals with periodontitis, a chronic inflammatory disease that has been previously associated with reduced BMD and systemic bone turnover [37]. Inflammatory mediators and pro-osteoclastogenic cytokines common in periodontitis may exacerbate hormonal and genetic susceptibilities, promoting skeletal fragility. Vitamin D exerts immunomodulatory effects that extend to bone remodeling, partly through the regulation of inflammatory cytokines such as IL-6, TNF-α, and modulation of RANKL expression [38]. Chronic elevations in PTH due to vitamin D deficiency may contribute to bone loss through pro-inflammatory pathways involving Th17 cells, TNF-α, and IL-17 [39]. Furthermore, previous studies have identified associations between periodontal status and VDR gene polymorphisms, reinforcing the complex interplay between immune, hormonal, and genetic factors in bone health [40–43].

Taken together, these findings support a broader framework in which biochemical, genetic, and inflammatory factors interact to modulate bone integrity. The consistent differences in PTH and T-scores across clinically relevant thresholds suggest that routine measurement of PTH, even in normocalcemic patients, may aid in the early identification of individuals at risk for bone loss. This may be particularly important in populations with coexisting inflammatory conditions or genetic predispositions.

Cyclical variations in PTH levels are expected within the normal physiological range: during the winter months, reduced sunlight exposure leads to lower serum 25(OH)D concentrations, resulting in a compensatory rise in PTH to maintain Ca homeostasis [44]. However, previous research has shown that, although VitD levels vary considerably with seasonal sun exposure, these fluctuations have only minor effects on PTH and Ca levels. In healthy individuals, Ca concentrations remain stable throughout the year, without clinically relevant variations. Therefore, seasonal variations in vitamin D do not influence clinical assessment or decision-making related to PTH and calcium metabolism [45,46].

A limitation of our study is the absence of biochemical and dietary covariates that may influence PTH levels independently of VitD status. Specifically, data on renal function (eGFR), albumin-corrected or ionized Ca, phosphate, magnesium, and dietary Ca intake were not systematically collected or available for the full cohort. These factors are known to modulate PTH secretion and may confound the attribution of elevated PTH levels solely to VitD deficiency-related SHPT. While our findings highlight a robust association between low 25(OH)D and elevated PTH concentrations, the lack of multivariable adjustment for these additional parameters could limit causal inference. Future prospective studies should incorporate these covariates into regression models and consider sensitivity analyses to better isolate the contribution of VitD deficiency to SHPT. Such approaches would further strengthen the mechanistic understanding and clinical interpretation of PTH thresholds in bone health risk stratification. Even more, future studies should aim to validate these findings in larger and more diverse cohorts, preferably through longitudinal designs. Investigating whether targeted interventions along the PTH-VitD axis (especially in genetically predisposed individuals) can prevent or mitigate skeletal complications would be a critical next step. Such research may ultimately inform personalized approaches to bone health, particularly in the context of systemic inflammatory diseases such as periodontitis.

5. Conclusions

In conclusion, our study supports a clear link between SHPT, VitD deficiency, and reduced BMD in patients with periodontitis. These findings underscore the need for careful monitoring of VitD and PTH levels in this population, while also highlighting the importance of further exploring the impact of genetic variants on bone metabolism. Integrating biochemical, densitometric, and genetic markers may help identify individuals at higher risk and guide personalized therapeutic approaches.

Moreover, these findings support a broader framework where biochemical, genetic, and skeletal parameters converge to modulate bone health. The high prevalence of periodontitis in this cohort further reinforces the clinical importance of systemic inflammation as a contributing factor in bone fragility. Given the subtle but consistent shifts observed in T-score and PTH across clinically relevant thresholds, routine assessment of PTH, even in normocalcemic patients, may help identify individuals at risk of bone loss earlier.

Future studies should aim to validate these findings longitudinally and assess whether targeted interventions on PTH or VitD pathways, especially in genetically predisposed individuals, can mitigate skeletal complications in inflammatory conditions such as periodontitis.

Author Contributions: Conceptualization, M.G. and M.F.S.; methodology, M.G. and B.F.; software, M.G. and M.S.; validation, M.G. and B.F.; formal analysis, M.G. and B.F.; investigation, M.G. and M.F.S.; resources, M.G. and M.F.S.; data curation, M.G.; writing—original draft preparation, M.G.; writing—review and editing, M.G., M.S. and F.P.; visualization, M.G., M.S., B.F., M.F.S., S.A. and F.P.; supervision, M.F.S.; project administration, M.F.S.; All authors have read and agreed to the published version of the manuscript.

Funding: This research received no external funding.

Institutional Review Board Statement: The study was conducted in accordance with the Declaration of Helsinki, and approved by the Ethic Committee of "Comitato Etico Regione Toscana, Area Vasta Centro" (protocol code 25008_oss, date of approval: 24 September 2024).

Informed Consent Statement: Informed consent was obtained from all subjects involved in the study.

Data Availability Statement: Data availability access of the database supporting the results and analysis in the article is available by request to the corresponding author. An official letter will be necessary because of institutional requirements.

Conflicts of Interest: The authors declare no conflicts of interest.

References

1. Brown, E.M.; Pollak, M.; Seidman, C.E.; Seidman, J.G.; Chou, Y.H.; Riccardi, D.; Hebert, S.C. Calcium-ion-sensing cell-surface receptors. *N. Engl. J. Med.* **1995**, *333*, 234–240. [CrossRef] [PubMed]
2. Murray, S.L.; Wolf, M. Calcium and Phosphate Disorders: Core Curriculum 2024. *Am. J. Kidney Dis.* **2024**, *8f3*, 241–256. [CrossRef] [PubMed]
3. Just, A.; Mallmann, R.T.; Grossmann, S.; Sleman, F.; Klugbauer, N. Two-pore channel protein TPC1 is a determining factor for the adaptation of proximal tubular phosphate handling. *Acta Physiol. Oxf.* **2023**, *237*, e13914. [CrossRef] [PubMed]
4. Murray, T.M.; Rao, L.G.; Divieti, P.; Bringhurst, F.R. Parathyroid hormone secretion and action: Evidence for discrete receptors for the carboxyl-terminal region and related biological actions of carboxyl- terminal ligands. *Endocr. Rev.* **2005**, *26*, 78–113. [CrossRef]
5. Nabeshima, Y. Discovery of alpha-Klotho unveiled new insights into calcium and phosphate homeostasis. *Proc. Jpn. Acad. Ser. B Phys. Biol. Sci.* **2009**, *85*, 125–141. [CrossRef]
6. Khundmiri, S.J.; Murray, R.D.; Lederer, E. PTH and Vitamin D. *Compr. Physiol.* **2016**, *6*, 561–601. [CrossRef]

7. Kumar, R.; Thompson, J.R. The regulation of parathyroid hormone secretion and synthesis. *J. Am. Soc. Nephrol.* **2011**, *22*, 216–224. [CrossRef]
8. Hemmati, H.R. Interactions of vitamin D with parathyroid glands; an updated mini-review. *J. Parathyr. Dis.* **2023**, *11*, e11228. [CrossRef]
9. Kowalski, G.J.; Buła, G.; Żądło, D.; Gawrychowska, A.; Gawrychowski, J. Primary hyperparathyroidism. *Endokrynol. Pol.* **2020**, *71*, 260–270. [CrossRef]
10. Institute of Medicine (US) Committee to Review Dietary Reference Intakes for Vitamin D and Calcium. *Dietary Reference Intakes for Calcium and Vitamin D*; Ross , A.C., Taylor , C.L., Yaktine , A.L., Del Valle , H.B., Eds.; National Academies Press: Washington, DC, USA, 2011. [PubMed]
11. Lips, P. Vitamin D deficiency and secondary hyperparathyroidism in the elderly: Consequences for bone loss and fractures and therapeutic implications. *Endocr. Rev.* **2001**, *22*, 477–501. [CrossRef]
12. van der Plas, W.Y.; Noltes, M.E.; van Ginhoven, T.M.; Kruijff, S. Secondary and Tertiary Hyperparathyroidism: A Narrative Review. *Scand. J. Surg.* **2020**, *109*, 271–278. [CrossRef]
13. Huang, J.C.; Sakata, T.; Pfleger, L.L.; Bencsik, M.; Halloran, B.P.; Bikle, D.D.; Nissenson, R.A. PTH differentially regulates expression of RANKL and OPG. *J. Bone Miner. Res.* **2004**, *19*, 235–244. [CrossRef] [PubMed]
14. Rhee, Y.; Lee, E.Y.; Lezcano, V.; Ronda, A.C.; Condon, K.W.; Allen, M.R.; Plotkin, L.I.; Bellido, T. Resorption controls bone anabolism driven by parathyroid hormone (PTH) receptor signaling in osteocytes. *J. Biol. Chem.* **2013**, *288*, 29809–29820. [CrossRef] [PubMed]
15. Holroyd, C.; Cooper, C.; Dennison, E. Epidemiology of osteoporosis. *Best. Pract. Res. Clin. Endocrinol. Metab.* **2008**, *22*, 671–685. [CrossRef] [PubMed]
16. Lombardi, G.; Di Somma, C.; Rubino, M.; Faggiano, A.; Vuolo, L.; Guerra, E.; Contaldi, P.; Savastano, S.; Colao, A. The roles of parathyroid hormone in bone remodeling: Prospects for novel therapeutics. *J. Endocrinol. Investig.* **2011**, *34*, 18–22.
17. Brandi, M.L.; Falchetti, A. Genetics of primary hyperparathyroidism. *Urol. Int.* **2004**, *72* (Suppl. 1), 11–16. [CrossRef]
18. Simonds, W.F. Clinical and Molecular Genetics of Primary Hyperparathyroidism. *Horm. Metab. Res.* **2020**, *52*, 578–587. [CrossRef]
19. Association, C.O. Diagnosis and treatment of osteoporotic fractures. *Orthop. Surg.* **2009**, *1*, 251–257. [CrossRef]
20. Brown, S.J.; Ruppe, M.D.; Tabatabai, L.S. The Parathyroid Gland and Heart Disease. *Methodist. Debakey Cardiovasc. J.* **2017**, *13*, 49–54. [CrossRef]
21. Lee, J.H.; O'Keefe, J.H.; Bell, D.; Hensrud, D.D.; Holick, M.F. Vitamin D deficiency an important, common, and easily treatable cardiovascular risk factor? *J. Am. Coll. Cardiol.* **2008**, *52*, 1949–1956. [CrossRef]
22. Cusano, N.E.; Cetani, F. Normocalcemic primary hyperparathyroidism. *Arch. Endocrinol. Metab.* **2022**, *66*, 666–677. [CrossRef] [PubMed]
23. Cuschieri, S. The STROBE guidelines. *Saudi J. Anaesth.* **2019**, *13*, S31–S34. [CrossRef] [PubMed]
24. Kanis, J.A. Assessment of fracture risk and its application to screening for postmenopausal osteoporosis: Synopsis of a WHO report. WHO Study Group. *Osteoporos. Int.* **1994**, *4*, 368–381. [CrossRef]
25. Gregson, C.L.; Armstrong, D.J.; Bowden, J.; Cooper, C.; Edwards, J.; Gittoes, N.J.L.; Harvey, N.; Kanis, J.; Leyland, S.; Low, R.; et al. UK clinical guideline for the prevention and treatment of osteoporosis. *Arch. Osteoporos.* **2022**, *17*, 58. [CrossRef]
26. Walker, M.D.; Cong, E.; Lee, J.A.; Kepley, A.; Zhang, C.; McMahon, D.J.; Silverberg, S.J. Vitamin D in Primary Hyperparathyroidism: Effects on Clinical, Biochemical, and Densitometric Presentation. *J. Clin. Endocrinol. Metab.* **2015**, *100*, 3443–3451. [CrossRef]
27. Kolcsar, M.; Szabó, L.; Dénes, O.M.; Gáll, Z. Assessment of Vitamin D Status in Primary Hyperparathyroidism Patients: A Retrospective Study. *Cureus* **2024**, *16*, e64988. [CrossRef]
28. Walker, M.D.; Bilezikian, J.P. Vitamin D and primary hyperparathyroidism: More insights into a complex relationship. *Endocrine* **2017**, *55*, 3–5. [CrossRef]
29. Tassone, F.; Gianotti, L.; Baffoni, C.; Visconti, G.; Pellegrino, M.; Cassibba, S.; Croce, C.G.; Magro, G.; Cesario, F.; Attanasio, R.; et al. Vitamin D status in primary hyperparathyroidism: A Southern European perspective. *Clin. Endocrinol.* **2013**, *79*, 784–790. [CrossRef]
30. Silva, B.C.; Bilezikian, J.P. Parathyroid hormone: Anabolic and catabolic actions on the skeleton. *Curr. Opin. Pharmacol.* **2015**, *22*, 41–50. [CrossRef]
31. Bioletto, F.; Barale, M.; Maiorino, F.; Pusterla, A.; Fraire, F.; Arvat, E.; Ghigo, E.; Procopio, M. Trabecular Bone Score as a Marker of Skeletal Fragility Across the Spectrum of Chronic Kidney Disease: A Systematic Review and Meta-analysis. *J. Clin. Endocrinol. Metab.* **2024**, *109*, e1534–e1543. [CrossRef] [PubMed]
32. di Filippo, L.; Bilezikian, J.P.; Canalis, E.; Terenzi, U.; Giustina, A. New insights into the vitamin D/PTH axis in endocrine-driven metabolic bone diseases. *Endocrine* **2024**, *85*, 1007–1019. [CrossRef]
33. Liu, Y.; Wang, W.; Yang, Y.; Deng, J.; Zhang, Z. Vitamin D and bone health: From physiological function to disease association. *Nutr. Metab.* **2025**, *22*, 113. [CrossRef]

34. Cusack, S.; Mølgaard, C.; Michaelsen, K.F.; Jakobsen, J.; Lamberg-Allardt, C.J.; Cashman, K.D. Vitamin D and estrogen receptor-alpha genotype and indices of bone mass and bone turnover in Danish girls. *J. Bone Miner. Metab.* **2006**, *24*, 329–336. [CrossRef]
35. Wang, J.; Shu, B.; Li, C.G.; Xie, X.W.; Liang, D.; Chen, B.L.; Lin, X.C.; Wei, X.; Wang, L.; Leng, X.Y.; et al. Polymorphisms of genes related to vitamin D metabolism and transportation and its relationship with the risk of osteoporosis: Protocol for a multicentre prospective cohort study in China. *BMJ Open* **2019**, *9*, e028084. [CrossRef]
36. Fang, Y.; van Meurs, J.B.; Arp, P.; van Leeuwen, J.P.; Hofman, A.; Pols, H.A.; Uitterlinden, A.G. Vitamin D binding protein genotype and osteoporosis. *Calcif. Tissue Int.* **2009**, *85*, 85–93. [CrossRef]
37. Garnero, P.; Vassy, V.; Bertholin, A.; Riou, J.P.; Delmas, P.D. Markers of bone turnover in hyperthyroidism and the effects of treatment. *J. Clin. Endocrinol. Metab.* **1994**, *78*, 955–959. [CrossRef]
38. Al-Daghri, N.M.; Aziz, I.; Yakout, S.; Aljohani, N.J.; Al-Saleh, Y.; Amer, O.E.; Sheshah, E.; Younis, G.Z.; Al-Badr, F.B.M. Inflammation as a contributing factor among postmenopausal Saudi women with osteoporosis. *Medicine* **2017**, *96*, e5780. [CrossRef] [PubMed]
39. Takayanagi, H. Osteoimmunology: Shared mechanisms and crosstalk between the immune and bone systems. *Nat. Rev. Immunol.* **2007**, *7*, 292–304. [CrossRef] [PubMed]
40. Cafiero, C.; Grippaudo, C.; Dell'Aquila, M.; Cimmino, P.; D'Addona, A.; De Angelis, P.; Ottaiano, M.P.; Costagliola, D.; Benincasa, G.; Micera, A.; et al. Association between Vitamin D Receptor Gene Polymorphisms and Periodontal Bacteria: A Clinical Pilot Study. *Biomolecules* **2022**, *12*, 833. [CrossRef] [PubMed]
41. Baldini, A.; Nota, A.; Fanti, E.; Martelli, F.S.; Ottomano, C.; Lippi, G. Association between periodontal disease and Interleukin-1β +3953 and vitamin D receptor Taq1 genetic polymorphisms in an Italian caucasian population. *Ann. Stomatol. Roma* **2013**, *4*, 191–195.
42. Martelli, F.S.; Mengoni, A.; Martelli, M.; Rosati, C.; Fanti, E. VDR TaqI polymorphism is associated with chronic periodontitis in Italian population. *Arch. Oral. Biol.* **2011**, *56*, 1494–1498. [CrossRef]
43. Martelli, F.S.; Martelli, M.; Rosati, C.; Fanti, E. Vitamin D: Relevance in dental practice. *Clin. Cases Miner. Bone Metab.* **2014**, *11*, 15–19. [CrossRef]
44. Piirainen, R.; Englund, E.; Henriksson, A.E. The impact of seasonal variation of 25-hydroxyvitamin D and parathyroid hormone on calcium levels. *Clin. Biochem.* **2016**, *49*, 850–853. [CrossRef]
45. Nevo-Shor, A.; Kogan, S.; Joshua, B.Z.; Bahat-Dinur, A.; Novack, V.; Fraenkel, M. Seasonal changes in serum calcium, PTH and vitamin D levels in patients with primary hyperparathyroidism. *Bone* **2016**, *89*, 59–63. [CrossRef]
46. Hegarty, V.; Woodhouse, P.; Khaw, K.-T. Seasonal Variation in 25-Hydroxyvitamin D and Parathyroid Hormone Concentrations in Healthy Elderly People. *Age Ageing* **1994**, *23*, 478–482. [CrossRef]

biomolecules

MDPI

Article

Exploring the Role of Ferroptosis in the Pathophysiology and Circadian Regulation of Restless Legs Syndrome

Maria Paola Mogavero [1,2,†], Giovanna Marchese [3,4,†], Giovanna Maria Ventola [3], Giuseppe Lanza [5,6], Oliviero Bruni [7], Luigi Ferini-Strambi [1,2] and Raffaele Ferri [5,*]

1 Vita-Salute San Raffaele University, 20132 Milan, Italy; paola_mogavero@libero.it (M.P.M.); ferinistrambi.luigi@hsr.it (L.F.-S.)
2 Sleep Disorders Center, Division of Neuroscience, San Raffaele Scientific Institute, 20127 Milan, Italy
3 Genomix4Life Srl, 84081 Baronissi, Italy; giovanna.marchese@genomix4life.com (G.M.); giovanna.ventola@genomix4life.com (G.M.V.)
4 Genome Research Center for Health-CRGS, 84081 Baronissi, Italy
5 Oasi Research Institute-IRCCS, 94018 Troina, Italy; glanza@oasi.en.it
6 Department of Surgery and Medical-Surgical Specialties, University of Catania, 95123 Catania, Italy
7 Department of Human Neuroscience, Sapienza University of Rome, 00185 Rome, Italy; oliviero.bruni@uniroma1.it
* Correspondence: rferri@oasi.en.it
† These authors contributed equally to this work.

Abstract

The study objectives were to investigate the role of ferroptosis, the mechanism linking iron accumulation, oxidative stress, and dopaminergic dysfunction, in restless legs syndrome (RLS), and to explore its connection with circadian regulation, a key feature of RLS and a known modulator of ferroptosis. We conducted pathway and gene expression analyses in 17 RLS patients and 39 controls, focusing on pathways related to ferroptosis, oxidative stress, iron metabolism, dopaminergic signaling, circadian rhythms, and immune responses. Enrichment analysis, differential gene expression, and cross-pathway gene overlaps were assessed. Ferroptosis and efferocytosis pathways were significantly upregulated in RLS, while oxidative phosphorylation, phosphatidylinositol signaling, PI3K-Akt, FoxO, and adipocytokine pathways were downregulated. The circadian rhythm pathway was markedly suppressed, with 12 circadian genes downregulated, suggesting that circadian disruption may drive ferroptosis activation. Decreased expression of protective pathways, including antioxidant responses and autophagy, was associated with increased iron accumulation, oxidative stress, and inflammation. Dopaminergic synapse genes were upregulated, possibly as a compensatory response to neuronal damage. Several genes overlapped across ferroptosis, circadian, and dopaminergic pathways, indicating a shared pathogenic mechanism. Our findings support a model in which circadian disruption promotes ferroptosis in RLS, contributing to iron overload, oxidative damage, and dopaminergic dysfunction. This pathogenic cascade may also enhance immune activation and inflammation. Circadian regulation and ferroptosis emerge as promising therapeutic targets in RLS. Further studies in larger cohorts are warranted to validate these mechanistic insights.

Keywords: ferroptosis; restless legs syndrome; circadian rhythms; oxidative stress; iron metabolism

1. Introduction

Restless legs syndrome (RLS) is a sensorimotor disorder whose symptoms, at any frequency or severity, affect approximately 5% to 10% of the general population in Western industrialized countries [1–3]. The condition is characterized by an irresistible urge to move the legs, often accompanied by unpleasant sensations. These symptoms typically begin or worsen during periods of rest or inactivity, especially in the evening or at night, and are partially or completely relieved by movement. Importantly, they are not attributable to other medical conditions [4].

RLS is clinically heterogeneous, with symptom expression varying by sex and age, which may also affect changes in time structure of periodic limb movements during sleep (PLMS) and the response to dopaminergic drugs [5–7]. However, a consistent and defining feature across phenotypes is its circadian pattern, with symptoms predominantly emerging in the evening and night [8].

Although the etiopathogenesis of RLS remains under investigation, current evidence points to the involvement of central nervous system structures, iron metabolism, and the dopaminergic system. Other neurotransmitter systems, including glutamatergic [9], GABAergic, and cholinergic pathways [8,10] have also been implicated. Notably, recent studies highlight the contribution of the mesolimbic dopaminergic pathway, as well as the basal ganglia, hippocampus, and amygdala, in RLS pathophysiology [11,12].

Several converging theories have been proposed to explain RLS pathogenesis, including central dopaminergic dysfunction, impaired brain iron homeostasis, increased glutamatergic transmission, and neuroinflammatory processes [8]. These mechanisms are not mutually exclusive and may act synergistically, with iron deficiency leading to altered dopamine synthesis and oxidative stress, further exacerbated by excitatory neurotransmission and immune activation.

Given the complex interplay of neural circuits and neurotransmitters involved, the origins of RLS circadianity are not fully understood. However, dopaminergic transmission is believed to play a central role. Symptom onset and worsening typically coincide with the circadian nadir of dopamine levels, while symptom relief occurs as dopamine levels normalize in the morning [8]. Moreover, this pattern is mirrored by PLMS, which occur in up to 80–90% of RLS patients [8]. The timing of PLMS, peaking from sleep onset to early morning, when dopamine levels are lowest, may help as a distinguishing feature of RLS compared to PLMS in other sleep disorders [13].

A recent transcriptomic analysis identified autophagy-related mechanisms, particularly ferroptosis, as key pathways in RLS [10]. Ferroptosis not only influences dopaminergic signaling but is also closely tied to iron metabolism [14–16], both of which are critical to RLS pathogenesis and treatment [17].

Despite these findings, the molecular basis for the circadian nature of RLS remains largely unresolved. Emerging evidence suggests that ferroptosis may itself be modulated by circadian genes [18], raising the possibility of a regulatory link between circadian molecular rhythms and the neurodegenerative processes observed in RLS. This intersection is particularly compelling, as it may explain how fluctuations in gene expression over the 24 h cycle influence dopaminergic vulnerability and symptom timing.

One emerging focus in RLS research is the role of the A11 dopaminergic nucleus, the primary source of descending dopaminergic input to the spinal cord [19]. Lesions or dysfunctions in the A11 region have been shown to induce RLS-like symptoms in animal models, suggesting its critical role in sensorimotor integration and dopaminergic control of spinal excitability [20]. While direct evidence of clock gene expression specifically in A11 neurons remains limited, recent findings demonstrate rhythmic expression of core circadian genes such as *PER1* and *BMAL1* in the posterior hypothalamus, which encompasses the

A11 region [21]. This suggests that A11 neurons may be influenced by local circadian gene activity and potentially act as a relay between central circadian networks and motor regulatory circuits. Such a dopaminergic-circadian interaction could provide a mechanistic bridge linking circadian timing and RLS symptomatology, particularly in relation to spinal excitability and the expression of PLMS.

The hypothalamus, a central node of circadian control, has also gained attention in RLS pathophysiology. Functional imaging and clinical studies suggest altered activity in hypothalamic regions involved in arousal and sleep regulation, including the suprachiasmatic nucleus (SCN) and lateral hypothalamus [12,19,22]. Dysregulation in these regions may influence both dopaminergic and orexinergic pathways, thereby contributing to the characteristic evening-worsening of RLS symptoms and their disruption of sleep. Furthermore, the SCN regulates peripheral clocks [23], which can modulate iron transport and metabolism, further linking circadian mechanisms with key pathological features of RLS.

Therefore, the aim of this study is to perform an in-depth transcriptomic analysis to investigate the interplay between circadian gene networks and ferroptosis-related pathways in RLS. Specifically, we seek to explore the role of ferroptosis, the mechanism linking iron accumulation, oxidative stress, and dopaminergic dysfunction, in the context of RLS, where these processes are also implicated. We further examine the connection between ferroptosis, RLS, and circadian regulation, given the established interplay between ferroptosis and circadian rhythms, and the fact that circadian disruption is a defining feature of RLS. By integrating high-resolution gene expression data, we aim to uncover novel molecular mechanisms driving RLS and identify potential targets for future therapeutic strategies.

2. Materials and Methods

Raw sequencing data for RLS patients and controls were obtained from datasets E-MTAB-13155 and E-MTAB-11326, respectively. A total of 17 RLS (13 women and 4 men, with an age range 24–76 years, mean age 55.8 years, a mean International RLS Study Group rating scale [24] of 21.1 and a mean disease duration of 6.6 years) were included in the analysis, along with 39 controls (22 women and 17 men, with an age range 23–92 years, mean age 62.9 years).

Two patients reported a family history of RLS. The diagnosis was established based on the criteria of the International RLS Study Group [4], using a semi-structured clinical interview designed to rigorously exclude conditions that may mimic RLS. Exclusion criteria included the presence of other sleep disorders, psychiatric, neurological, cardiovascular, or neurodegenerative conditions, neurodevelopmental delay, use of central nervous system-active medications within the year prior to the study, or any pharmacological treatment within three weeks before the polysomnographic recording. Participants with an apnea–hypopnea index greater than 10 events per hour of sleep were also excluded. Control subjects were free from medication and had no history of sleep, neurological, psychiatric, or physical disorders. Written informed consent was obtained in accordance with the Declaration of Helsinki, and the study protocol was approved by the Ethics Committee of the Oasi Research Institute.

The RNA-seq pipeline followed was previously described in 2024 by Mogavero et al. [10]. For all samples, from both controls and patients, venous blood was drawn in the morning following a 12 h fast. Immediately after collection, the samples were processed using Ficoll-Paque density gradient centrifugation (Ficoll-Paque PLUS-GE Healthcare Life Sciences, Piscataway, NJ, USA), and peripheral blood mononuclear cells (PBMCs) were stored at −80 °C until RNA extraction. RNA was isolated using TRIzol reagent (TRIzol Reagent, Invitrogen Life Technologies, Carlsbad, CA, USA) according to the manufacturer's instructions. All samples were handled, stored, and processed using standardized procedures.

RNA extraction was performed simultaneously for samples within the same group. The extracted RNA was stored at −80 °C until further analysis. A DNase treatment was applied after RNA isolation to ensure sample purity and enable accurate quantification.

Before proceeding with the protocol, RNA yield and quality were assessed using a NanoDrop One spectrophotometer (NanoDrop Technologies LLC, Wilmington, DE, USA) and a TapeStation 4200 (Agilent Technologies, 5301 Stevens Creek Blvd, Santa Clara, CA, USA), respectively. Specifically, the 260/280 absorbance ratio, as measured by NanoDrop, ranged from 1.8 to 2.0, while the RNA Integrity Number (RIN), as determined by the TapeStation, ranged from 6 to 8.

RNA extraction, library preparation, and sequencing were all performed using a standardized protocol, with consistent procedures applied within the same laboratory. Specifically, indexed libraries for all samples were prepared using the Illumina Stranded mRNA Prep Kit (Illumina, San Diego, CA, USA) according to the manufacturer's instructions. Following mRNA enrichment and fragmentation, cDNA synthesis, adapter ligation, and PCR amplification were carried out. Library concentrations were measured using the TapeStation 4200 (Agilent Technologies) and the Qubit 4 Fluorometer (Thermo Fisher Scientific, Waltham, MA, USA). Equimolar amounts of the indexed libraries were pooled and sequenced on the Illumina platform using a 2 × 75 bp paired-end read configuration.

Gene expression quantification was performed using the featureCounts tool, aligning reads to the GRCh38 human reference genome and GENCODE Version 37 annotations, via the STAR aligner with standard parameters. Quantification of expressed genes for each sample was computed using the featureCounts algorithm.

Data normalization was conducted in Rversion 4.4.2 using negative binomial generalized linear models, implemented through the Bioconductor DESeq2 version 1.49age. Only genes expressed in ≥25% of samples were included in the analysis. This model incorporates both the mean expression level and a gene-specific dispersion parameter, allowing for robust estimation of differential expression between conditions.

The DESeq2 package applies the Benjamini–Hochberg (BH) correction for multiple testing to calculate adjusted *p*-values (*p*adj), using default parameters. The BH correction controls the false discovery rate (FDR), which is the expected proportion of false positives among the results deemed statistically significant.

Genes were considered differentially expressed if they exhibited a fold change ≥ 1.50 or ≤ -1.50 ($|FC| \geq 1.50$) and an adjusted *p*-value (*p*adj) ≤ 0.01.

For functional analysis, pathway enrichment was conducted using the pathfindR version 2.4.2in R, leveraging the Kyoto Encyclopedia of Genes and Genomes (KEGG) database. This approach enabled the identification of biological pathways significantly associated with the observed gene expression changes.

Principal Component Analysis (PCA) was performed to assess overall variability among samples and to detect potential outliers or group-specific clustering patterns. PCA plots, along with volcano plots, used to visualize the distribution of regulated genes in terms of statistical significance and magnitude of fold change, were generated using the ggplot2 version 3.5.1 in R. Heatmaps of the differentially expressed genes were generated using the ComplexHeatmap version 2.14.0

3. Results

The first step involved assessing sample distribution using PCA, which revealed a clear separation between RLS and control groups along the first two principal components, indicating systematic transcriptomic differences (Figure 1).

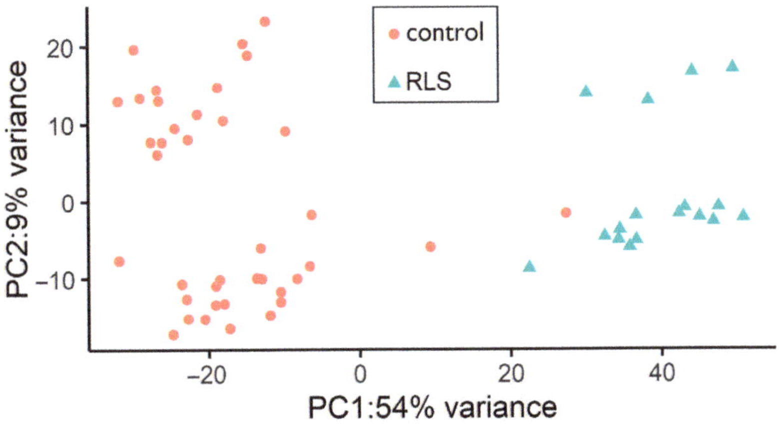

Figure 1. Principal Component Analysis (PCA) of RLS and control samples showing distinct clustering of the two groups based on the first principal component (PC1) and second principal component (PC2), explaining 54% of the variance.

We detected 23,331 expressed genes across all samples (expressed in at least 25% of RLS and control samples). Differential expression analysis identified 10,185 differentially expressed genes ($padj \leq 0.01$). Among these, 4008 genes were significantly upregulated ($padj \leq 0.01$ and fold-change ≥ 1.5), and 3413 were significantly downregulated ($padj \leq 0.01$ and fold-change ≤ -1.5) in RLS samples compared to controls (Supplementary Table S1). The volcano plot illustrates the distribution of differentially expressed genes between groups (Figure 2).

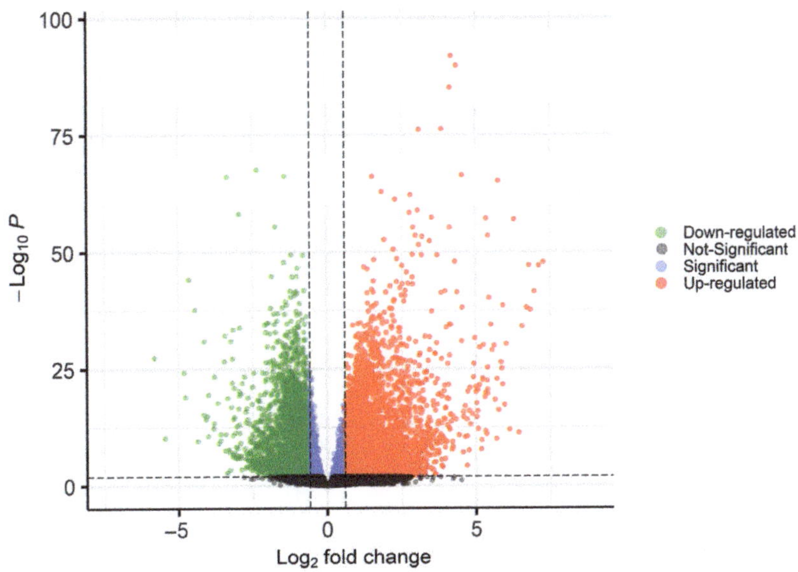

Figure 2. Volcano plot of differentially expressed genes between RLS and control samples. Upregulated genes are shown in red, downregulated genes in green, non-significant genes ($padj > 0.01$) in gray, and genes significant only for $padj \leq 0.01$ in blue.

Pathway enrichment analysis using pathfindR revealed 228 enriched KEGG pathways. We focused our analysis on 12 pathways of particular relevance, selected for their functional connections to ferroptosis, oxidative phosphorylation, and circadian clock genes, three critical biological axes involved in the regulation of oxidative stress and neuronal survival, particularly within the dopaminergic system.

All selected pathways were significantly enriched ($p$adj < 0.05), with fold enrichment values ranging from 1.09 for Oxidative Phosphorylation to 2.89 for Circadian Rhythm, as shown in Table 1.

Table 1. List of 12 KEGG pathways enriched in RLS versus controls, with fold enrichment values, statistical significance (lowest and highest p-values), and the corresponding sets of upregulated and downregulated genes for each pathway.

ID	Term Description	Fold Change	Lowest p	Highest p	Up-Regulated	Down-Regulated
hsa04151	PI3K-Akt signaling pathway	1.39	3.30×10^{-18}	2.89×10^{-14}	*PIK3CB, PIK3R3, PIK3R1, EGF, ITGA5, LPAR2, FGF22, PDGFA, GNG11, SYNE1, COL1A1, FGF1, MAP2K2, MAP3K5, COL4A6, LAMA2, LAMB1*	*SPP1, MMP9, ITGA5, COL1A1, COL4A6, LAMB1, LAMA2, LAMA4, COL4A5, LAMB2, COL4A4, LAMC2, COL4A3, COL4A2*
hsa04068	FoxO signaling pathway	1.95	1.12×10^{-15}	3.34×10^{-8}	*MAP3K5, FOXO4, MAP2K2, BAD, PIK3CB, PIK3R3, PIK3R1, GABARAPL1, CALM3, GABARAP, CALM2, CALM1*	*MAP3K5, PIK3CB, PIK3R3, PIK3R1, SOX4, IRS1, BAD, GABARAPL1, GABARAP, CALM3, CALM2, CALM1*
hsa04140	Autophagy—animal	1.73	2.00×10^{-13}	1.16×10^{-13}	*IRS1, PIK3CB, PIK3R3, PIK3R1, MAP3K5, ATG2B, ATG4C, ATG7, RB1CC1, STX17, GABARAPL1, GABARAP, CALM3, CALM2, CALM1*	*IRS1, PIK3CB, PIK3R3, PIK3R1, MAP3K5, ATG2B, ATG4C, ATG7, RB1CC1, STX17, GABARAPL1, GABARAP, CALM3, CALM2, CALM1*
hsa04728	Dopaminergic synapse	1.36	3.84×10^{-9}	1.09×10^{-4}	*MAOA, VMAT2, GNAI1, GNG5, GNG3, GNG4, GNAQ, CAMK2B, CAMK2A, GRIN1, GRIA1, GRID1, GABRA2, CALM2, CALM1*	*SLC6A3, LEP, CALM1, CALM2, CAMK2A, CAMK2B, GNG5, GNG4, GNG3, GNAI1*
hsa04920	Adipocytokine signaling pathway	1.29	3.45×10^{-8}	1.04×10^{-2}	*CRPRA, ACSL1, LEP, POMC, ADIPOR1, STK11, RXRA, SLC2A1, TNFSF4, CAMKK2*	*NR1H3, PTPN11, SOCS3, IRS1, RXRA, PRKAA1, PRKAA2, LEP, RXRG, AKT3, MAPK8, STAT1, STAT3*
hsa00190	Oxidative phosphorylation	1.09	7.27×10^{-6}	1.01×10^{-5}	*NUDFB8, COX6C, TTC19, ATP6V2, ATP6V1B2, ATP6V1E1, COX7A2, ATP6V1G2*	*NUDFB6, NUDFB4, NDUFA1, NUDFB9, NUDFB5, NDUFA2, COX6C, ATP6V1E1, ATP6V1B2*
hsa04148	Efferocytosis	1.32	3.10×10^{-5}	1.30×10^{-3}	*AGER, GRK6, ROCK1, ERBB4, MAP3K5, BNIP3L, ITGA5, SYNE1, MAP2K2, GABARAPL1, PIK3CB, PIK3R1, PIK3R3, COL4A6, LAMA2, LAMB1*	*METRNL, OXTR, ITGA1, S1PR1, CD47, HVCN1, LRP1, VAV1, TUBB2B, CSF1R, GRK6, BNIP3L, MAP3K5, SYNE1, AGER*

Table 1. *Cont.*

ID	Term Description	Fold Change	Lowest *p*	Highest *p*	Up-Regulated	Down-Regulated
hsa05208	Chemical carcinogenesis—reactive oxygen species	1.23	1.05×10^{-6}	1.04×10^{-5}	*PIK3CB, MAP3K5, MAP2K2, FGF1, CYBA, NCF1, NCF2, NDUFA2, COX6C, PRDX2, PRDX6, PRDX1, MAPKAPK2, MAPKAPK3, KEAP1, ASMT, SOD2*	*PIK3CB, MAP3K5, MAP2K2, FGF1, CYBA, NCFA, NCF2, PRDX2, PRDX1, PRDX6, MAPKAPK2, MAPKAPK3, KEAP1, ASMT, SOD2*
hsa04710	Circadian rhythm	2.89	3.91×10^{-16}	4.62×10^{-3}	*DBP, BHLHE41*	*RORA, PER1, PER2, PER3, CLOCK*
hsa04070	Phosphatidylinositol signaling system	1.35	6.61×10^{-8}	3.73×10^{-2}	*PTEN, CALM1, CALM2, PIK3R1, PIK3R3, DGKZ, GRIN1, CALM3, CALML4, CAMK2A, CAMK2B, CAMK2G, MAP2K2, MAP3K5*	*PIK3CB, MAP3K5, MAP2K2, GRIN1, CALM1, CALM2, CALM3, CAMK2A, CAMK2B, CAMK2G*
hsa04713	Circadian entrainment	1.29	4.36×10^{-7}	4.85×10^{-3}	*GRIN1, CALM1, CALM2, CAMK2A, CAMK2B, CAMK2G, MAP2K2, MAP3K5*	*GRIN1, CALM1, CALM2, CAMK2A, CAMK2B, CAMK2G*
hsa04216	Ferroptosis	1.57	2.45×10^{-8}	4.60×10^{-2}	*TP53, SLC1A4, PINK1, SLC7A11, ACSL4*	*TP53, SLC1A4, PINK1, SLC7A11, ACSL4*

Among these, the ferroptosis pathway (hsa04216) may represent a particularly relevant mechanism potentially implicated in RLS, given its role in iron-mediated oxidative damage and its potential contribution to disease pathophysiology.

As ferroptosis is closely linked to oxidative stress and redox regulation, we extended our analysis to pathways such as oxidative phosphorylation (hsa00190) and chemical carcinogenesis—reactive oxygen species (ROS) (hsa05208), both critical for understanding oxidative stress-mediated injury.

Oxidative stress and iron dysregulation appear to intersect with circadian rhythm control (hsa04710, hsa04713), influencing cellular energy metabolism and susceptibility to ferroptosis. Disruptions in circadian rhythms may impair antioxidant defense systems, promoting ROS accumulation and cell death.

Additional pathways, FoxO signaling (hsa04068), PI3K-Akt (hsa04151), autophagy (hsa04140), and adipocytokine signaling (hsa04920), were also analyzed for their roles in stress responses, cell survival, and inflammation, all of which are relevant to ferroptosis susceptibility.

The phosphatidylinositol signaling system (hsa04070) is highlighted for its involvement in regulating iron metabolism and oxidative damage responses. Autophagy (hsa04140) plays a protective role by removing excess iron and damaged cellular components, while the dopaminergic synapse pathway (hsa04728) is crucial for brain function and is notably affected in RLS. Dysregulation of iron metabolism in dopaminergic neurons may increase susceptibility to ferroptosis activation, thereby contributing to the progression and exacerbation of neurodegenerative processes. Additionally, the efferocytosis pathway (hsa04148), the mechanism by which immune cells clear apoptotic or damaged cells, appears to play a protective role by mitigating oxidative damage. (Figure 3, Table 1).

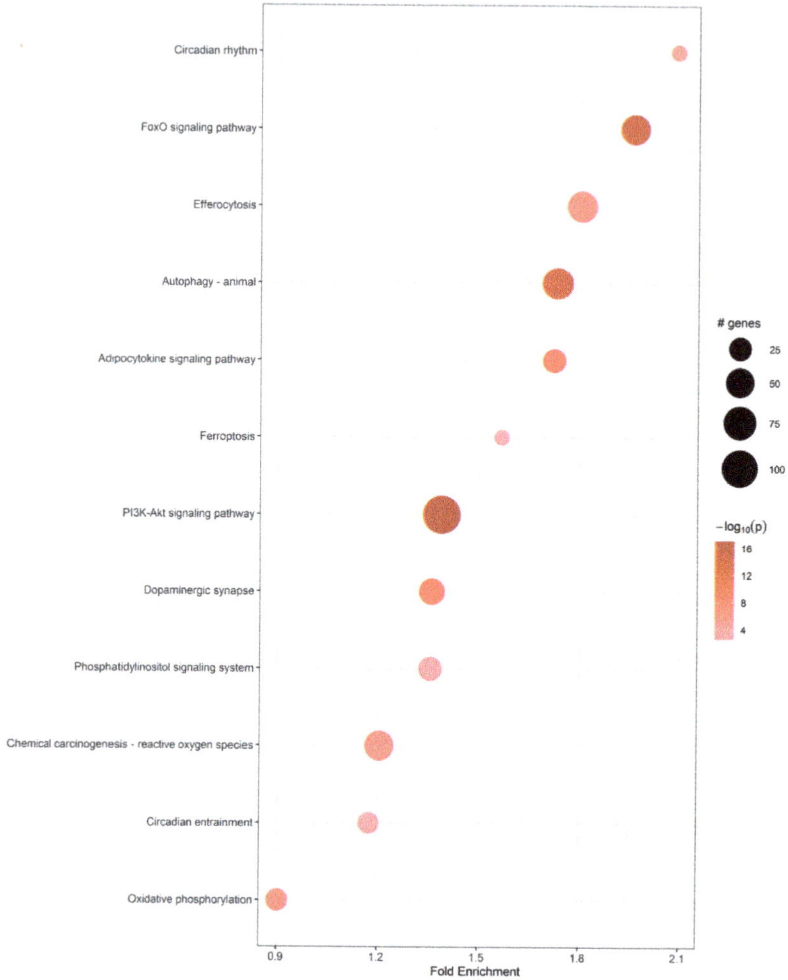

Figure 3. Summary of pathway enrichment results. Graphical representation of 12 selected enriched pathways based on differentially expressed genes.

Shared genes among these pathways suggest a common pathogenic thread. Figure 4 displays a histogram of upregulated (red) and downregulated (green) genes for each biological pathway. In the Chemical Carcinogenesis–ROS pathway, there was a predominance of downregulated genes (32) versus upregulated ones (20), suggesting a suppressed oxidative stress response. Although this may reflect an adaptive mechanism to limit harmful oxidative pathways, it could also impair the cell's ability to neutralize ROS, favoring sustained oxidative stress and the initiation of ferroptosis.

The Circadian Rhythm pathway shows a strong imbalance, with only 2 upregulated and 12 downregulated genes, a marked downregulation of the genes involved in this pathway. The concurrent enrichment of pathways related to energy metabolism, the circadian clock, and ferroptosis suggests the existence of an integrated biological network in which oxidative stress, iron dysregulation, and disruption of circadian rhythms converge to increase neuronal vulnerability.

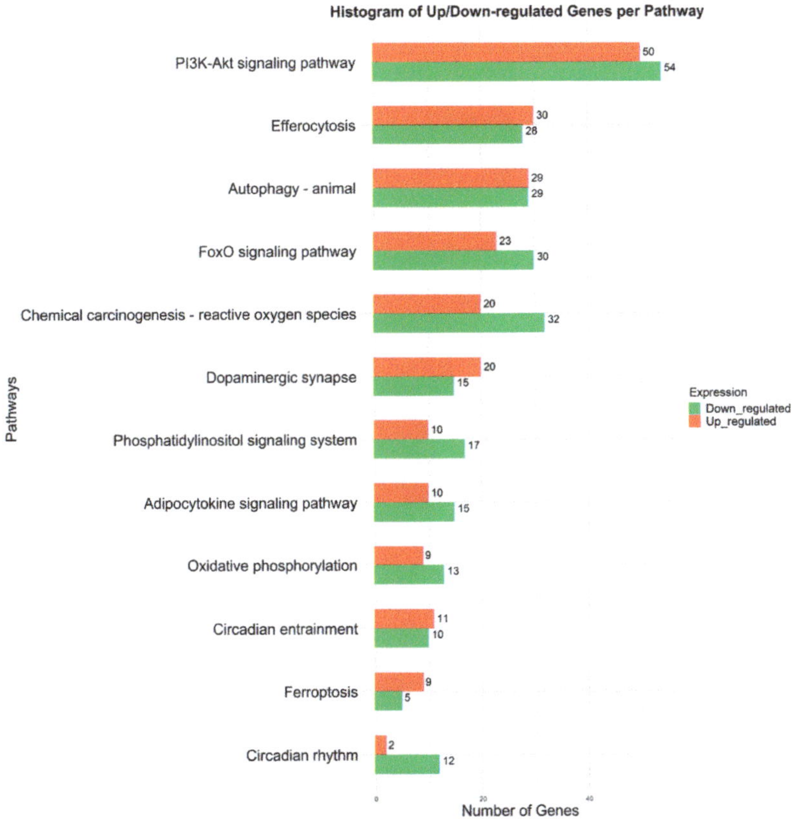

Figure 4. Histogram showing the number of upregulated (red) and downregulated (green) genes contributing to the enrichment of the 12 selected biological pathways.

Downregulation of core circadian clock genes (*PER1-3*, *CLOCK*, *RORA*), which are essential for maintaining the rhythmicity of antioxidant defenses, may impair cellular homeostasis and heighten susceptibility to ferroptosis.

This network appears to be further modulated by key regulatory pathways, including FoxO signaling, PI3K-Akt, and autophagy, which orchestrate cellular stress responses and the clearance of damaged organelles and proteins.

Finally, the involvement of the dopaminergic synapse pathway suggests potential alterations in synaptic function and neuronal clearance mechanisms, both of which are critical to the pathophysiology of RLS.

In Figure 5, we show a network representation of pathway interactions, illustrating how gene expression changes are integrated across pathways. This visualization highlights the interplay between biological processes and the complexity of signaling alterations in RLS.

In Figure 6A–E, we focus on key pathways including ferroptosis (hsa04216), circadian entrainment (hsa04713), circadian rhythm (hsa04710), dopaminergic synapse (hsa04728), and efferocytosis (hsa04148), respectively, analyzing gene expression variations between controls and RLS patients.

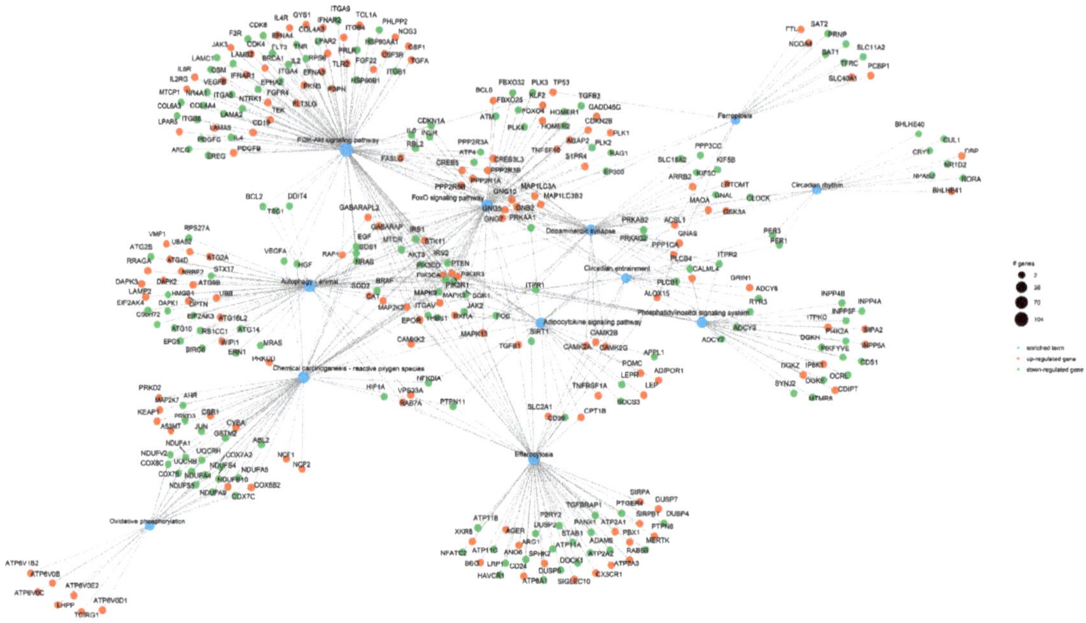

Figure 5. Network diagram illustrating the relationships between upregulated and downregulated genes and the 12 selected enriched KEGG pathways. Each node represents a KEGG pathway; red and green nodes indicate pathways with predominantly upregulated or downregulated gene sets, respectively. The complexity of the network reflects the multifactorial nature of RLS pathophysiology and the high degree of gene overlap among pathways. While dense, the full network was retained to preserve the biological detail and interconnectedness revealed by the transcriptomic data.

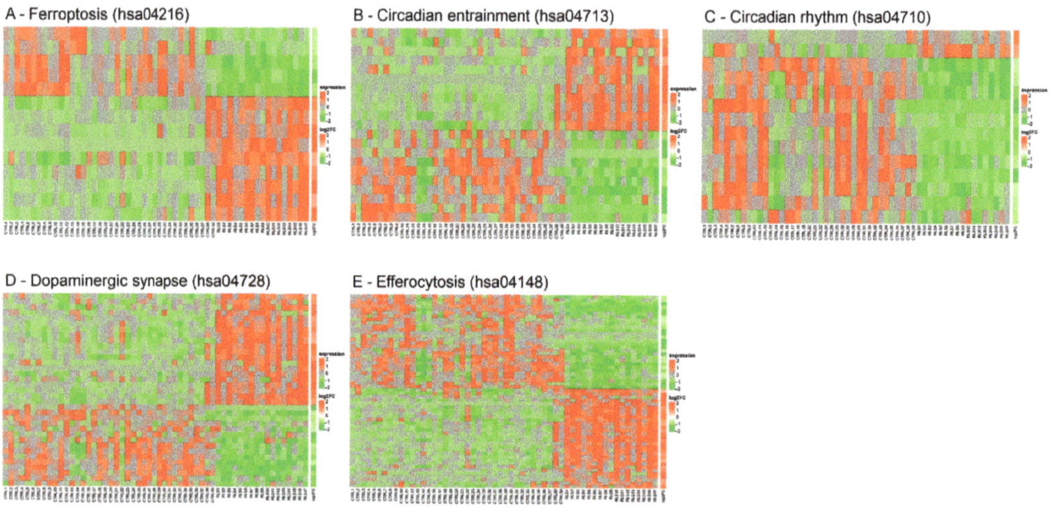

Figure 6. Heatmap of (**A**) ferroptosis-related gene expression; (**B**) circadian entrainment pathway gene expression; (**C**) circadian rhythm pathway gene expression; (**D**) dopaminergic synapse pathway gene expression; and (**E**) efferocytosis pathway gene expression, across all RLS and controls (CTRL). samples.

Hierarchical clustering analysis reveals consistent and statistically significant expression differences ($padj \leq 0.01$ and $|FC| \geq 1.5$). RLS samples exhibited a clear upregulation pattern (red) compared to controls (green/neutral), particularly for ferroptosis-related genes. This heatmap strongly supports the hypothesis that ferroptosis pathways are activated or dysregulated in RLS.

4. Discussion

Our research investigated the role of ferroptosis in RLS, a mechanism recently implicated in the disorder [10], also considering its crucial involvement in inflammation and oxidative stress [14,15], both of which have been increasingly linked to RLS [25,26]. Moreover, ferroptosis is known to contribute to intracellular dysregulation of iron metabolism [14] and to impact dopaminergic metabolism [18], both of which are highly relevant to the pathogenesis of RLS [8,27].

Given that circadian dysregulation is a hallmark of RLS [4], we also explored potential connections between ferroptosis-related mechanisms (oxidative phosphorylation, phosphatidylinositol signaling, FoxO, PI3K-Akt and adipocytokine pathways, autophagy, efferocytosis), circadian regulation, and the dopaminergic system. Our analyses revealed that multiple genes are shared across these pathways, indicating a common underlying thread.

Our findings demonstrated a significant difference in the expression of the analyzed pathways, as well as in the regulation of transcripts associated with these biological processes, between RLS patients and controls. Each pathway exhibited high fold enrichment, with circadian rhythms showing the greatest enrichment and involving numerous genes. Importantly, we highlighted the predominance of either up- or down-regulated genes within each pathway, providing insights into the genetic mechanisms underpinning circadian dysfunction and its related biological processes in RLS.

Specifically, we observed a marked predominance of down-regulated genes within the Chemical Carcinogenesis–ROS pathway in RLS, suggesting a potential suppression of the cellular oxidative stress response. This suppression could lead to the accumulation of oxidative damage, creating a favorable environment for the activation of ferroptosis, a process highly sensitive to increased ROS levels and redox imbalance [15]. Indeed, ferroptosis was markedly upregulated in RLS. This process appears further exacerbated by the downregulation of FoxO, PI3K-Akt, and phosphatidylinositol signaling pathways, all of which are crucial for oxidative stress response, cell survival, autophagy, and cellular repair mechanisms [28,29]. Additionally, downregulation of oxidative phosphorylation and adipocytokine signaling pathways may also critically contribute to increased oxidative stress and inflammation, thereby further promoting ferroptosis and neuronal damage [30,31].

Among these mechanisms, phosphatidylinositol signaling is also involved in iron metabolism regulation [30], suggesting it might directly influence ferroptosis and further contribute to a vicious cycle of dysregulation of iron metabolism.

It is well established that disruption of brain iron homeostasis is a fundamental pathological feature of RLS, with consequences for both dopaminergic and glutamatergic dysfunction [27]. However, the mechanisms driving dysregulation of iron metabolism remain unclear. Our group was the first to report a pronounced upregulation of ferroptosis in RLS [10], and this study provides a more in-depth exploration of that finding. Ferroptosis activation promotes intracellular dysregulation of iron metabolism [31], offering a key explanation for previous observations and the onset of inflammatory mechanisms in RLS [25], while also suggesting therapeutic potential for ferroptosis inhibitors in the disorder.

Of note, the upregulation of efferocytosis identified in our study, a crucial immune-modulatory process by which macrophages and other immune cells clear apoptotic or damaged cells [32], reinforces growing evidence implicating inflammatory, immune, and infectious mechanisms in RLS [10,25], consistent with autopsy studies in the disorder [33].

Another critical finding of our study is the downregulation of the PI3K-Akt pathway in RLS, a signaling cascade known to regulate neuroinflammation, neurogenesis, synaptic plasticity, and neurotransmission [29]. Impairment of this pathway could thus contribute to the dysfunction of all these processes, which are known to be altered in RLS [10,34], and may also be linked to the frequent comorbidity of depression in RLS patients [35].

Importantly, no difference in autophagy regulation was observed, with an equal number of up- and downregulated genes, suggesting that the protective effect of autophagy, key in eliminating excess iron and damaged cellular components, may be insufficient to counteract ferroptosis in RLS [36], thereby facilitating further dysregulation of iron metabolism, oxidative stress, and inflammation.

Another major goal of our study was to investigate the interplay between ferroptosis, circadian rhythms, and dopaminergic transmission; again, we found that multiple genes are shared among these pathways. Namely, we observed a strongly disrupted circadian rhythm pathway, with only two upregulated genes versus twelve downregulated ones, suggesting a marked suppression of the circadian system. Since circadian regulation governs many metabolic and antioxidant processes [37], its disruption may indirectly facilitate ferroptosis by impairing protective systems such as glutathione production and lipid peroxide detoxification.

Therefore, the dysfunction of the molecular clock may act as a key modulator of cellular vulnerability to ferroptosis [38,39] by amplifying oxidative stress imbalance. Investigation of RLS circadian rhythmicity in relation to circadian markers remains a crucial yet largely unexplored research area [40]. Despite the recognized importance of genetic factors in RLS and the diagnostic relevance of its circadian characteristics [8], studies to date have mainly focused on possible links between *CLOCK* genes and dopaminergic transmission [40].

It is important to note that the causal relationship between circadian dysregulation and ferroptosis activation remains to be clarified. While our findings suggest that circadian disruption may promote ferroptotic vulnerability, the inverse scenario is also plausible: ferroptosis-induced oxidative damage and neuroinflammation, particularly in dopaminergic pathways, may in turn impair circadian gene expression. The interaction may thus be bidirectional or self-reinforcing. Future longitudinal and experimental studies will be crucial to dissect these complex temporal dynamics and clarify the sequence of events in RLS pathophysiology.

Recent evidence suggests that circadian genes may also regulate ferroptosis [18], with *BMAL1* emerging as a key regulator of antioxidant systems suppressing ferroptosis [41]. In our study, *CRY1*, *PER1*, *PER3*, *CLOCK*, and *RORA* were all downregulated, also *BMAL1* (alias *ARNTL*) was downregulated, but with fold change -1.48 and $p\text{adj} = 2.59 \times 10^{-6}$, indicating that circadian suppression might represent a primary trigger for ferroptosis activation, intracellular dysregulation of iron metabolism, and inflammation in RLS.

Furthermore, recent studies in mouse models have shown that hypothalamic A11 dopaminergic neurons, known to be implicated in RLS pathogenesis [8], may be influenced by local circadian gene activity and could act as a relay between central circadian networks and motor regulation circuits [21]. This might explain the upregulation of dopaminergic synapses observed in our study as a compensatory mechanism for circadian rhythm disruption, even though increased synaptic transmission does not necessarily imply enhanced dopaminergic output, which is known to be impaired in RLS, likely due to ferroptosis-induced neuronal damage. Further research on this topic is necessary to confirm this hypothesis. Similarly, upregulation of circadian entrainment might reflect compensatory responses to circadian gene downregulation.

The circadian modulation of dopamine production has long been associated with RLS symptoms and motor manifestations such as PLMS [8]; however, this does not fully explain the link between circadian rhythms and other pathogenetic factors like iron deficiency, inflammation, or neurotransmitter network alterations [8].

In this context, it would be intriguing to investigate the potential correlation between daily and seasonal variations in RLS symptoms [8,42] and periodic, including seasonal, regulation of circadian genes. Innate circannual timing is an evolutionarily conserved trait present in many species, including humans, and contributes to seasonal variations in cellular function across different geographical regions [43]. At the molecular level, epigenetically regulated chromatin remodeling in pituitary cells of the hypothalamus is thought to drive seasonal oscillations in the transcriptional activity of specific circadian timer genes, mediating the shift between summer and winter phenotypes [43]. Further research into circannual rhythms may provide valuable insights into their contribution to human physiology and disease, including RLS, and may help explain both the circadian pattern of symptoms and the geographical variability in disease prevalence.

In this context, it is intriguing to consider recent hypotheses suggesting involvement of additional biological pathways in RLS, including orexinergic signaling [44] and the Calcitonin Gene-Related Peptide [45], both linked to hypothalamic structures and subject to circadian regulation [46].

This study has some limitations. First, the relatively small sample size (17 RLS patients and 39 controls) may limit the generalizability of the findings. However, this limitation was mitigated by stringent criteria for differential gene expression (adjusted p-value ≤ 0.01 and $|\text{fold-change}| \geq 1.5$) and by the use of robust pathway enrichment analyses. An additional limitation is the unequal group size (17 RLS patients vs. 39 controls), which may introduce potential bias. However, all samples were processed using standardized protocols, and differential expression analysis was conducted using DESeq2, a method specifically designed to handle unbalanced groups through rigorous normalization and dispersion modeling. Furthermore, gene expression data were obtained from peripheral blood rather than neural tissue, which may not fully reflect central nervous system processes. Nonetheless, blood-based transcriptomic profiling has been increasingly recognized as a valid proxy for identifying systemic molecular alterations relevant to neurological disorders. Another limitation of our study is the lack of standardized assessment for subclinical anxiety and depression symptoms, which are common comorbidities in RLS. Although major psychiatric disorders were part of the exclusion criteria, the absence of formal rating scales may have limited our ability to account for milder affective symptoms that could potentially influence gene expression. Finally, while the study establishes associations between circadian gene expression, ferroptosis, and RLS, causal relationships cannot be inferred from transcriptomic data alone. Future studies with larger cohorts, longitudinal designs, and functional validation will be necessary to confirm and expand upon these findings.

5. Conclusions

In conclusion, our study offers a comprehensive framework of RLS, suggesting that downregulation of circadian genes may drive ferroptosis overexpression, leading to intracellular dysregulation of iron metabolism, neuronal damage, dopaminergic dysfunction, and heightened oxidative stress, inflammation, and immune activation, thus establishing a vicious pathogenic cycle (Figure 7). Nevertheless, further studies with larger cohorts are needed to validate these findings and to clarify the complex role of circadian genes and ferroptosis in RLS pathogenesis.

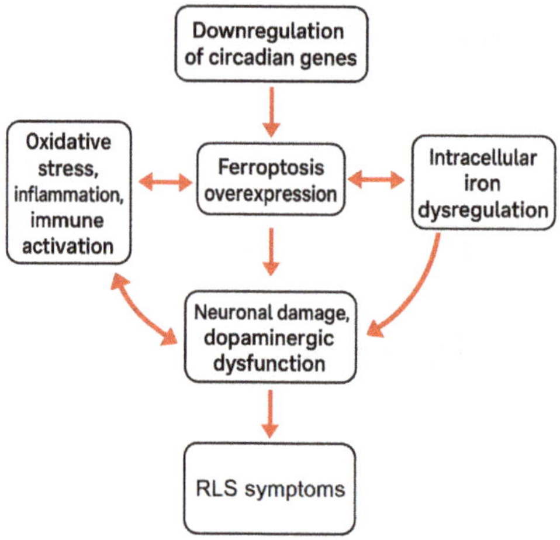

Figure 7. Schematic representation of the proposed pathogenic cycle in RLS. The diagram illustrates how the downregulation of circadian genes may initiate the overexpression of ferroptosis pathways. This, in turn, leads to intracellular iron accumulation, resulting in neuronal damage and dopaminergic dysfunction. These alterations promote increased oxidative stress, inflammation, and immune system activation, which further amplify ferroptosis processes. The cycle represents a self-perpetuating loop that may contribute to the neurobiological and clinical manifestations of RLS.

Such research may also help determine whether ferroptosis could serve as a novel therapeutic target, as has been proposed in other conditions, including Parkinson's disease and other neurodegenerative disorders [47].

Supplementary Materials: The following supporting information can be downloaded at: https://www.mdpi.com/article/10.3390/biom15081184/s1. Table S1: List of differentially expressed genes (DEGs) identified between Restless Legs Syndrome (RLS) patients and control subjects. The table includes gene symbols, log2 fold change (log2FC), adjusted *p*-values (*p*adj), and expression direction (upregulated or downregulated) based on RNA-seq analysis. Genes were considered differentially expressed if $padj \leq 0.01$ and $|log2FC| \geq 1.5$. This dataset forms the basis for subsequent pathway enrichment and gene network analyses presented in the main text.

Author Contributions: Conceptualization, M.P.M. and G.M.; methodology, G.M. and G.M.V.; investigation, M.P.M., G.M. and G.M.V.; resources, R.F.; writing—original draft preparation, M.P.M., G.M., G.M.V. and R.F.; writing—review and editing, M.P.M., G.M., G.M.V., G.L., O.B., L.F.-S. and R.F.; supervision, R.F.; funding acquisition, R.F. All authors have read and agreed to the published version of the manuscript.

Funding: This study was partially supported by a grant from the Italian Ministry of Health "Ricerca Corrente" (RC n. 2793745) (R. Ferri).

Institutional Review Board Statement: The study was conducted in accordance with the Declaration of Helsinki and approved by the Ethics Committee "Comitato Etico Locale IRCCS Oasi Maria SS." (approval code: CEL-IRCCS OASI/18-04-2024/04, date of approval: 18 April 2024).

Informed Consent Statement: Written informed consent was obtained from the patients participating in this study.

Data Availability Statement: Raw sequencing data for RLS patients and controls were obtained from datasets E-MTAB-13155 and E-MTAB-11326, respectively. No new raw data were generated during the conduct of this new study.

Conflicts of Interest: Authors Giovanna Marchese and Giovanna Maria Ventola were employed by the company Genomix4Life Srl. The remaining authors declare that the research was conducted in the absence of any commercial or financial relationships that could be construed as a potential conflict of interest.

Abbreviations

The following abbreviations are used in this manuscript:

FC	Fold Change
FDR	False Discovery Rate
KEGG	Kyoto Encyclopedia of Genes and Genomes
PBMCs	Peripheral Blood Mononuclear Cells
PCA	Principal Component Analysis
PLMS	Periodic Limb Movements during Sleep
RIN	RNA Integrity Number
RLS	Restless Legs Syndrome
RNA	Ribonucleic Acid
ROS	Reactive Oxygen Species
SCN	Suprachiasmatic Nucleus

References

1. Koo, B.B. Restless Leg Syndrome Across the Globe: Epidemiology of the Restless Legs Syndrome/Willis-Ekbom Disease. *Sleep Med. Clin.* **2015**, *10*, 189–205. [CrossRef] [PubMed]
2. Ulfberg, J.; Nystrom, B.; Carter, N.; Edling, C. Prevalence of restless legs syndrome among men aged 18 to 64 years: An association with somatic disease and neuropsychiatric symptoms. *Mov. Disord. Off. J. Mov. Disord. Soc.* **2001**, *16*, 1159–1163. [CrossRef]
3. Ulfberg, J.; Nystrom, B.; Carter, N.; Edling, C. Restless Legs Syndrome among working-aged women. *Eur. Neurol.* **2001**, *46*, 17–19. [CrossRef]
4. Allen, R.P.; Picchietti, D.L.; Garcia-Borreguero, D.; Ondo, W.G.; Walters, A.S.; Winkelman, J.W.; Zucconi, M.; Ferri, R.; Trenkwalder, C.; Lee, H.B.; et al. Restless legs syndrome/Willis-Ekbom disease diagnostic criteria: Updated International Restless Legs Syndrome Study Group (IRLSSG) consensus criteria—History, rationale, description, and significance. *Sleep Med.* **2014**, *15*, 860–873. [CrossRef] [PubMed]
5. Mogavero, M.P.; DelRosso, L.M.; Lanza, G.; Lanuzza, B.; Bruni, O.; Strambi, L.F.; Ferri, R. Changes in time structure of periodic leg movements during sleep in restless legs syndrome: Effects of sex and age. *Sleep Med.* **2024**, *115*, 137–144. [CrossRef]
6. Mogavero, M.P.; Lanza, G.; DelRosso, L.M.; Lanuzza, B.; Bruni, O.; Ferini Strambi, L.; Ferri, R. Exploring sex differences in periodic leg movements during sleep across the lifespan of patients with restless legs syndrome. *Sleep Med.* **2024**, *122*, 253–257. [CrossRef] [PubMed]
7. Mogavero, M.P.; Antelmi, E.; Lanza, G.; Marelli, S.; Castelnuovo, A.; Tinazzi, M.; DelRosso, L.M.; Silvestri, R.; Ferri, R.; Ferini Strambi, L. Sex-based disparities in dopamine agonist response in patients with restless legs syndrome. *J. Sleep Res.* **2024**, *34*, e14311. [CrossRef]
8. Manconi, M.; Garcia-Borreguero, D.; Schormair, B.; Videnovic, A.; Berger, K.; Ferri, R.; Dauvilliers, Y. Restless legs syndrome. *Nat. Rev. Dis. Primers* **2021**, *7*, 80. [CrossRef] [PubMed]
9. Trenkwalder, C.; Allen, R.; Hogl, B.; Clemens, S.; Patton, S.; Schormair, B.; Winkelmann, J. Comorbidities, treatment, and pathophysiology in restless legs syndrome. *Lancet Neurol.* **2018**, *17*, 994–1005. [CrossRef]
10. Mogavero, M.P.; Salemi, M.; Lanza, G.; Rinaldi, A.; Marchese, G.; Ravo, M.; Salluzzo, M.G.; Antoci, A.; DelRosso, L.M.; Bruni, O.; et al. Unveiling the pathophysiology of restless legs syndrome through transcriptome analysis. *iScience* **2024**, *27*, 109568. [CrossRef]
11. Mogavero, M.P.; Mezzapesa, D.M.; Savarese, M.; DelRosso, L.M.; Lanza, G.; Ferri, R. Morphological analysis of the brain subcortical gray structures in restless legs syndrome. *Sleep Med.* **2021**, *88*, 74–80. [CrossRef]
12. Rizzo, G.; Plazzi, G. Neuroimaging Applications in Restless Legs Syndrome. *Int. Rev. Neurobiol.* **2018**, *143*, 31–64. [CrossRef]

13. Ferri, R.; Zucconi, M.; Manconi, M.; Bruni, O.; Ferini-Strambi, L.; Vandi, S.; Montagna, P.; Mignot, E.; Plazzi, G. Different periodicity and time structure of leg movements during sleep in narcolepsy/cataplexy and restless legs syndrome. *Sleep* **2006**, *29*, 1587–1594. [CrossRef]
14. Dixon, S.J.; Lemberg, K.M.; Lamprecht, M.R.; Skouta, R.; Zaitsev, E.M.; Gleason, C.E.; Patel, D.N.; Bauer, A.J.; Cantley, A.M.; Yang, W.S.; et al. Ferroptosis: An iron-dependent form of nonapoptotic cell death. *Cell* **2012**, *149*, 1060–1072. [CrossRef] [PubMed]
15. Stockwell, B.R.; Friedmann Angeli, J.P.; Bayir, H.; Bush, A.I.; Conrad, M.; Dixon, S.J.; Fulda, S.; Gascon, S.; Hatzios, S.K.; Kagan, V.E.; et al. Ferroptosis: A Regulated Cell Death Nexus Linking Metabolism, Redox Biology, and Disease. *Cell* **2017**, *171*, 273–285. [CrossRef] [PubMed]
16. Morris, G.; Berk, M.; Carvalho, A.F.; Maes, M.; Walker, A.J.; Puri, B.K. Why should neuroscientists worry about iron? The emerging role of ferroptosis in the pathophysiology of neuroprogressive diseases. *Behav. Brain Res.* **2018**, *341*, 154–175. [CrossRef] [PubMed]
17. Winkelman, J.W.; Berkowski, J.A.; DelRosso, L.M.; Koo, B.B.; Scharf, M.T.; Sharon, D.; Zak, R.S.; Kazmi, U.; Falck-Ytter, Y.; Shelgikar, A.V.; et al. Treatment of restless legs syndrome and periodic limb movement disorder: An American Academy of Sleep Medicine clinical practice guideline. *J. Clin. Sleep Med.* **2024**, *21*, 137–152. [CrossRef]
18. Wang, X.; Wang, M.; Zhi, H.; Li, J.; Guo, D. REV-ERBα inhibitor rescues MPTP/MPP^+-induced ferroptosis of dopaminergic neuron through regulating FASN/SCD1 signaling pathway. *Heliyon* **2024**, *10*, e40388. [CrossRef]
19. Clemens, S.; Rye, D.; Hochman, S. Restless legs syndrome: Revisiting the dopamine hypothesis from the spinal cord perspective. *Neurology* **2006**, *67*, 125–130. [CrossRef]
20. Ondo, W.G.; He, Y.; Rajasekaran, S.; Le, W.D. Clinical correlates of 6-hydroxydopamine injections into A11 dopaminergic neurons in rats: A possible model for restless legs syndrome. *Mov. Disord. Off. J. Mov. Disord. Soc.* **2000**, *15*, 154–158. [CrossRef]
21. Pina-Leyva, C.; Lara-Lozano, M.; Rodriguez-Sanchez, M.; Vidal-Cantu, G.C.; Barrientos Zavalza, E.; Jimenez-Estrada, I.; Delgado-Lezama, R.; Rodriguez-Sosa, L.; Granados-Soto, V.; Gonzalez-Barrios, J.A.; et al. Hypothalamic A11 Nuclei Regulate the Circadian Rhythm of Spinal Mechanonociception through Dopamine Receptors and Clock Gene Expression. *Life* **2022**, *12*, 1411. [CrossRef]
22. Schilling, C.; Schredl, M.; Strobl, P.; Deuschle, M. Restless legs syndrome: Evidence for nocturnal hypothalamic-pituitary-adrenal system activation. *Mov. Disord. Off. J. Mov. Disord. Soc.* **2010**, *25*, 1047–1052. [CrossRef]
23. Damiola, F.; Le Minh, N.; Preitner, N.; Kornmann, B.; Fleury-Olela, F.; Schibler, U. Restricted feeding uncouples circadian oscillators in peripheral tissues from the central pacemaker in the suprachiasmatic nucleus. *Genes Dev.* **2000**, *14*, 2950–2961. [CrossRef] [PubMed]
24. Walters, A.S.; LeBrocq, C.; Dhar, A.; Hening, W.; Rosen, R.; Allen, R.P.; Trenkwalder, C. Validation of the International Restless Legs Syndrome Study Group rating scale for restless legs syndrome. *Sleep Med.* **2003**, *4*, 121–132. [CrossRef]
25. Jimenez-Jimenez, F.J.; Alonso-Navarro, H.; Garcia-Martin, E.; Agundez, J.A.G. Inflammatory factors and restless legs syndrome: A systematic review and meta-analysis. *Sleep Med. Rev.* **2023**, *68*, 101744. [CrossRef] [PubMed]
26. Gottlieb, D.J.; Somers, V.K.; Punjabi, N.M.; Winkelman, J.W. Restless legs syndrome and cardiovascular disease: A research roadmap. *Sleep Med.* **2017**, *31*, 10–17. [CrossRef] [PubMed]
27. Earley, C.J.; Jones, B.C.; Ferre, S. Brain-iron deficiency models of restless legs syndrome. *Exp. Neurol.* **2022**, *356*, 114158. [CrossRef]
28. Link, W.; Ferreira, B.I. FOXO Transcription Factors: A Brief Overview. *Methods Mol. Biol.* **2025**, *2871*, 1–8. [CrossRef]
29. Guo, N.; Wang, X.; Xu, M.; Bai, J.; Yu, H.; Le, Z. PI3K/AKT signaling pathway: Molecular mechanisms and therapeutic potential in depression. *Pharmacol. Res.* **2024**, *206*, 107300. [CrossRef]
30. Dixon, S.J.; Olzmann, J.A. The cell biology of ferroptosis. *Nat. Rev. Mol. Cell Biol.* **2024**, *25*, 424–442. [CrossRef]
31. Tang, D.; Chen, X.; Kang, R.; Kroemer, G. Ferroptosis: Molecular mechanisms and health implications. *Cell Res.* **2021**, *31*, 107–125. [CrossRef] [PubMed]
32. Ma, Y.; Jiang, T.; Zhu, X.; Xu, Y.; Wan, K.; Zhang, T.; Xie, M. Efferocytosis in dendritic cells: An overlooked immunoregulatory process. *Front. Immunol.* **2024**, *15*, 1415573. [CrossRef] [PubMed]
33. Walters, A.S.; Paueksakon, P.; Adler, C.H.; Moussouttas, M.; Weinstock, L.B.; Spruyt, K.; Bagai, K. Restless Legs Syndrome Shows Increased Silent Postmortem Cerebral Microvascular Disease with Gliosis. *J. Am. Heart Assoc.* **2021**, *10*, e019627. [CrossRef]
34. Schormair, B.; Zhao, C.; Bell, S.; Didriksen, M.; Nawaz, M.S.; Schandra, N.; Stefani, A.; Hogl, B.; Dauvilliers, Y.; Bachmann, C.G.; et al. Genome-wide meta-analyses of restless legs syndrome yield insights into genetic architecture, disease biology and risk prediction. *Nat. Genet.* **2024**, *56*, 1090–1099. [CrossRef]
35. Para, K.S.; Chow, C.A.; Nalamada, K.; Kakade, V.M.; Chilakamarri, P.; Louis, E.D.; Koo, B.B. Suicidal thought and behavior in individuals with restless legs syndrome. *Sleep Med.* **2019**, *54*, 1–7. [CrossRef]
36. Zhu, Y.; Fujimaki, M.; Rubinsztein, D.C. Autophagy-dependent versus autophagy-independent ferroptosis. *Trends Cell Biol.* **2025**. [CrossRef] [PubMed]
37. Bass, J.; Takahashi, J.S. Circadian integration of metabolism and energetics. *Science* **2010**, *330*, 1349–1354. [CrossRef]

38. Ying-Hao, P.; Yu-Shan, Y.; Song-Yi, C.; Hua, J.; Peng, Y.; Xiao-Hu, C. Time of day-dependent alterations of ferroptosis in LPS-induced myocardial injury via Bmal-1/AKT/Nrf2 in rat and H9c2 cell. *Heliyon* **2024**, *10*, e37088. [CrossRef]
39. Xu, S.; Tang, Q.; Du, H.; Xie, J.; He, R.; Wang, R.; Sun, Q. Mechanism of Circadian Regulation in Ferroptosis of the BMAL1/NRF2 Pathway in Renal Ischemia-Reperfusion. *Biomedicines* **2025**, *13*, 1375. [CrossRef]
40. Walters, A.S.; Zee, P.C. Why the worsening at rest and worsening at night criteria for Restless Legs Syndrome are listed separately: Review of the circadian literature on RLS and suggestions for future directions. *Front. Neurol.* **2023**, *14*, 1153273. [CrossRef]
41. Liu, Y.; Wang, Y.; Liu, J.; Kang, R.; Tang, D. The circadian clock protects against ferroptosis-induced sterile inflammation. *Biochem. Biophys. Res. Commun.* **2020**, *525*, 620–625. [CrossRef] [PubMed]
42. Liguori, C.; Holzknecht, E.; Placidi, F.; Izzi, F.; Mercuri, N.B.; Hogl, B.; Stefani, A. Seasonality of restless legs syndrome: Symptom variability in winter and summer times. *Sleep Med.* **2020**, *66*, 10–14. [CrossRef] [PubMed]
43. Lincoln, G. A brief history of circannual time. *J. Neuroendocrinol.* **2019**, *31*, e12694. [CrossRef]
44. Garcia-Borreguero, D.; Aragon, A.G.; Moncada, B.; Romero, S.; Granizo, J.J.; Quintas, S.; Castillo, M. Treatment of Sleep, Motor and Sensory Symptoms with the Orexin Antagonist Suvorexant in Adults with Idiopathic Restless Legs Syndrome: A Randomized Double-Blind Crossover Proof-of-Concept Study. *CNS Drugs* **2024**, *38*, 45–54. [CrossRef]
45. Mogavero, M.P.; Fowowe, M.; Sanni, A.; Goli, M.; Lanza, G.; L'Episcopo, F.; Ferini-Strambi, L.; Mechref, Y.; Ferri, R. Evidence of Involvement of the Calcitonin Gene-Related Peptide in Restless Legs Syndrome. *Mov. Disord. Off. J. Mov. Disord. Soc.* **2025**, *40*, 1148–1159. [CrossRef]
46. Imai, N. Molecular and Cellular Neurobiology of Circadian and Circannual Rhythms in Migraine: A Narrative Review. *Int. J. Mol. Sci.* **2023**, *24*, 10092. [CrossRef]
47. Ryan, S.K.; Ugalde, C.L.; Rolland, A.S.; Skidmore, J.; Devos, D.; Hammond, T.R. Therapeutic inhibition of ferroptosis in neurodegenerative disease. *Trends Pharmacol. Sci.* **2023**, *44*, 674–688. [CrossRef] [PubMed]

biomolecules

MDPI

Article

Field-Effect Transistor Based on Nanocrystalline Graphite for DNA Immobilization

Bianca Adiaconita, Eugen Chiriac *, Tiberiu Burinaru, Catalin Marculescu, Cristina Pachiu, Oana Brincoveanu, Octavian Simionescu and Marioara Avram *

National Institute for Research and Development in Microtechnologies—IMT Bucharest, 126A Erou Iancu, Nicolae, 077190 Voluntari, Ilfov, Romania; bianca.adiaconita@imt.ro (B.A.); tiberiu.burinaru@imt.ro (T.B.); octavian.simionescu@imt.ro (O.S.)

* Correspondence: eugen.chiriac@imt.ro (E.C.); marioara.avram@imt.ro (M.A.)

Abstract: In recent years, field-effect transistors (FETs) based on graphene have attracted significant interest due to their unique electrical properties and their potential for biosensing and molecular detection applications. This study uses FETs with a nanocrystalline graphite (NCG) channel to detect DNA nucleobases. The exceptional electronic properties of NCG, and its high surface area, enable strong π–π stacking interactions with DNA nucleobases, promoting efficient adsorption and stabilization of the biomolecules. The direct attachment of nucleobases to the NCG channel leads to substantial changes in the device's electrical characteristics, which can be measured in real time to assess DNA binding and sequence recognition. This method enables highly sensitive, label-free DNA detection, opening up new possibilities for rapid genetic analysis and diagnostics. Understanding the interactions between DNA nucleobases and graphene-based materials is crucial for advancing genetic research and biotechnology, paving the way for more accurate and efficient diagnostic tools.

Keywords: field-effect transistor; graphene-related material; nucleobase; Dirac point; mobility; DNA technologies

1. Introduction

DNA is the foundation material of biological heredity because it carries the genetic information used in controlling all biological processes [1,2]. The detection of abnormal concentrations of nucleic acids can be used for evaluating the occurrence and progress of different pathological conditions. Therefore, nucleic acid biomarkers could be employed for the early detection of disease. They can also be employed for lab-on-a-chip (LOC), point-of-care detection (POC), precision medicine, and personalized therapy. There are different types of nucleic acids present in body fluids, such as cell-free DNA (cfDNA) and circulating tumor DNA (ctDNA). Of the two, ctDNA is the most widely searched tumor-related element in clinical applications to detect minimal residual disease, disease recurrence, and targetable genomic alterations, and to detect and monitor the emergence of resistance mechanisms [3,4]. Nucleic acids can also be used for rapid pathogen detection, which is an emerging issue in the clinical, environmental, and food industry sectors. Multidrug resistant bacteria like Enterococcus faecium, Staphylococcus aureus, Klebsiella pneumoniae, Acinetobacter baumannii, Pseudomonas aeruginosa, Enterobacter species (ESKAPE), and some foodborne pathogens such as Listeria monocytogenes, Salmonella spp., and Campylobacter spp. are responsible for 700,000 deaths per year globally and have a significant economic impact [5]. Additionally, the recent COVID-19 pandemic has demonstrated that there is an urgent need for fast, accurate, and cheap diagnosis methods [6]. It has been estimated that health issues caused by ESKAPE organisms generated 55 billion USD in excess direct and societal costs per year in the United States (US). In the European Union (EU) and European Economic Area (EEA) countries, healthcare costs sum up to 1.1–1.5 billion EUR yearly [7]. The global economic impact of COVID-19 was estimated to be between 77 billion and 2.7 trillion USD in 2019 [8].

In the last few years, field-effect transistors (FETs) have attracted much attention in the area of bio-detection due to their sensitivity in detecting molecular interactions. The FET-based biosensor relies on a biomolecular recognition event at the FET gate [9–12]. During specific bio-recognition interactions, the electric charge distribution changes the surface density of the electric charge carriers and, therefore, changes the conductivity of the source–drain channel [10–12]. Graphene is considered an ideal material for the construction of FET biosensors because the bandgap can be tuned by modifying the surface. In a typical graphene-based FET architecture, graphene is deposited/transferred on a Si substrate with a 300 nm SiO_2 layer. The doped Si substrate acts as a back gate, which induces a charge density at the surface and hence changes the Fermi energy level in the graphene layer [13]. So far, several reports have demonstrated that graphene-based FETs with electrolytic top gates can be effectively used for charged molecule detection [14–17]. Compared with electrochemical signal readout, graphene-based FETs utilize electrical sensing to exploit the change in resistivity, mobility, and electric charge distribution when nucleic acids adsorb on the surface of the source–drain channel of the FET, and their small or even nanoscale dimensions allow them to provide better sensing abilities. Thus, these properties make graphene-based FETs a promising instrument for interface studies. The development of the graphene–nucleic acid nano–bio interface is straightforward and easy to tailor due to the self-assembly properties of DNA molecules and the detection specificity that single-stranded DNA (ssDNA) or single-stranded RNA (ssRNA) molecules offer [9,18–25].

Theoretical studies indicate that DNA molecules attach to the graphene surface through $\pi-\pi$ stacking interactions [26,27]. The calculated binding energies are almost identical for adenine, thymine, and cytosine. The hierarchy of binding energies is as follows: G > A ≈ T ≈ C > U (G—guanine, A—adenine, T—thymine, C—cytosine, U—uracil). The authors found that the calculated binding energies were correlated with the polarizabilities of the nucleobases, making polarizability the dominant source of interaction between nucleobases and graphene. Single-stranded DNA possesses a much higher binding capacity to graphene than double-stranded DNA. Shorter strands of DNA adsorb faster and bind more tightly to graphene's surface [26,27]. However, these theoretical studies assumed perfectly flat graphene layers, which differ from the real behavior of graphene layers that naturally tend to roll to achieve structural stability. The appearance of curvatures in the graphene layer can significantly alter the electronic properties and the binding capacity of nucleobases. Theoretical calculations that investigate the binding capacity of nucleobases to curved graphene, such as carbon nanotubes (CNTs), have been performed using high-precision first principle calculations [28]. The results show that the binding energy increases as the curvature decreases and reaches a maximum for planar graphene. However, these studies are based on CNT structures, not on the nonplanarity of graphene. Another study has reported that nucleobases have a stronger binding affinity to graphene with larger radii of curvature compared with the smaller CNTs, as large-size graphene exhibits significant curvature, as observed by quantum chemical calculations and atomic force microscopy (AFM) measurements [29]. It has been reported that the interaction energy increases systematically as the size of the system expands [30]. The curvature may provide additional stability for nucleobase binding on the graphene surface. Studies also show that ssDNA binds to graphene mainly through π–π attachment interactions [31–34]. Zhao et al. performed molecular dynamics (MD) simulations to investigate the self-assembly of double-stranded DNA segments on graphene's surface [35]. They observed that DNA can form two distinct types of assembly. Both of them use π–π stacking as the main binding force onto the graphene layer.

As nucleic acids have a charged phosphate backbone, graphene-based FETs are an ideal tool in nucleic acid/graphene interface property sensing, interfacial studies, and interface-based sensing applications [14–17,36]. DNA probes, aptamers, or CRISPR/Cas9 molecules have been used as bio-receptors for LOC devices [37–39]. Additionally, as DNA molecules can bind directly to graphene, DNA origami molecules can be directly attached to it to develop new types of biosensors or drug delivery platforms [40–42]. Wasfi et al. developed an FET device based on graphite oxide decorated with a trimetallic nanocluster of gold, silver, and platinum for real-time nucleic acid detection with a 1.28 nM limit of detection (LOD) [43]. Ganguli et al. demonstrated that crumpled gFETs can be used to detect E. coli DNA down to zeptomolar (zM) concentrations [44]. Chen et al. fabricated a DNA-functionalized graphene field-effect transistor for quantitation of vascular endothelial growth factor, with a 3.24 pg/mL LOD [24]. Hwang et al. used deformed monolayer graphene channels for the detection of nucleic acids with a detection sensitivity down to 600 zM in buffer and 20 aM in human serum sample [45].

The transfer characteristic curves of BioFET devices, particularly those utilizing graphene and graphene-related field-effect transistors (GFETs), are crucial for understanding their biosensing applications. Studies indicate that DNA immobilization significantly affects the electrical characteristics of these devices and highlight the importance of reducing defects and preserving graphene's properties to maintain stable transfer characteristics during DNA hybridization detection [46]. Similarly, Jia and Ju used simulations to demonstrate how ssDNA immobilization and hybridization affect GFETs' performance [47]. They found that DNA's negative charge alters the Dirac point and modulates conductivity, supporting the device's role in biosensing. The charge accumulation method enhances signal detection by directly transferring charges, improving signal-to-noise ratios [48]. Additionally, Selvarajan et al. modeled GFET transfer characteristics, confirming that doping and environmental factors influence electrical responses, which is vital for optimizing DNA detection [49].

As there are no published works that describe this, we propose to study the interaction of ssDNA probes with nanocrystalline graphite (NCG). In the current paper, we propose to use NCG for the sensing area of an FET device. NCG is an innovative 3D carbonic nanomaterial with excellent electric properties. Additionally, the FETs used in the current paper are a two-gate FET with a back gate and a front gate. By understanding and characterizing the attachment of the DNA samples and how it influences the behavior of the FETs, we hope to further advance the fabrication of NCG-based LOC- and POC-type devices for disease diagnosis and pathogen detection.

This study represents a groundbreaking advancement in the field, as it is the first to report the direct interaction between guanine, adenine, cytosine, and thymine, with NCG and its significant impact on the electrical conductivity of NCG. The NCG–guanine complex demonstrates the highest conductivity among the nucleobases tested, marking a significant leap in understanding the role of nucleobases in modifying graphene-based materials' electronic properties. This pioneering research opens new avenues for the development of guanine-based biosensors, providing insights that could shape the future of bioelectronics and biosensing technologies.

2. Materials and Methods

The cleaning of FET devices, deposition of single-stranded DNA probes, incubation, and washing were carried out at room temperature in a laminar flow hood within an ISO6 cleanroom (class 1000) to minimize surface contamination.

The source–drain channel surface characteristics were evaluated after the immobilization of nucleobases through energy dispersive X-ray spectroscopy (EDX/EDS), at an acceleration voltage of 15 kV, using the Nova NanoSEM 630 system (FEI Company, Hillsboro, OR, USA), which is equipped with an EDX detector (EDAX TEAM™, AMETEK, Inc., Berwyn Pennsylvania, USA).

Structural characterization was performed using Raman spectroscopy, which provides information regarding the shape and type of chemical bonds. Raman spectra were acquired using a Witec Raman spectrometer (Alpha-SNOM 300 S, WiTec GmbH, Germany), utilizing a 532 nm diode-pumped solid-state laser with a maximum power of 145 mW. The incident laser spot size was approximately 1.0 μm, with two objectives attached (50× and 100×) to a Thorlabs MY100X-806 microscope.

Electrical characterization was carried out using the semiconductor characterization system (DC) with the Wafer Probing Station-4200-SCS/C/Keithley Easyprobe EP6/Suss MicroTec.

Single-stranded DNA samples were purchased from Integrated DNA Technologies. We used ssDNA with a length of 20 bases of the same type with 100 μM concentration. We named them poly G (GGG GGG GGG GGG GGG GGG GG), poly A (AAA AAA AAA AAA AAA AAA AA), poly C (CCC CCC CCC CCC CCC CCC CC), and poly T (TTT TTT TTT TTT TTT TTT TT).

Fabrication of the Nanocrystalline Graphite-Based Field-Effect Transistor for Direct Immobilization of Single-Stranded DNA Probes

The NCG FET was fabricated using conventional photolithography on a 4-inch p-doped Si wafer. The wafers were cleaned in Piranha solution by immersion. The FET dimension was 4.7 mm^2, with a sensing area of 100 μm × 100 μm.

The fabrication process began with p-doped silicon wafers, which serve as the substrate, cleaned by immersion in Piranha solution for 30 min. The Si wafers were subjected to a thermal oxidation process in an oxidation furnace at a temperature of 900 °C for 340 min. The NCG channel between the source and drain was synthesized via plasma-enhanced chemical vapor deposition (PECVD) at a substrate temperature of ≈ 900 °C, with a growth time of 75 min, in a gas flow of CH_4 (60 sccm) and H_2 (75 sccm). The NCG film is deposited on the full wafer surface and patterned via photolithography. A thick positive photoresist (AZ4562) is used to transfer the pattern from the photomask, as the NCG film is etched via reactive ion etching (RIE) in an O_2 plasma. The removal of the photoresist after RIE takes place in acetone and isopropyl alcohol. The source and drain electrodes are fabricated/patterned via the lift-off process: a negative photoresist (LOR 5A) and a positive photoresist (AZ1518) are spin-coated onto the substrate, followed by UV exposure through a specific photomask. A thin film of Cr/Au (10/300 nm) is then deposited, and the lift-off process is carried out in acetone with ultrasonic treatment to remove both the metal and the photoresist, resulting in the formation of the source–drain electrodes. The front gate fabrication is also accomplished via a lift-off procedure: after the deposition of two sacrificial photoresists, followed by alignment and exposure to a photomask with the front gate design, a thin aluminum oxide (Al_2O_3) film, with a thickness of 10 nm, is deposited at 120 °C via atomic layer deposition (ALD). This film serves to isolate the gate contact. This is followed by the deposition of a 10 nm thick Cr layer and a 300 nm thick Au layer, with subsequent electrode patterning using the lift-off technique. The final step is dicing the wafers into individual sensors, according to the device geometry.

Figure 1a,b display a schematic representation of the functionalized NCG-based FET and an optical microscopy image of the FET device, respectively, while Figure 1c showcases the experimental setup for the electrical characterizations. The microfabrication process is more extensively presented in a previous article [50]. Figure 1d represents the transfer characteristic curves of an NCG-FET after the direct immobilization of the different DNA nucleobases (Poly G, poly T, poly A, and poly C). The immobilization of different nucleobases shifts the transfer curves, affecting the drain current (I_d) and shifting the Dirac point—Poly G (Red) and poly T (Blue) show relatively similar shifts. These shifts occur due to the varying charge distribution and dipole moment of the DNA nucleobases interacting with the graphene surface, affecting the charge carrier concentration and transport properties of the NCG-FET. The shift direction and magnitude depend on the specific nucleobase's electronic properties and its interaction with the graphene-based sensing layer.

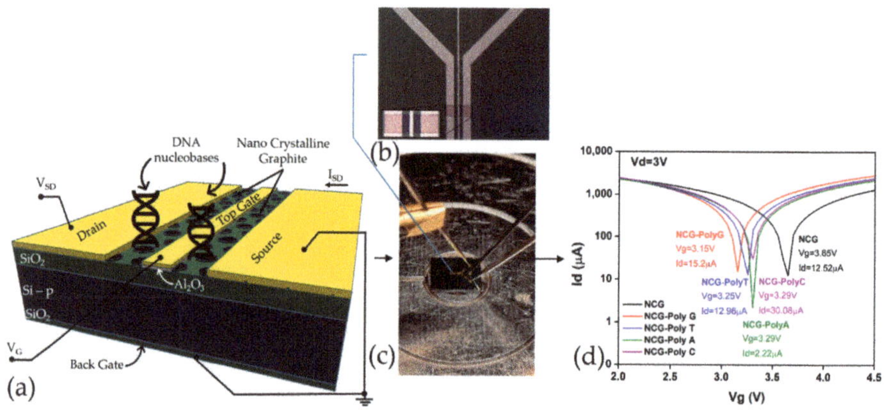

Figure 1. (**a**) Schematic representation of the NCG-based FET device depicting the NCG channel and the 4 electrodes. (**b**) Optical microscopy image of the NCG-FET at 5× magnification with an enhanced view of the NCG channel (inset). (**c**) Experimental setup for the electrical characterizations. (**d**) I_d/V_g plot for different nucleotide at V_d = 3 V.

3. Results

Understanding the nucleobase–graphene interactions is essential for advancements in genetic research and biotechnology, as they play a critical role in maintaining genetic accuracy and participating in various biological processes such as energy metabolism and cell signaling. Direct detection of these nucleobases on NCG surfaces represents a significant step toward the development of advanced biosensors. Due to its superior electrical properties, NCG provides an optimal platform for such applications. While existing literature primarily reports theoretical data on the direct interaction between graphene and nucleotide bases, our study presents novel experimental findings, demonstrating promising results for the direct detection of DNA. When integrated into an FET, this approach enables rapid and highly sensitive nucleobase detection, paving the way for high-performance biosensors with applications in genetics, medical diagnostics, and biotechnology.

Direct Adsorption of Nucleobase on NCG

Based on the theory that DNA interacts with the graphene surface through various mechanisms, such as π–π attachment, hydrogen bonding, electrostatic interactions, van der

Waals forces, and hydrophobic interactions, forming spherical particles in the case of poly A and poly C, while poly T and poly G form a network on the surface, the main mechanism we applied in the preliminary studies is free adsorption.

For the attachment of single-stranded DNA probes on the GFET, we aimed to functionalize the NCG channel non-covalently by utilizing π–π interactions between the graphene layers and the nucleotides. We chose this method to avoid altering the structure and properties of the graphene-based material. The hydrophobic interaction between the DNA bases and NCG demonstrated excellent nucleotide base adsorption capabilities without compromising their structural integrity.

First, each sample is cleaned with 3 μL of isopropyl alcohol. Then, the samples are dried at room temperature in a laminar flow hood for 10 min, and the ssDNA specimens are incubated. The immobilization process involves the π–π interaction between the NCG surfaces and the four nucleotide bases (G, A, T, C), each containing 20 nucleotides. We added 5 μL of a 100 μM solution—poly G/poly A/poly T/poly C—onto the surface of the NCG channel. A droplet formed above the channel, and the FET devices were kept at room temperature overnight in the hood. After incubation, the samples were cleaned three times with the specific nucleotide buffer IDTE: 1× TE solution, pH = 8, certified nuclease-free, and stored at room temperature to dry.

The electrical response of the NCG-based FETs for the four types of nucleotides is presented in Figure 2. The transfer characteristics, where the gate voltage (V_g) is swept in the interval [−5 V; 5 V]; the drain voltage is biased at −3 V, 1 V, 2 V, and 3 V; and the drain current is measured, are displayed in Figure 2a,c,e,g. Based on the presented transfer characteristics, the mobility is computed in each case and is showcased in Figure 2b,d,f,h.

The detection of DNA nucleobases using NCG-FETs is achieved by measuring the electrical response following the adsorption of ssDNA samples onto the NCG source–drain channels. The adsorption of nucleobases onto the FET channels induces a leftward shift of the Dirac point. This leftward shift is attributed to n-type doping of NCG, caused by π–π stacking interactions between NCG and the electron-rich nucleobases in DNA molecules. The corresponding values of the Dirac point and its respective shift after ssDNA adsorption are presented for each of the four nucleobases in Table 1. Depending on the applied gate voltage, the Dirac point shifts to the left by 0.45 V and 0.9 V in the case of guanine, by 0.1 V to 0.4 V in the case of thymine, by 0.1 V to 0.46 V in the case of adenine, and by 0.1 V and 0.45 V in the case of cytosine. The Dirac point shift—an indicator of sensitivity—is significant upon nucleobases adsorption compared with the reference NCG-FET.

Table 1. Analysis of Dirac point shift of the NCG-FETs at different drain potentials, after guanine, thymine, adenine, and cytosine adsorption.

		V_d = 1 V	V_d = 2 V	V_d = 3 V	V_d = −3 V
	V_{DP}-NCG1 (V)	1.5	2.7	4.09	−4.05
Guanine	V_{DP}-NCG1 poly G (V)	1.05	2.09	3.15	−3.15
	ΔV_{DP} (V)	**−0.45**	**−0.61**	**−0.94**	**0.9**
	V_{DP}-NCG4 (V)	1.15	2.4	3.65	−3.6
Thymine	V_{DP}-NCG4 poly T (V)	1.05	2.15	3.25	−3.2
	ΔV_{DP} (V)	**−0.1**	**−0.25**	**−0.4**	**0.4**
	V_{DP}-NCG8 (V)	1.2	2.4	3.75	−3.7
Adenine	V_{DP}-NCG8 poly A (V)	1.1	2.2	3.29	−3.29
	ΔV_{DP} (V)	**−0.1**	**−0.2**	**−0.46**	**0.41**
	V_{DP}-NCG12 (V)	1.2	2.45	3.7	−3.7
Cytosine	V_{DP}-NCG12 poly C (V)	1.1	2.2	3.3	−3.25
	ΔV_{DP} (V)	**−0.1**	**−0.25**	**−0.4**	**0.45**

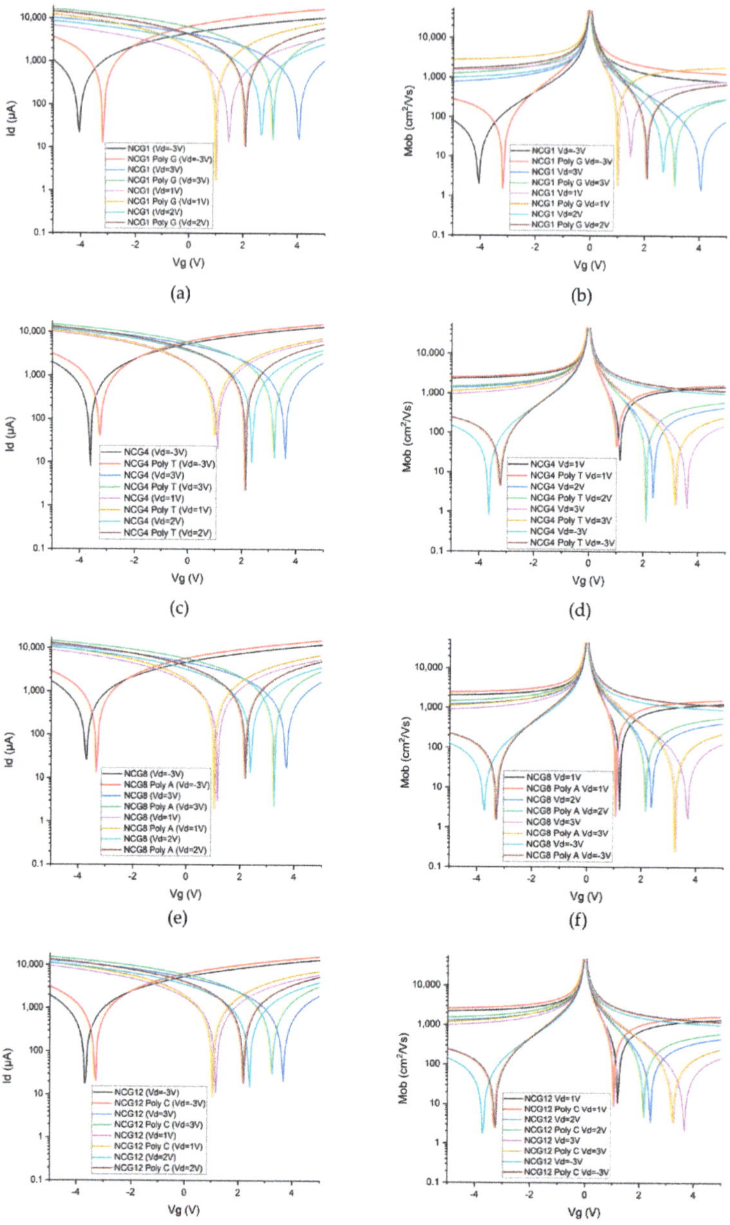

Figure 2. I–V characteristics of the NCG-FETs before and after (**a**) guanine, (**c**) thymine, (**e**) adenine, and (**g**) cytosine adsorption/direct interaction. Mobility of the NCG-FETs before and after (**b**) guanine, (**d**) thymine, (**f**) adenine, and (**h**) cytosine adsorption at a drain potential of (top-left) 1 V, (top-right) 2 V, (bottom-left) 3 V, and (bottom-right) −3 V.

From an EDAX perspective (Table 2), the adsorption of guanine, thymine, adenine, and cytosine onto the NCG channel is confirmed by the presence of carbon (C), oxygen (O), and nitrogen (N). Guanine exhibits significant charge transfer at the NCG interface, associated with high interaction energy due to π-π interactions playing a crucial role, as well as the presence of both amino ($-NH_2$) and carbonyl ($=O_6$) groups, which can interact with the π-electron cloud at the NCG surface. For thymine, the nitrogen signal intensity is lower than that observed for guanine, which is expected given that guanine is a stronger nitrogenous base. Conversely, an increase in oxygen presence is observed, as thymine contains two oxygen atoms in its chemical structure, whereas guanine has only one. Adenine shows a more balanced composition, with a notable increase in nitrogen, while in the case of cytosine, the nitrogen presence is lower, which is likely due to both the lower number of nitrogen atoms in cytosine (three nitrogen atoms) and the fact that cytosine is a weaker base.

Table 2. Elemental distribution extracted from EDAX measurements of the NCG FETs' channel after poly G, poly T, poly A, and poly C adsorption.

	Atoms	Weight %	Atomic %	Net. Int	Error %
	C	34.47	41.17	408.10	5.28
Poly G	N	0.64	0.65	2.92	68.46
	O	64.89	58.18	641.85	6.56
	C	20.44	25.49	535.7	11.12
Poly T	N	0.06	0.07	1	99.99
	O	79.54	74.4	2326.10	7.17
	C	37.54	44.45	319.37	5.21
Poly A	N	0.21	0.22	0.64	0.0007
	O	62.62	55.34	421.42	6.90
	C	33.47	40.12	248.06	5.69
Poly C	N	0.17	0.17	0.49	90.67
	O	66.36	59.71	423.55	6.74

The presence of nucleobases can either enhance or attenuate NCG-specific Raman peaks (Figure 3), which is expected due to electrostatic interactions and hydrogen bond formation. The characteristic NCG Raman peaks, such as D (≈1350 cm^{-1}), G (≈1580 cm^{-1}), 2D (≈2680 cm^{-1}), and D + D′ (≈2980 cm^{-1}), may experience frequency shifts in the presence of nucleobases. These shifts indicate changes in the electronic structure or graphene lattice dynamics.

The Raman spectra acquired from the NCG-FETs after poly G, poly A, and poly C adsorption reveal the appearance of a 1152 cm^{-1} peak, while after poly T the new peak appears at 1143 cm^{-1}. This peak is frequently associated with specific vibrational modes of chemical bonds in organic molecules, including nucleobases. This peak may be related to C–N bond vibrations common in nucleobase structures, C–C bond vibrations, especially within the aromatic ring, or C–H stretching modes. Additionally, D-band doubling is observed, which may indicate changes in the electronic structure. When the ssDNA is adsorbed onto the NCG surface, it influences the lattice vibrations, leading to D-band splitting. Initially, the D-band appeared at 1336 cm^{-1} for the control sample for poly G, poly T, and poly A, and at 1332 cm^{-1} for the control sample for poly C. After ssDNA adsorption, the D band splits into two peaks at 1332 cm^{-1} and 1349 cm^{-1} for poly G and poly C, 1328 cm^{-1} and 1349 cm^{-1} for poly T, and 1328 cm^{-1} and 1349 cm^{-1} for poly A. The G-band, initially detected at 1594 cm^{-1} in the pristine NCG channels, shifts to 1578 cm^{-1} after ssDNA attachment for all nucleobases. Peaks observed at 1489 cm^{-1} and 1529 cm^{-1} are associated with N–H bond vibrations and other vibrational modes of amino groups. Furthermore, N–H stretching vibrations appear at approximately 3125 cm^{-1} and 3248 cm^{-1}.

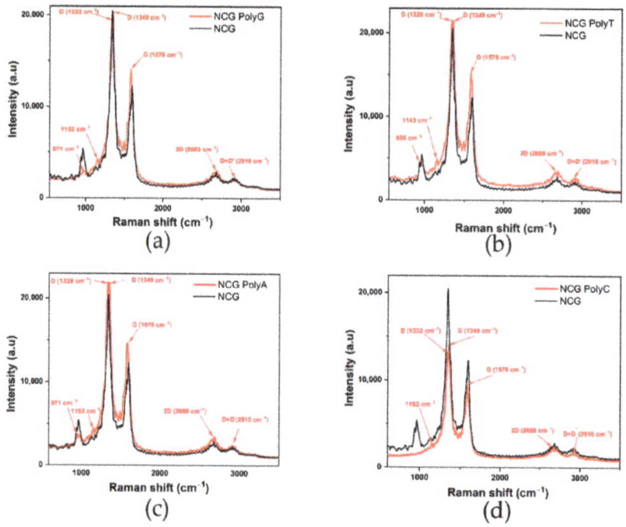

Figure 3. Comparative Raman spectra of the NCG-FETs before and after (**a**) poly G, (**b**) poly T, (**c**) poly A, and (**d**) poly C adsorption.

The results from our experimental study support the conclusions drawn in previous theoretical studies from the literature, which suggest that the presence of guanine significantly enhances the electrical conductivity of pure graphene. Theoretical models indicate that this enhancement is due to guanine's ability to donate electrons to graphene, thus facilitating n doping and promoting easier electron flow. Furthermore, the NCG–guanine complex exhibits the highest conductivity among the nucleobases tested, confirming its superior electrical properties as predicted by theoretical analyses.

4. Discussion

The comparison of the four nucleobases (adenine, thymine, guanine, and cytosine) adsorbed onto the NCG-FETs reveals distinct differences in their interaction with the NCG surface. Figure 4 represents the transfer characteristic curves of the NCG-FET before and after the direct immobilization of different DNA nucleobases (poly G, poly T, poly A, and poly C) at a 3 V drain potential, which displayed the highest shift in the Dirac point. Poly G and poly T show relatively similar drain currents at the Dirac point. These shifts occur due to the varying charge distribution and dipole moment of the DNA nucleobases interacting with the NCG surface, affecting the charge carrier concentration and transport properties of the NCG-FET. The shift direction and magnitude depend on the specific nucleobase's electronic properties and its interaction with the NCG-based sensing layer.

Both adenine and guanine, which are purine bases, exhibit the strongest interactions with NCG, resulting in significant n-doping effects. The Dirac point shift, a key indicator of doping, is most pronounced for these two bases, with shifts reaching up to 0.46 V for adenine and even greater for guanine. This is attributed to the electron-rich nature of purines, particularly their nitrogen atoms, which participate in π–π stacking interactions with the NCG surface. These strong interactions lead to significant changes in the electronic structure of NCG, as evidenced by the pronounced Dirac point shifts and enhanced electron flow through the NCG upon adsorption.

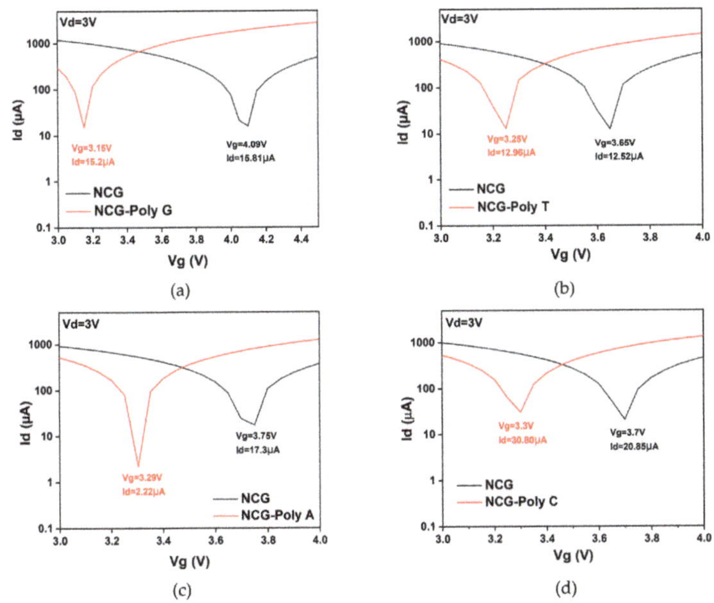

Figure 4. Transfer characteristic curves of the NCG-FET before and after (**a**) guanine, (**b**) thymine, (**c**) adenine, and (**d**) cytosine adsorption. Drain current (I_d) vs. gate potential (V_g) plot for the different nucleotides at a fixed drain potential of $V_d = 3$ V.

In contrast, thymine and cytosine, the pyrimidine bases, induce weaker electrical interactions with the NCG surface. The Dirac point shift in these bases is more moderate, ranging from 0.1 V to 0.4 V for thymine and up to 0.45 V for cytosine. These shifts indicate a weaker n-doping effect compared with purines, as these bases have fewer nitrogen atoms and exhibit less electron-donating behavior. The interactions are still detectable, but they do not induce a significant change in the NCG electronic structure. The electronic sensitivity of NCG in nucleobase adsorption studies correlates directly with the size of the Dirac point shift. Purine bases induce more significant shifts in the Dirac point, making them more effective in modulating the conductivity of NCG, which is crucial for biosensing applications.

These findings indicate that FET devices can effectively differentiate DNA nucleobases based on their distinct electronic interactions, providing a pathway for highly selective and sensitive biosensing applications.

The transconductance behavior (Figure 5) of the NCG-FET devices shows notable changes after DNA nucleobase immobilization, indicating interactions between the nucleobases and the NCG channel. The variations in transconductance suggest that different nucleobases influence charge transport differently, likely due to their distinct electronic properties, indicating specific interactions between each nucleobase and the NCG material. DFT studies show that poly G has the lowest ionization potential (≈7.75 eV), meaning it donates electrons more easily than other nucleobases. This matches the higher transconductance observed experimentally NCG-poly G [51].

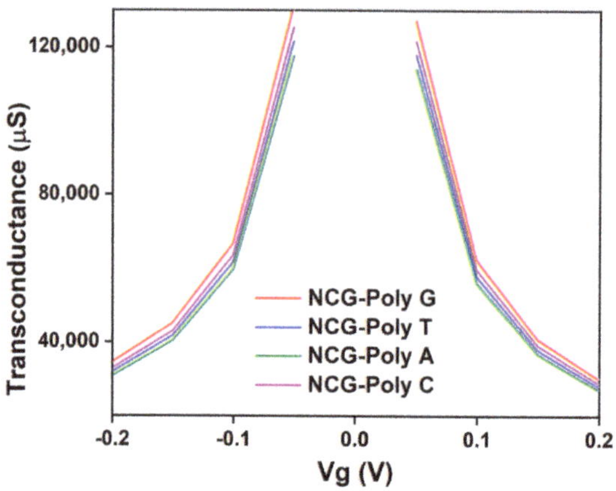

Figure 5. Transconductance of the NCG-FET device after DNA nucleobases immobilization on the NCG channel.

Poly G exhibits the strongest interaction with the NCG surface, as it contains a high percentage of oxygen (≈64.89 wt%), contributing to significant π–π stacking interactions with NCG. The high intensity of oxygen-related peaks in the EDAX analysis (641.85) further suggests a robust interaction with the NCG surface. This is also supported by its substantial shift in the electrical characteristics of the device, which indicates strong doping effects and charge transfer to the top graphenic layers. Poly T, while containing a high percentage of oxygen (≈79.54 wt%), has a lower overall carbon content (≈20.44 wt%), resulting in weaker π–π stacking interactions compared with guanine. Its higher error margin (11.12%) for carbon and low nitrogen content (≈0.06 wt%) indicate a less pronounced interaction with the NCG surface, which is reflected in the more moderate changes in the device's electrical properties. Adenine shows a more balanced composition of carbon (≈37.54 wt%) and oxygen (≈62.62 wt%) but with a lower nitrogen content (≈0.21 wt%), which results in moderate interactions with the NCG channel. The device shows notable changes in electrical behavior, though not as strong as poly G, indicating less effective electron donation to the NCG surface. Cytosine, with a composition similar to guanine (≈33.47 wt% carbon and ≈66.36 wt% oxygen), exhibits moderate interactions with the NCG surface. However, the nitrogen content (≈0.17 wt%) is slightly lower than in guanine, leading to a weaker doping effect. This results in less pronounced modifications in the electrical properties, as seen in the relatively lower net intensity values for carbon and nitrogen (248.06 and 0.49, respectively). These data confirm that guanine induces the most significant modification in the electronic properties of the NCG channel, while thymine and cytosine show weaker interactions, highlighting the important role of nitrogen and oxygen content in modulating the performance of graphene-based biosensors.

The Raman spectra (Figure 6) also confirm these findings, showing D-band splitting and G-band shifts in all cases, with the purine bases exhibiting more pronounced shifts compared with the pyrimidine bases. These results emphasize the stronger π–π stacking interactions and n-doping effects in adenine and guanine, while thymine and cytosine cause less pronounced structural changes in the graphene lattice.

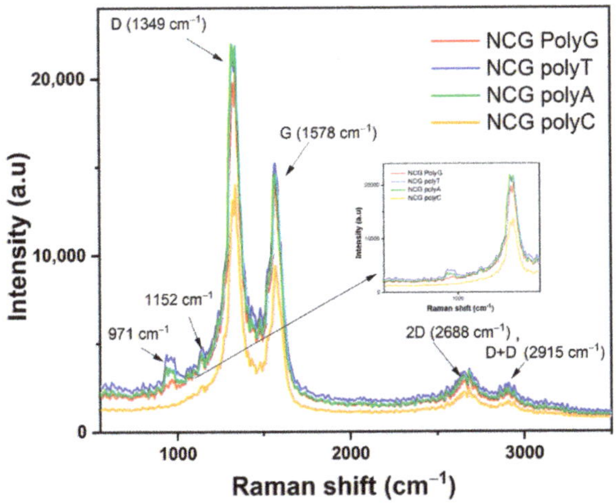

Figure 6. Raman spectra after direct immobilization of DNA nucleobases.

5. Conclusions

Each DNA nucleobase contributes differently to the detection sensitivity of the NCG-FET devices based on its interaction with the NCG channel. The differences in interaction, observed through changes in the Dirac point, transconductance, and Raman spectra, underline the ability of NCG-FETs to differentiate between nucleobases.

Guanine induces the most significant changes in the NCG-FET response, with a pronounced shift in the Dirac point and significant modifications in the Raman spectra and transconductance. Its electron-rich nature, due to the presence of nitrogen atoms, makes it the most effective nucleobase for modulating the electronic properties of the NCG channel. This strong interaction increases the sensitivity of the NCG-FET, allowing for the precise detection of guanine in DNA sequences. Therefore, guanine is the most influential base for improving detection sensitivity in NCG-FET-based biosensors.

Adenine also induces measurable changes in the NCG-FET, though less dramatic than guanine. The Dirac point shift and variations in transconductance are moderate, which suggests a weaker interaction with the NCG channel. However, these changes are still sufficient to contribute to the detection of adenine in a DNA sequence, though with lower sensitivity when compared with guanine. Adenine provides a valuable but less impactful contribution to the detection capabilities of the NCG-FET.

Both thymine and cytosine show relatively weak interactions with the NCG channel, resulting in smaller shifts in the Dirac point and less significant changes in transconductance and Raman spectra. These weaker interactions indicate that thymine and cytosine have a limited impact on the electrical properties of the NCG-FET, leading to lower sensitivity in detecting these bases. However, they still contribute to the overall ability of the NCG-FET to differentiate between nucleobases, though their influence on detection sensitivity is weaker compared with guanine and adenine.

Theoretically, the binding energies of the nucleotide bases are influenced by their electronic structures, with purine bases (guanine and adenine) typically exhibiting stronger binding energies due to their electron-donating properties, which facilitate more robust interactions with graphene or other sensing materials.

In summary, all four nucleobases contribute to the modulation of the NCG channel in FETs. Guanine stands out as having the biggest influence on NCG's electrical properties. Its strong electronic interaction with the NCG channel makes it a key component for precise and sensitive DNA detection. Adenine, thymine, and cytosine also play roles in detection, but their weaker interactions result in less pronounced effects on the NCG-FET's electrical properties, contributing to lower sensitivity for detecting these bases.

Current findings pave the way for integrating NCG into FET biosensors. Each nucleobase has its own electric signature and its influence on the electric properties of NCG. By understanding how each of them interacts with the NCG substrate, different types of sensors can be further developed with applications in DNA-based diagnosis, single-nucleotide polymorphism or mutation detection, and even DNA sequencing. Furthermore, artificial intelligence can be trained to analyze the electric signals generated by an NCG-FET sensor so that heterogeneous DNA sequence detection or DNA sequencing is enabled. NCG can also be coupled with CRISPR/Cas molecules to develop new types of biosensors or delivery vectors.

Author Contributions: Conceptualization, M.A. and T.B.; methodology, B.A., M.A., T.B. and E.C.; validation, M.A.; formal analysis, B.A., E.C., T.B., C.M. and O.S.; investigation, B.A., E.C., T.B., C.P. and O.B.; resources, M.A.; writing—original draft preparation, B.A., E.C., T.B. and C.M.; writing—review and editing, B.A., M.A., E.C. and O.S.; supervision, M.A.; project administration, M.A.; funding acquisition, M.A. All authors have read and agreed to the published version of the manuscript.

Funding: This work was funded from the project "National Platform for Semiconductor Technologies", contract no. G 2024-85828/390008/27.11.2024, SMIS code 304244, co-funded by the European Regional Development Fund under the Program for Intelligent Growth, Digitization, and Financial Instruments. This research was funded as well by Romanian Ministry of Research, PNCDI-UEFISCDI, in the "Core program from the Research, Development and Innovation National Plan 2022–2027", conducted with MCID aid, contract no. 8 N/2023, project no. 2307.

Institutional Review Board Statement: Not applicable.

Informed Consent Statement: Not applicable.

Data Availability Statement: The data presented in this study are available on request from the corresponding author.

Acknowledgments: The authors greatly acknowledge the support of Adrian Albu, for the electrical characterizations, and of Cătălin Pârvulescu, for the photolithography processes.

Conflicts of Interest: The authors declare no conflicts of interest. The funders had no role in the design of the study; in the collection, analyses, or interpretation of data; in the writing of the manuscript; or in the decision to publish the results.

References

1. Leslie, A. Pray Discovery of DNA Structure and Function: Watson and Crick. *Nat. Educ.* **2008**, *1*, 100.
2. Dwivedi, S.D.; Yadav, S.D.; Sahu, D.; Singh, D.; Singh, M.R. Revolutionizing Genetic Diagnostics: Innovative Techniques for Inherited Disease Detection. *Gene Rep.* **2024**, *36*, 101963. [CrossRef]
3. Cisneros-Villanueva, M.; Hidalgo-Pérez, L.; Rios-Romero, M.; Cedro-Tanda, A.; Ruiz-Villavicencio, C.A.; Page, K.; Hastings, R.; Fernandez-Garcia, D.; Allsopp, R.; Fonseca-Montaño, M.A.; et al. Cell-Free DNA Analysis in Current Cancer Clinical Trials: A Review. *Br. J. Cancer* **2022**, *126*, 391–400. [CrossRef] [PubMed]
4. Fernandes, M.G.O.; Cruz-Martins, N.; Machado, J.C.; Costa, J.L.; Hespanhol, V. The Value of Cell-Free Circulating Tumour DNA Profiling in Advanced Non-Small Cell Lung Cancer (NSCLC) Management. *Cancer Cell Int.* **2021**, *21*, 675. [CrossRef] [PubMed]
5. Vidic, J.; Manzano, M. Electrochemical Biosensors for Rapid Pathogen Detection. *Curr. Opin. Electrochem.* **2021**, *29*, 100750. [CrossRef]
6. Kevadiya, B.D.; Machhi, J.; Herskovitz, J.; Oleynikov, M.D.; Blomberg, W.R.; Bajwa, N.; Soni, D.; Das, S.; Hasan, M.; Patel, M.; et al. Diagnostics for SARS-CoV-2 Infections. *Nat. Mater.* **2021**, *20*, 593–605. [CrossRef]
7. Zhen, X.; Lundborg, C.S.; Sun, X.; Hu, X.; Dong, H. Economic Burden of Antibiotic Resistance in ESKAPE Organisms: A Systematic Review. *Antimicrob. Resist. Infect. Control* **2019**, *8*, 137. [CrossRef]

8. Faramarzi, A.; Norouzi, S.; Dehdarirad, H.; Aghlmand, S.; Yusefzadeh, H.; Javan-Noughabi, J. The Global Economic Burden of COVID-19 Disease: A Comprehensive Systematic Review and Meta-Analysis. *Syst. Rev.* **2024**, *13*, 68. [CrossRef]
9. Murugan, P.; Nagarajan, R.D.; Shetty, B.H.; Govindasamy, M.; Sundramoorthy, A.K. Recent Trends in the Applications of Thermally Expanded Graphite for Energy Storage and Sensors—A Review. *Nanoscale Adv.* **2021**, *3*, 6294–6309. [CrossRef]
10. Wu, H.; Wu, J.; Liu, Y.; Wang, H.; Zou, P. Silver Nanoclusters-Based Fluorescent Biosensing Strategy for Determination of Mucin 1: Combination of Exonuclease I-Assisted Target Recycling and Graphene Oxide-Assisted Hybridization Chain Reaction. *Anal. Chim. Acta* **2020**, *1129*, 40–48. [CrossRef]
11. He, Y.; Xiong, L.-H.; Xing, X.-J.; Tang, H.-W.; Pang, D.-W. An Ultra-High Sensitive Platform for Fluorescence Detection of Micrococcal Nuclease Based on Graphene Oxide. *Biosens. Bioelectron.* **2013**, *42*, 467–473. [CrossRef] [PubMed]
12. Bahri, M.; Shi, B.; Djebbi, K.; Elaguech, M.A.; Zhou, D.; Ben Ali, M.; Tlili, C.; Wang, D. Toward Clean and Crackless Polymer-Assisted Transfer of CVD-Grown Graphene and Its Recent Advances in GFET-Based Biosensors. *Mater. Today Chem.* **2021**, *22*, 100578. [CrossRef]
13. Seo, G.; Lee, G.; Kim, M.J.; Baek, S.-H.; Choi, M.; Ku, K.B.; Lee, C.-S.; Jun, S.; Park, D.; Kim, H.G.; et al. Rapid Detection of COVID-19 Causative Virus (SARS-CoV-2) in Human Nasopharyngeal Swab Specimens Using Field-Effect Transistor-Based Biosensor. *ACS Nano* **2020**, *14*, 5135–5142. [CrossRef]
14. Mensah, K.; Cissé, I.; Pierret, A.; Rosticher, M.; Palomo, J.; Morfin, P.; Plaçais, B.; Bockelmann, U. DNA Hybridization Measured with Graphene Transistor Arrays. *Adv. Healthc. Mater.* **2020**, *9*, 2000260. [CrossRef]
15. Khan, N.I.; Mousazadehkasin, M.; Ghosh, S.; Tsavalas, J.G.; Song, E. An Integrated Microfluidic Platform for Selective and Real-Time Detection of Thrombin Biomarkers Using a Graphene FET. *Analyst* **2020**, *145*, 4494–4503. [CrossRef]
16. Tu, J.; Gan, Y.; Liang, T.; Hu, Q.; Wang, Q.; Ren, T.; Sun, Q.; Wan, H.; Wang, P. Graphene FET Array Biosensor Based on SsDNA Aptamer for Ultrasensitive Hg^{2+} Detection in Environmental Pollutants. *Front. Chem.* **2018**, *6*, 333. [CrossRef] [PubMed]
17. Laxmi, L.; Mahapatra, B.; Krishna, R.V.; Patel, P.K. A Review of Recent Advancements in Graphene Based Field-Effect Transistor Biosensors. *AIP Conf. Proc.* **2021**, *2327*, 20011. [CrossRef]
18. Sun, J.; Chen, W.; Jia, K.; Li, S.; Jia, P.; Wang, W.; Song, Z.; Zhao, X.; Mao, Y.; Chen, S. Progress on the Microwave-Assisted Recycling of Spent Lithium Battery Graphite. *Processes* **2023**, *11*, 1451. [CrossRef]
19. Kawaguchi, N.; Shibata, K.; Mizoguchi, T. Possible New Graphite Intercalation Compounds for Superconductors and Charge Density Wave Materials: Systematic Simulations with Various Intercalants Using a van Der Waals Density Functional Method. *J. Phys. Chem. C* **2023**, *127*, 9833–9843. [CrossRef]
20. Suvarnaphaet, P.; Pechprasarn, S. Graphene-Based Materials for Biosensors: A Review. *Sensors* **2017**, *17*, 2161. [CrossRef]
21. Green, N.S.; Norton, M.L. Interactions of DNA with Graphene and Sensing Applications of Graphene Field-Effect Transistor Devices: A Review. *Anal. Chim. Acta* **2015**, *853*, 127–142. [CrossRef] [PubMed]
22. Jahromi, A.K.; Shieh, H.; Low, K.; Tasnim, N.; Najjaran, H.; Hoorfar, M. Experimental Comparison of Direct and Indirect Aptamer-Based Biochemical Functionalization of Electrolyte-Gated Graphene Field-Effect Transistors for Biosensing Applications. *Anal. Chim. Acta* **2022**, *1222*, 340177. [CrossRef] [PubMed]
23. Wang, Y.; Qu, H.; Wang, R.; Dong, B.; Zheng, L. Label-Free Biosensing of Mercury (II) in Milk Using an Aptamer-Gated Graphene Field-Effect Transistor. *J. Electroanal. Chem.* **2022**, *904*, 115931. [CrossRef]
24. Chen, L.; Li, G.; Yang, A.; Wu, J.; Yan, F.; Ju, H. A DNA-Functionalized Graphene Field-Effect Transistor for Quantitation of Vascular Endothelial Growth Factor. *Sens. Actuators B Chem.* **2022**, *351*, 130964. [CrossRef]
25. Campos, R.; Borme, J.; Guerreiro, J.R.; Machado, G.J.; Cerqueira, M.F.; Petrovykh, D.Y.; Alpuim, P. Attomolar Label-Free Detection of DNA Hybridization with Electrolyte-Gated Graphene Field-Effect Transistors. *ACS Sens.* **2019**, *4*, 286–293. [CrossRef] [PubMed]
26. Gao, Y.; Wang, Y. Interplay of Graphene–DNA Interactions: Unveiling Sensing Potential of Graphene Materials. *Appl. Phys. Rev.* **2024**, *11*, 11306. [CrossRef]
27. Tang, L.; Wang, Y.; Li, J. The Graphene/Nucleic Acid Nanobiointerface. *Chem. Soc. Rev.* **2015**, *44*, 6954–6980. [CrossRef]
28. Umadevi, D.; Sastry, G.N. Quantum Mechanical Study of Physisorption of Nucleobases on Carbon Materials: Graphene versus Carbon Nanotubes. *J. Phys. Chem. Lett.* **2011**, *2*, 1572–1576. [CrossRef]
29. Panigrahi, S.; Bhattacharya, A.; Banerjee, S.; Bhattacharyya, D. Interaction of Nucleobases with Wrinkled Graphene Surface: Dispersion Corrected DFT and AFM Studies. *J. Phys. Chem. C* **2012**, *116*, 4374–4379. [CrossRef]
30. Vijay, D.; Sastry, G.N. Exploring the Size Dependence of Cyclic and Acyclic π-Systems on Cation–π Binding. *Phys. Chem. Chem. Phys.* **2008**, *10*, 582–590. [CrossRef]
31. Wang, L.; Wu, A.; Wei, G. Graphene-Based Aptasensors: From Molecule–Interface Interactions to Sensor Design and Biomedical Diagnostics. *Analyst* **2018**, *143*, 1526–1543. [CrossRef] [PubMed]
32. Liu, B.; Salgado, S.; Maheshwari, V.; Liu, J. DNA Adsorbed on Graphene and Graphene Oxide: Fundamental Interactions, Desorption and Applications. *Curr. Opin. Colloid Interface Sci.* **2016**, *26*, 41–49. [CrossRef]
33. Manna, A.K.; Pati, S.K. Theoretical Understanding of Single-Stranded DNA Assisted Dispersion of Graphene. *J. Mater. Chem. B* **2013**, *1*, 91–100. [CrossRef]

34. Zeng, S.; Chen, L.; Wang, Y.; Chen, J. Exploration on the Mechanism of DNA Adsorption on Graphene and Graphene Oxide via Molecular Simulations. *J. Phys. D. Appl. Phys.* **2015**, *48*, 275402. [CrossRef]
35. Zhao, X. Self-Assembly of DNA Segments on Graphene and Carbon Nanotube Arrays in Aqueous Solution: A Molecular Simulation Study. *J. Phys. Chem. C* **2011**, *115*, 6181–6189. [CrossRef]
36. Hao, R.; Liu, L.; Yuan, J.; Wu, L.; Lei, S. Recent Advances in Field Effect Transistor Biosensors: Designing Strategies and Applications for Sensitive Assay. *Biosensors* **2023**, *13*, 426. [CrossRef] [PubMed]
37. Dkhar, D.S.; Kumari, R.; Malode, S.J.; Shetti, N.P.; Chandra, P. Integrated Lab-on-a-Chip Devices: Fabrication Methodologies, Transduction System for Sensing Purposes. *J. Pharm. Biomed. Anal.* **2023**, *223*, 115120. [CrossRef]
38. Thaisa-nga, R.; Chaumpluk, P. Aptamer Based Simple Plastic Lab-on-a-Chip for Rapid on-site Monitoring of Paraquat Herbicide Residue in Soils. *Mater. Today Proc.* **2023**, *75*, 124–131. [CrossRef]
39. Razavi, Z.; Soltani, M.; Pazoki-Toroudi, H.; Chen, P. CRISPR-Microfluidics Nexus: Advancing Biomedical Applications for Understanding and Detection. *Sens. Actuators A Phys.* **2024**, *376*, 115625. [CrossRef]
40. Khakpour, E.; Salehi, S.; Naghib, S.M.; Ghorbanzadeh, S.; Zhang, W. Graphene-Based Nanomaterials for Stimuli-Sensitive Controlled Delivery of Therapeutic Molecules. *Front. Bioeng. Biotechnol.* **2023**, *11*, 1129768. [CrossRef]
41. Postigo, A.; Marcuello, C.; Verstraeten, W.; Sarasa, S.; Walther, T.; Lostao, A.; Göpfrich, K.; del Barrio, J.; Hernández-Ainsa, S. Folding and Functionalizing DNA Origami: A Versatile Approach Using a Reactive Polyamine. *J. Am. Chem. Soc.* **2025**, *147*, 3919–3924. [CrossRef] [PubMed]
42. Weiden, J.; Bastings, M.M.C. DNA Origami Nanostructures for Controlled Therapeutic Drug Delivery. *Curr. Opin. Colloid Interface Sci.* **2021**, *52*, 101411. [CrossRef]
43. Wasfi, A.; Awwad, F.; Qamhieh, N.; Iratni, R.; Ayesh, A.I. Real-Time Nucleic Acid Detection via Field-Effect Transistor Sensors Based on Graphite Oxide Decorated with Trimetallic Nanocluster of Gold, Silver, and Platinum. *New J. Phys.* **2021**, *23*, 103041. [CrossRef]
44. Ganguli, A.; Faramarzi, V.; Mostafa, A.; Hwang, M.T.; You, S.; Bashir, R. High Sensitivity Graphene Field Effect Transistor-Based Detection of DNA Amplification. *Adv. Funct. Mater.* **2020**, *30*, 2001031. [CrossRef]
45. Hwang, M.T.; Heiranian, M.; Kim, Y.; You, S.; Leem, J.; Taqieddin, A.; Faramarzi, V.; Jing, Y.; Park, I.; van der Zande, A.M.; et al. Ultrasensitive Detection of Nucleic Acids Using Deformed Graphene Channel Field Effect Biosensors. *Nat. Commun.* **2020**, *11*, 1543. [CrossRef] [PubMed]
46. Sung, D.; Koo, J. A Review of BioFET's Basic Principles and Materials for Biomedical Applications. *Biomed. Eng. Lett.* **2021**, *11*, 85–96. [CrossRef] [PubMed]
47. Jia, Y.; Ju, C. Sentaurus Based Modeling and Simulation for GFET's Characteristic for SsDNA Immobilization and Hybridization. *J. Semicond.* **2016**, *37*, 14005. [CrossRef]
48. Shin, D.-Y.; Lee, H.; Kim, S. Improving the Accuracy of Capacitance-to-Frequency Converter by Accumulating Residual Charges. *IEEE Trans. Instrum. Meas.* **2011**, *60*, 3950–3955. [CrossRef]
49. Selvarajan, R.S.; Hamzah, A.A.; Majlis, B.Y. Transfer Characteristics of Graphene Based Field Effect Transistor (GFET) for Biosensing Application. In Proceedings of the 2017 IEEE Regional Symposium on Micro and Nanoelectronics (RSM), Batu Ferringhi, Malaysia, 23–25 August 2017; pp. 88–91. [CrossRef]
50. Simionescu, O.-G.; Avram, A.; Adiaconiţă, B.; Preda, P.; Pârvulescu, C.; Năstase, F.; Chiriac, E.; Avram, M. Field-Effect Transistors Based on Single-Layer Graphene and Graphene-Derived Materials. *Micromachines* **2023**, *14*, 1096. [CrossRef]
51. Mohammadi-Manesh, E.; Mir-Mahdevar, M. Adsorption Behavior of Guanine, Adenine, Thymine, and Cytosine Nucleobases of DNA on Zinc Oxide-Graphene Nanosensor: A DFT Study. *Synth. Met.* **2020**, *267*, 116486. [CrossRef]

biomolecules

MDPI

Article

UPLC-MS/MS High-Risk Screening for Sphingolipidoses Using Dried Urine Spots

Tristan Martineau, Bruno Maranda and Christiane Auray-Blais *

Division of Medical Genetics, Department of Pediatrics, Faculty of Medicine and Health Sciences, Université de Sherbrooke, Centre de Recherche-CHUS, 3001, 12th Avenue North, Sherbrooke, QC J1H 5N4, Canada; tristan.martineau@usherbrooke.ca (T.M.); bruno.maranda@usherbrooke.ca (B.M.)

* Correspondence: christiane.auray-blais@usherbrooke.ca; Tel.: +1-819-346-1110 (ext. 14706); Fax: +1-819-564-5217

Abstract: Background: Early detection of sphingolipidoses is crucial to prevent irreversible complications and improve patient outcomes. The use of urine samples dried on filter paper (DUS) is a non-invasive strategy that simplifies the collection, storage, and shipping of samples compared to using liquid urine specimens. Objectives: (1) Develop and validate a multiplex ultra-performance liquid chromatography–tandem mass spectrometry (UPLC-MS/MS) methodology using DUS to quantify twenty-one lysosphingolipids normalized to creatinine for eight different sphingolipidoses. (2) Establish normal reference values to evaluate the clinical utility of the methodology. Methods: Samples were eluted from a 5 cm filter paper disk (~1 mL of urine) and extracted on Oasis MCX solid-phase extraction cartridges prior to injection in the UPLC-MS/MS system. Results: Urinary lysosphingolipids were stable on DUS at −80 °C and −30 °C for 117 days, at 21.5 °C and 4 °C for at least 26 days, and at 35 °C for 3 days. Globotriaosylsphingosine, glucosylsphingosine, and their analogs were elevated in patients with Fabry disease and Gaucher disease, respectively, compared to controls (p-value < 0.0001). The analysis of related analog profiles suggests a better overall reliability in detecting patients early, especially for Fabry patients. Conclusions: This approach is feasible and might be useful for the early detection, monitoring, and follow-up of patients with sphingolipidoses.

Keywords: tandem mass spectrometry; urine dried on filter paper; lysosomal storage disorders; lysosphingolipids; sphingolipidoses; Fabry disease; Gaucher disease

1. Introduction

Lysosomal storage disorders (LSDs) regroup over 70 inborn errors of metabolism characterized by the excess systemic accumulation of metabolites due to mutated genes encoding lysosomal enzymes, transporters, or enzyme cofactors [1].

Sphingolipidoses, an LSD subgroup, represent several rare diseases caused by a dysfunctional pathway of sphingolipid catabolism. Sphingolipids form an essential lipid class regarding the cellular structure, homeostasis, adhesion, signaling, senescence, development, and death. Their chemical structure is based on a sphingosine backbone comprised of a N-acylated fatty acid with a variable chain length (isoforms) or modifications on the sphingosine moiety (analogs) combined with a carbohydrate group (cerebrosides and gangliosides) or hydrogen (ceramide) in the C-1 hydroxyl [2–8].

The accumulation of a specific sphingolipid leads to different diseases, such as metachromatic leukodystrophy, Gaucher disease, Krabbe disease, Fabry disease, GM1 and GM2 gangliosidoses, and Niemann–Pick disease. They are autosomal recessive genetic disorders, except for Fabry disease, which is X-linked [9].

Gaucher disease (OMIM no. 230800) has an average incidence of 1 in 50,000 births but is more prevalent in some populations, such as the Ashkenazi Jewish population (1 in 800). It is caused by mutations in the *GBA1* gene (chromosome 1 [1q21]), leading to the deficiency of the glucocerebrosidase enzyme (EC: 4.2.1.25), which is essential to the catabolism of glucosylceramide [10]. The cellular accumulation of metabolites, including glucosylceramide, glucosylsphingosine and related analogs in several tissues (bone marrow, spleen, and liver) and biological fluids (urine and plasma) will cause cellular dysfunctions and organ involvement [6,7,11]. There are three major Gaucher disease clinical types (1, 2, 3) and two variant phenotype forms (perinatal-lethal and cardiovascular). Type 1 is the most common and is characterized by several clinical manifestations, such as bone involvement, hepatosplenomegaly, anemia, thrombocytopenia, and lung disease. Compared to type 2 (acute; infantile) and type 3 (subacute; juvenile), Gaucher disease type 1 is non-neuronopathic [10].

Krabbe disease (OMIM no. 245200) is caused by mutations in the *GALC* gene and is characterized by progressive demyelination and the presence of globoid cells [12]. It is caused by the deficiency of galactosylceramidase (EC: 3.2.1.46), which causes the accumulation of galactosylceramide and galactosylsphingosine and leads to the degeneration of oligodendrocytes. These lipids are important for the integrity of the myelin, kidney, and epithelial cells of the small intestine and the colon [12,13]. Elevated levels of galactosylsphingosine in dried blood spots were previously detected [14].

Fabry disease (OMIM no. 301500) is a multisystemic X-linked sphingolipidosis caused by mutations in the *GLA* gene affecting the α-galactosidase A enzyme (EC: 3.2.1.22) activity, which leads to the accumulation of globotriaosylceramide (Gb_3), globotriaosylsphingosine (lyso-Gb_3), and their analogs/isoforms in biological fluids and tissues [5,8,15]. The severity and signs/symptoms are quite variable. The main features are acroparesthesia, angiokeratomas, corneal opacity, cornea verticillata, digestive tract issues and severe complications such as kidney disease, heart failure and stroke. Fabry females may also develop the disease, albeit generally with less severe symptoms compared to male patients [16].

Niemann–Pick disease types A and B (type A: OMIM no. 257200; type B: OMIM no. 607616) are characterized by the deficiency of acid sphingomyelinase (EC: 3.1.4.12), which leads to lipid storage accumulation and to visceral or neurovisceral manifestations. Intermediate forms (types A and B) are caused by mutations in the *SMPD1* gene. Niemann–Pick disease type C (OMIM no. 257220) is a less severe form resulting from mutations in the *NPC1* or *NPC2* gene and affecting the cholesterol trafficking without altering the enzyme activity of acid sphingomyelinase [17]. The clinical manifestation spectrum largely depends on the disease form, but patients may suffer from common neurological signs and symptoms such as gaze palsy, cerebellar ataxia, dysarthria, cataplexy, seizures, dystonia, and progressive dementia [17].

Metachromatic leukodystrophy (OMIM no. 250100) is the most common leukodystrophy, with a prevalence of 1 in 40,000–160,000 worldwide. Due to the deficiency of the lysosomal enzyme arylsulfatase A (EC: 3.1.6.1), sulfatides and sphingolipids accumulate in affected cells and damage the myelin sheath of the central and peripheral nervous systems, causing progressive motor and cognitive involvement [18].

GM1 gangliosidosis (OMIM no. 230500) is caused by mutations in the *GLB1* gene (chromosome 3p21.33), leading to the deficiency of β-galactosidase activity (EC: 3.2.1.23), while GM2 gangliosidosis (Tay–Sachs disease: OMIM no. 272800; Sandhoff disease: OMIM no. 268800) is characterized by mutations in the *HEXA* or *HEXB* gene, respectively, leading to the deficiency of β-hexosaminidase A and B (EC: 3.2.1.52). GM1 gangliosidosis is distinguished by the systemic and central nervous system neurodegeneration phenotype with a severity spectrum categorized as type I (infantile), type II (late infantile), and type III (juvenile). GM2 gangliosidosis is characterized only by central nervous system involvement [19].

There is no treatment available to cure these diseases. However, several approaches were developed for patients affected by sphingolipidoses to reduce or stop disease progression: enzyme replacement therapy (ERT) for Gaucher disease [20] and Fabry disease [21], substrate reduction therapy (SRT) for Gaucher disease [20] and Niemann–Pick C [22], pharmacological chaperone therapy for Gaucher disease [20] and Fabry disease [21], gene therapy for Fabry disease [21,23,24], Gaucher disease [20] and metachromatic leukodystrophy [25], and allogeneic hematopoietic stem cell transplantation for Gaucher disease [20], Krabbe disease [26], and metachromatic leukodystrophy [18]. There is no FDA-approved therapy option for GM1 and GM2 gangliosidoses. Genome editing by CRISPR-Cas9 is an emerging therapy currently in progress and seems promising in terms of curing sphingolipidoses [27]. Biomarker monitoring is important at various stages of therapy development to evaluate the safety and the therapeutic efficiency [28].

Early detection of these disorders using biomarkers is important to treat affected patients in a window of action before the development of irreversible organ damage [29,30]. Newborn screening programs (NBSs) were developed to reduce the diagnostic delays experienced by LSD patients [31,32]. Currently, the gold standard is to measure specific enzyme activities from dried blood spots (DBSs), followed by confirmation tests such as clinical evaluations, enzyme activity measurements in leukocytes/lymphocytes, biomarker quantitation, and mutation analyses [31–33]. However, analytical challenges remain. The analytical accuracy when using DBS can be affected by the limited volume of blood available on the filter paper and the hematocrit effect [34]. Also, patients with late-onset forms may have residual enzyme activity that is not correlated with the genotype and phenotype [31,35,36]. A high number of patients with variants of uncertain significance (VUS) are detected in NBSs [31,36]. It has been reported that among 248,616 newborns screened in the NBS program of Northeast Italy, 22 of 126 confirmed cases were asymptomatic with VUS (3 Gaucher disease, 7 Pompe disease, 10 Fabry disease, 2 mucopolysaccharidosis type I) [31]. Furthermore, even if NBS programs allow early detection of patients, the most appropriate course of action for the follow-up, management, and treatment of these patients still needs to be confirmed [31]. Guidelines for clinicians are updated frequently. Recent expert consensus guidelines recommend the monitoring of specific biomarkers, such as lyso-Gb_3 instead of Gb_3 in plasma, for the evaluation of the pharmacodynamic response and treatment outcomes in patients with the classic form of Fabry disease. Nevertheless, detection of late-onset forms or female Fabry patients is still challenging [37].

Lysosphingolipids, which represent deacylated sphingolipids generated by the acid ceramidase from intra-lysosomal sphingolipid accumulations, are interesting biomarkers for the follow-up and monitoring of LSDs [38,39]. As mentioned above, analog forms of primary lysosphingolipids were found in plasma and urine samples from Fabry disease [5, 8] and Gaucher disease patients [6,7]. In urine samples, the analog levels may be more elevated than the primary lysosphingolipid itself. More studies analyzing larger cohorts of Gaucher and Fabry disease patients are needed to assess the clinical benefits of analyzing the analog profiles. Still, previous studies have shown the advantages of the analog profile analysis in Fabry patients with late-onset mutations [40,41].

Several analytical MS methods were developed to analyze different lysosphingolipids in tissues and biological fluids, especially in plasma samples [42–44]. Some methodologies were developed to analyze the lysosphingolipid profiles using DBSs [45,46]. Some analytical MS methods were developed using urine samples from patients with metachromatic leukodystrophy [47], GM1 gangliosidosis [48], GM2 gangliosidosis [49], and Niemann–Pick disease [50]. We have previously shown that LC-MS sensitivity can ensure the detection and measurement of sphingolipids and lysosphingolipids in urine, including in Gaucher disease and Fabry disease [51,52]. We have already developed a multiplex UPLC-MS/MS methodology to measure lysosphingolipids in liquid urine specimens for the detection of the aforementioned sphingolipidoses, including Krabbe disease [53]. Compared to plasma, it remains true that urine sample collection is non-invasive for patients and can easily be performed at home, but the number of studies using urine samples for analysing lysosphingolipids is limited. Also, home collection of liquid urine samples from the pediatric population can still be tedious, according to many parents [54].

Therefore, a urine filter paper collection approach can be readily used. Existing programs, such as the Provincial Neonatal Urine Screening Program in Sherbrooke, Quebec, Canada, have proven that using urine dried on filter paper allows the detection of inborn errors of metabolism (IEM) in newborns [55]. Until now, more than 3.7 million babies have been screened using DUSs collected at home by parents, with a voluntary compliance ranging from 84–93%. Also, the program has highlighted that it is possible to easily collect samples and analyze 5 cm DUS disks. More recently, a mass spectrometry approach was used to evaluate the feasibility of screening IEM using DUS specimens as part of a sensitive, specific, non-invasive, and low-cost strategy [56]. Furthermore, the DUS approach was previously used to collect LSD patient samples: mucopolysaccharidoses [57], Pompe disease [58], Fabry disease [59], and MLD [60]. Finally, considering that the concept of "hospital at home" and virtual medicine is more accessible than before [61], using DUSs to collect samples from patients may simplify the process, reduce the cost, and facilitate sample shipping by regular mail to the analytical laboratory [62,63].

Considering these previous studies, there is an incentive to develop a new multiplex non-invasive high-risk screening MS approach for the early detection, monitoring, and follow-up of sphingolipidosis-affected pediatric patients, as well as some adult patients, using DUS instead of liquid urine. Therefore, our main objectives were twofold: (1) to develop and validate a multiplex quantitative UPLC-MS/MS methodology using DUS to quantify the levels of 21 lysosphingolipids and their related analogs normalized to creatinine for the detection of eight different sphingolipidoses: metachromatic leukodystrophy, Gaucher disease, Krabbe disease, Fabry disease, GM1 gangliosidosis, GM2 gangliosidosis (Tay–Sachs and Sandhoff diseases) and Niemann–Pick disease; and (2) to establish normal reference values to evaluate the clinical utility of this assay for high-risk screening, monitoring, and follow-up of patients.

2. Materials and Methods

2.1. Ethics Approval

This research project followed the principles outlined in the Declaration of Helsinki for all human experimental investigations and was approved by the Research Ethics Board at the Centre intégré universitaire de santé et de services sociaux de l'Estrie—Centre hospitalier universitaire de Sherbrooke under project ID 2023-4955, which was an extended study of project ID 2021-3435. Urine samples were obtained from consenting participants as part of this latter study [53].

2.2. Urine Sample Collection

Urine samples selected from healthy controls (n = 59), patients diagnosed with specific sphingolipidoses (n = 119), and pathological controls (patients suffering from LSDs other than sphingolipidoses (n = 21)) were used. The available demographic details, such as the treatment status, sex, age, and mutation of participants, are summarized in Tables S1 and S2.

Briefly, the patients recruited for this study according to each disease were as follows: Fabry disease, 97 patients (untreated males (UFMs): n = 19; treated males (TFMs): n = 22); untreated females (UFFs): n = 33; treated females (TFFs): n = 23); Gaucher disease, 14 patients (untreated (UG): n = 9; treated (TG): n = 5); GM1 gangliosidosis: two treated patients; metachromatic leukodystrophy (MLD), four untreated patients; and Niemann–Pick disease type C, one treated patient. Healthy control samples were obtained from healthy participants with no incidental findings (<18 years: n = 26; ≥18 years: n = 34). Samples from patients suffering from other LSDs were used as pathological controls: MPS type I (n = 4); MPS type II (n = 5), MPS type III (n = 2), MPS type IV (n = 6), MPS type VI (n = 2), MPS type VII (n = 1), and Pompe disease (n = 2).

Liquid urine specimens were vortexed for 30 s, and one mL was deposited on a 5 cm Whatman 903 filter paper disk (CF10, GE Healthcare, Chicago, IL, USA). The disks were flat-dried for 4 h at room temperature, then transferred at −30 °C until the UPLC-MS/MS analysis. For the MLD patients, no liquid urine specimens were available and only a quarter of a 5 cm DUS was available per patient for the extraction.

2.3. Reagents

LC/MS-grade water (H_2O), methanol (MeOH), American Chemical Society-grade ammonium formate, o-phosphoric acid (H_3PO_4, 85%), and ammonium hydroxide (NH_4OH, 28–30% purity) were purchased from Fisher Scientific (Hampton, NH, USA). Formic acid (FA, 99+%) was obtained from Thermo Fisher Scientific (Waltham, MA, USA) and LC-MS-grade acetonitrile, ReagentPlus-grade dimethyl sulfoxide (DMSO) (≥99.5%), ammonium formate (Amm. Form.) (≥99.995% trace metals basis) and creatinine (≥98%) were obtained from Millipore Sigma (Burlington, MA, USA). Glucosylsphingosine (d18:1) (psychosine; GluSph), galactosylsphingosine (d18:1) (GalSph), globotriaosylsphingosine (d18:1) (lyso-Gb_3), lyso-sulfatide (d18:1) (lyso-Sulf), lyso-monosialoganglioside GM1 (d18:1) (lyso-GM1), lyso-monosialoganglioside GM2 (d18:1) (lyso-GM2), $^{13}C_6$-glucosylsphingosine (d18:1) ($^{13}C_6$-GluSph), D-erythro sphingosylphosphorylcholine (d18:1) (D-erythro-lysosphingomyelin; lyso-SM) and N-glycinated lyso-sulfatide (d18:1) (N-Gly-lyso-Sulf) were purchased from Cayman Chemical Company (Ann Arbor, MI, USA). D_9-sphingosylphosphorylcholine (d18:1) (d_9-lyso-SM) was purchased from Avanti Polar Lipids (Alabaster, AL, USA). The d_3-creatinine standard (99.8 atom %D) was purchased from CDN Isotopes (Pointe-Claire, QC, Canada). N-glycinated globotriaosylsphingosine (d18:1) (N-Gly-lyso-Gb_3) was synthesized and purified in our laboratory [51]

2.4. Calibration Standards and Quality-Control Preparation

Spiked (S-) and urine (U-) in-house quality controls (QCs) (lower limit of quantification level (LLOQC), low level (LQC) at 3xLLOQC, medium level (MQC) at the midrange of the calibrators and high level (HQC) at the higher range of the calibrators) were used to evaluate the effectiveness and robustness of the method, as described in the Section 2.7 Method Validation. The S-QCs and calibrators were prepared on the day of the analysis, while the U-QCs were prepared before the analysis and stored with the working solutions and urine matrices in the freezer (−30 °C) until analysis. The concentration of the QCs and calibrator working solutions are shown in Table S3.

For creatinine, the S-QCs were freshly prepared before the extraction by spiking 100 µL of working solutions, prepared in water, uniformly on a 5 cm blank filter paper disk, which was flat-dried at room temperature (21.5 °C) for 30 min. Blank filter paper disks (without urine) were used because the urine matrix is not exempt from endogenous creatinine. The U-QCs were made from a pool of urine samples from five healthy controls (three males and two females) to an endogenous creatinine concentration at 15 mmol/L, then diluted with water to obtain the same final concentration ranges as the S-QCs, before depositing 1 mL on a 5 cm blank filter paper disk flat-dried at room temperature (21.5 °C) for at least 4 h. For the creatinine calibrators, eight working solutions prepared in water were spiked as the creatinine S-QCs.

For lysosphingolipids, the S-QCs were prepared similarly to the S-QCs for creatinine, but the working solutions were prepared in 80:20 MeOH:DMSO, sonicated 5 min prior to use and spiked on 5 cm filter paper disks containing 1 mL of urine with an endogenous creatinine level at 7 mmol/L, already dried. The U-QCs were made by spiking the same liquid urine pool used to prepare the creatinine U-MQCs with the same multiplex lysosphingolipid working solutions used for the S-QCs before depositing 1 mL on a 5 cm blank filter paper, as mentioned previously for the creatinine U-QCs. For the lysosphingolipid calibrators, eight working solutions prepared in 80:20 MeOH:DMSO were spiked as the lysosphingolipids S-QCs.

2.5. Extraction of Urine from Filter Paper Disks

Prior to the extraction, the samples, QCs, and calibrators were thawed at room temperature (21.5 °C). A volume of 100 µL of multiplex lysosphingolipid internal standard working solution, containing 50 nmol/L of N-Gly-lyso-Sulf, $^{13}C_6$-GluSph, N-Gly-lyso-Gb_3 and d_9-lyso-SM in 80:20 MeOH:DMSO, and 100 µL of creatinine internal standard working solution, containing 10 mmol/L of d_3-creatinine in water, was deposited on each filter paper and flat-dried at room temperature for 30 min. Afterwards, each 5 cm filter paper disk was folded in half and deposited in a 20 mL glass vial with the fold on top. To elute the compounds, 3 mL of 80:20 MeOH:H_2O was added to the glass vial, covered with parafilm M PM-996 (Bemis Co., Neenah, WI, USA), and agitated for 30 min at 300 rpm on a model G2 gyratory shaker (New Brunswick Scientific Co., Edison, NJ, USA). Then, 1 mL of eluate was transferred to a 13 × 100 mm borosilicate test tube containing 100 µL of H_3PO_4 2% in water. For each acidified eluate, the preparations were loaded on mixed-mode strong cation-exchange cartridges (Oasis MCX, 1 cc, 30 mg, Waters Corp., Milford, MA, USA) previously conditioned with 1200 µL of MeOH followed by 1200 µL of H_3PO_4 2% in water. Each cartridge was washed successively with 1200 µL of water containing 2% FA and 1200 µL of 50:50 MeOH:H_2O containing 0.2% FA, then transferred in a clean 13 × 100 mm borosilicate test tube to eluate the compounds by gravity with 1200 µL of 2% NH_4OH in MeOH. The collected eluates were dried under a nitrogen flow for about 60 min. The prepared samples were resuspended in 100 µL of 60:40 mobile phase A:B (82.2:17.8 ACN:H_2O + 5mM Amm. Form. + 0.3% FA). The reconstituted samples were transferred to a 300 µL borosilicate insert fitted in a 2 mL borosilicate capped vial to be injected into the UPLC-MS/MS system.

2.6. UPLC-MS/MS Analysis of Lysosphingolipids and Creatinine

An Acquity UPLC I-Class Plus combined with a triple quadrupole Xevo TQ-XS (MS/MS) from Waters Corporation (Milford, MA, USA) was used to analyze the targeted molecules. An HILIC column approach with a Halo HILIC column (90 Å, 2.1 × 50 mm, 1.6 µm, Advanced Materials Technology, Wilmington DE, USA) combined with an Acquity Column In-Line filter (2.1 mm nut, Frit 0.2 µm, Waters Corp., Milford, MA, USA) was used to separate the lysosphingolipids according to the interaction of their polar moiety with the UPLC column. This chromatographic approach was successfully used in our laboratory in previous studies on sphingolipids [53]. The UPLC-MS/MS parameters are described in Table 1. Briefly, the system was operated in the selected reaction monitoring (SRM) mode with positive electrospray ionization (ESI+). Using this tandem mode, one precursor ion specific to each studied lysosphingolipid was selected in the first quadrupole, and fragmented in the collision cell, followed by the selection of a specific and sensitive fragment ion in a second quadrupole. The SRM transitions are described in Table 2. The M + 1 peak was used for the detection of creatinine and its internal standard to avoid saturating the MS detector signal.

2.7. Method Validation

The method validation was performed for research purposes according to the guideline recommendations from the FDA guidance for industry [64]. All the S-QCs levels were used to evaluate the assays from three intradays (n = 5) and one interday (n = 3) for accuracy (%bias) and precision (%CV). The U-LQCs and U-HQCs were used to evaluate the sample stability (n = 3) up to 4 months (117 days) for lysosphingolipids and 1 month (26 days) for creatinine at different temperatures: high temperature (35 °C), room temperature (21.5 °C), 4 °C, −30 °C, and −80 °C. The effects of 3 and 5 freeze–thaw cycles (n = 3) were also assessed for all the biomarkers with the U-LQCs and U-HQCs. The autosampler stability (10 °C) of the extracted U-LQCs and U-HQCs was evaluated for 24 h (n = 3). Biomarkers were considered stable in samples when the accuracy (bias nominal%) was ≤15% at each concentration level within the precision range of the replicates.

The adsorption of the metabolites to polypropylene tubes and glass tubes was not investigated because no significant impact was observed in our previous study using liquid urine samples [53]. The limit of detection (LOD) and the lower limit of quantification (LLOQ) were, respectively, established as 3 and 10 times the standard deviation of the analyte concentration after 10 injections of S-LLOQ.

Table 1. UPLC and MS/MS parameters used for the analysis of the targeted lysosphingolipids and creatinine.

UPLC Parameters		MS/MS Parameters	
Column	**Halo® HILIC 90 Å (USP: L3)**	**Ionization Mode**	**ESI+**
ID × Length	2.1 × 50 mm	Acquisition Mode	SRM
Particle Size	1.6 µm	Capillary Voltage	3.0 kV
Column Temperature	40 °C	Source Offset Voltage	90 V
Autosampler Temperature	10 °C	Source Temperature	150 °C
Flow Rate	0.8 mL/min	Desolvation Temperature	650 °C
Injection Volume	3 µL	Desolvation Gas Flow	1100 L/h
Weak Wash	1800 µL ACN	Cone Gas Flow	150 (L/h)
Strong Wash	600 µL 90/10 (H_2O/ACN)	Collision Gas Flow	0.15 (mL/min)
Injection Mode	PLNO	Nebuliser Gas Flow	5.00 Bar
Liquid Chromatographic Method			
Time (min)	**Mobile Phase A**	**Mobile Phase B**	**Gradient types**
Initial	100%	0%	1 (Isocratic)
0.50	100%	0%	1 (Isocratic)
1.00	92%	8%	6 (Linear)
8.00	92%	8%	1 (Isocratic)
8.50	66%	34%	6 (Linear)
10.50	50%	50%	6 (Linear)
11.00	50%	50%	1 (Isocratic)
12.00	0%	100%	1 (Isocratic)
13.00	100%	0%	1 (Isocratic)

ID: internal diameter; PLNO: partial loop with needle overfill with fixed loop; UPLC: ultra-high-performance liquid chromatography; MS/MS: tandem mass spectrometry; HILIC: hydrophilic interaction chromatography; mobile phase A: (97/2.7/0.3 (ACN/H_2O/FA) + 5 mm Amm. Form.); mobile phase B: (60/39.7/0.3 (ACN/H_2O/FA) + 5 mm Amm. Form.); SRM: selected reaction monitoring; ACN: acetonitrile; H_2O: water; ESI: electrospray ionisation.

The criteria for the extraction recovery and matrix effect (nominal bias%) were assessed based on Matuszewski et al.'s approach [65]. Three urine matrices with different levels of endogenous creatinine (1.5, 7, and 15 mM) were used to represent the impact of the matrix on the results in different spiked QC ranges (L, M, and H). Pre- and post-spiked extracted matrices were used to evaluate the compound recovery, while post-spiked extracted matrices with spiked blank results were compared to evaluate the ion suppression or enhancement. Internal standards were added in the final step before injection on the instrument. The detailed approach for these assays is described in Protocol S-1.

The sample dilution impact was investigated by comparing the measured concentrations of sphingolipids normalized to creatinine in the U-MQCs of different filter paper sizes (n = 5): 5 cm filter paper disk, half of a 5 cm filter paper disk and a quarter of a 5 cm filter paper disk. The carry-over impact was evaluated by analyzing a blank sample after the injection of the most concentrated calibrator, HQCs, and every 20 injections. The selectivity was evaluated by analyzing a blank sample, zero calibrators, and six healthy controls without internal standards to verify if they were free of interferences at each SRM transition and the chromatographic retention times of the biomarkers and the internal standards. Finally, the LOD, LLOQ, and specificity were the only parameters validated for the analogs of GluSph and lyso-Gb_3, considering that commercial standards are not available for these molecules. These tests were performed with diluted samples obtained from an untreated Fabry male and untreated Gaucher patient with a confirmed diagnosis excreting related analogs.

Table 2. Selected reaction monitoring (SRM) transitions used to analyze the targeted lysosphingolipids and creatinine. The SRM transitions were separated into two detection windows according to their retention times to optimize the number of datapoints collected across the chromatographic peaks.

Compounds	Disease Targeted	SRM Transitions $[M+H]^+$	Dwell Time	Cone	CE
Creatinine (M + 1)	n/a	m/z 115.07 > 45.05	0.01 s	30 V	5 V
d_3-creatinine (M + 1)	ISTD	m/z 118.09 > 48.75	0.01 s	30 V	5 V
Lyso-Sulf	MLD	m/z 462.34 > 282.28	0.01 s	30 V	20 V
N-Gly-lyso-Sulf	ISTD	m/z 519.36 > 264.27	0.01 s	30 V	20 V
GalSph	Krabbe	m/z 462.34 > 282.28	0.01 s	30 V	20 V
GluSph	Gaucher	m/z 462.34 > 282.28	0.01 s	30 V	20 V
GluSph analog (−28)	Gaucher	m/z 434.31 > 254.25	0.01 s	30 V	20 V
GluSph analog (−26)	Gaucher	m/z 436.33 > 238.25	0.01 s	30 V	20 V
GluSph analog (−12)	Gaucher	m/z 450.31 > 252.23	0.01 s	30 V	20 V
GluSph analog (+2)	Gaucher	m/z 464.36 > 266.28	0.01 s	30 V	20 V
GluSph analog (+14)	Gaucher	m/z 476.32 > 278.25	0.01 s	30 V	20 V
GluSph analog (+16)	Gaucher	m/z 478.34 > 280.26	0.01 s	30 V	20 V
GluSph analog (+30)	Gaucher	m/z 492.32 > 294.24	0.01 s	30 V	20 V
GluSph analog (+32)	Gaucher	m/z 494.33 > 296.26	0.01 s	30 V	20 V
$^{13}C_6$-GluSph	ISTD	m/z 468.36 > 282.28	0.01 s	30 V	20 V
Lyso-SM	Niemann-Pick	m/z 465.35 > 184.07	0.01 s	30 V	21 V
D_9-Lyso-SM	ISTD	m/z 474.40 > 193.13	0.01 s	30 V	21 V
Lyso-Gb_3 analog (−28)	Fabry	m/z 758.42 > 254.25	0.01 s	30 V	35 V
Lyso-Gb_3 analog (−12)	Fabry	m/z 774.41 > 252.23	0.01 s	30 V	35 V
Lyso-Gb_3 analog (−2)	Fabry	m/z 784.43 > 280.26	0.01 s	30 V	35 V
Lyso-Gb_3	Fabry	m/z 786.45 > 282.28	0.01 s	30 V	35 V
Lyso-Gb_3 analog (+14)	Fabry	m/z 800.43 > 278.25	0.01 s	30 V	35 V
Lyso-Gb_3 analog (+16)	Fabry	m/z 802.44 > 280.26	0.01 s	30 V	35 V
Lyso-Gb_3 analog (+34)	Fabry	m/z 820.45 > 334.30	0.01 s	30 V	35 V
Lyso-Gb_3 analog (+50)	Fabry	m/z 836.45 > 350.29	0.01 s	30 V	35 V
Lyso-GM2	GM2	m/z 1118.57 > 204.09	0.01 s	30 V	42 V
Lyso-GM1	GM1	m/z 1280.62 > 204.09	0.01 s	30 V	50 V
N-Gly-lyso-Gb_3	ISTD	m/z 843.47 > 264.47	0.01 s	30 V	48 V

SRM: selected reaction monitoring; ISTD: internal standard; $[M+H]^+$: protonated ion; CE: collision energy.

2.8. Statistical Analyses

The MS data generated were processed with MassLynx-TargetLynx V4.2 SCN 1040 software (Waters, Milford, MA, USA). The calibration curve parameters for all the compounds were defined as a linear curve with the point of origin excluded and with a $1/x$ weighting. Statistical analysis was performed using GraphPad Prism 10.2.2. software (Dotmatics, Boston, MA, USA). Non-parametric tests were chosen because of the small group size and the non-normal distribution of the results analyzed. Differences between the healthy controls, pathological controls, other sphingolipidoses, and specific patient groups were determined by comparing the group medians with the Kruskal–Wallis test with the post hoc Dunn's multiple comparisons test. For the healthy control group, the Spearman test was used to evaluate if there was a correlation between the biomarker levels and the participant age, and the Mann–Whitney U test was used to evaluate if the biomarker levels were associated with the sex of the participant. The area under the receiver operating characteristic (ROC) curve (AUC) was calculated to evaluate the overall diagnostic accuracy of the putative biomarkers and related analogs in the Fabry and Gaucher disease untreated groups compared to the healthy controls (n = 59). The ROC curve represents the true-positive rate (sensitivity: (true positives/(true positives + false negatives)) on the y-axis versus the false-positive rate (1 − (specificity: true negatives/(true negatives + false positives)) on the x-axis, as a function of the cut-off value tested. The optimal cut-off according to the Youden index (sensitivity + specificity − 1) was evaluated. The diagnostic reliability of the assay regarding the other sphingolipidoses could not be evaluated due to the limited number of patients. However, the normal reference values were determined by calculating the 95th percentile of the results from the biomarker levels of the healthy control specimens.

3. Results and Discussion

3.1. Chromatographic Separation of Lyso-Sulf, GluSph and GalSph

In our previous studies, it was already shown that chromatographic HILIC approaches were efficient in separating structural isomers, such as GluSph and GalSph, which allowed the differentiation between Gaucher disease and Krabbe disease biomarkers and ensured a reliable quantification [53]. It was observed that with positive electrospray ionization (ESI+), the lyso-Sulf precursor ion lost its H_2SO_4 group by in-source fragmentation and had the same mass-to-charge (m/z) ratio as GluSph and GalSph (m/z 462.34). Even if another study selected the m/z 542 precursor ion in ESI+ for the quantification of lyso-Sulf in plasma [66], we observed, by MS infusion (50:50 MeOH:H_2O + 10 mM ammonium formate), that the in-source fragmentation ion was ten times more sensitive than the m/z 542 precursor ion in our specific setting. Also, lyso-Sulf was analyzed in negative electrospray ionization (ESI−) in our previous study [53]. However, we used a more recent triple quadrupole MS for this study, and with our optimized parameters, at molar equivalence, the precursor ion at m/z 462.34 was three times more abundant in ESI+ than analyzing m/z 540.48 using ESI− according to the signal-to-noise ratio (S/N) (Figure 1). For these reasons, the lyso-Sulf precursor ion at m/z 462.34 was selected.

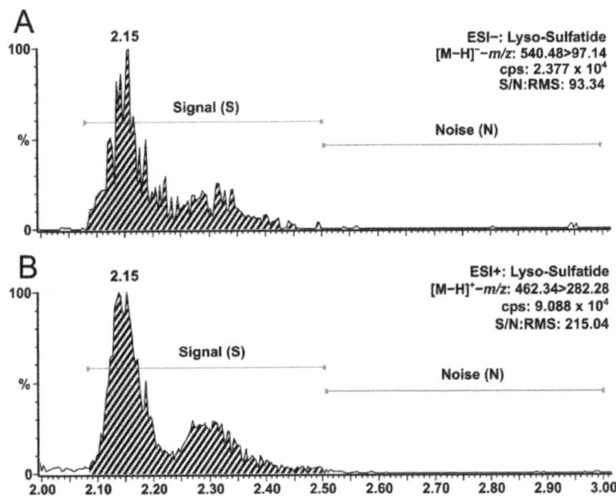

Figure 1. Chromatograms of lyso-Sulf using: (**A**) negative electrospray ionization (ESI−); and (**B**) positive electrospray ionization (ESI+) at 1 nmol/L. Red lines show chromatogram sections to calculate the signal-to-noise ratio (S/N) and the root mean square (RMS) of lyso-Sulf with the MassLynx software. Cps: counts per second.

Considering that lyso-Sulf was not retained with the previous chromatography strategy developed [53], a new chromatographic method was devised to optimize the lyso-Sulf retention time while still allowing a good separation of GluSph, and GalSph. The HPLC column previously used (Halo® HILIC 2.7 (4.6 × 150 mm, 2.7 µm, Advanced Materials Technology, Wilmington, DE, USA) was replaced by the equivalent particles of a UPLC column to improve the chromatographic resolution and separation efficiency (Halo® HILIC 90 Å (2.1 × 50 mm, 1.6 µm, Advanced Materials Technology, DE, USA), with a slight modification of the mobile phase A to ensure the retention of lyso-Sulf. Also, the strong wash and the injection mode were modified to avoid carry-over in-between injections using the different UPLC system. All the other compounds targeted in this study were still clearly separated by the new UPLC-MS/MS method developed, as shown in Figure 2. Furthermore, analogs of lyso-Gb_3 and GluSph were analyzed, and their chromatographic separation is shown in Figure 3. The urinary analogs of lyso-Gb_3 (−28, −12, −2, +14, +16, +34 and +50) in Fabry disease patients and of GluSph (−28, −26, −12, +2, +14, +16, +30, and +32) in Gaucher disease patients were discovered in previous metabolomic studies in urine [5,7]. For Fabry disease, analogs +16, +34, and +50 had a positive association with the left ventricular mass index in Fabry patients from Taipei carrying the IVS4 + 919G>A mutation (cardiac variant) [40]. For Gaucher disease, larger cohorts are needed to understand the clinical significance of GluSph-related analogs in urine [52].

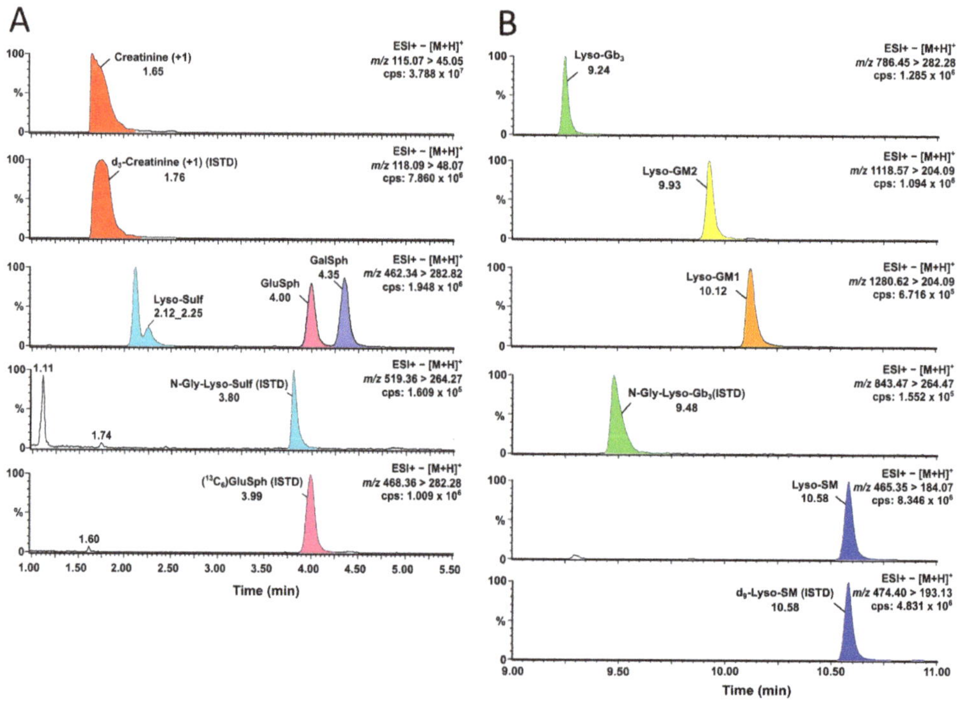

Figure 2. Chromatograms of the principal biomarkers analyzed by UPLC-MS/MS. (**A**) Extracted ion currents from SRM transitions of creatinine, lyso-Sulf, GluSph, GalSph and their internal standards; and (**B**) extracted ion currents from SRM transitions of lyso-Gb_3, lyso-GM2, lyso-GM1, lyso-SM and their internal standards. Cps: counts per second; %: relative abundance; ESI+: positive electrospray ionization; m/z: mass-to-charge ratio; ISTD: internal standard.

3.2. Method Validation Results

The intra- and interday accuracy, and the intra- and interday precision assays, are summarized in Table S4, while the LOD, LLOQ and linearity assays for sphingolipids and creatinine are shown in Table S5. The matrix effect and extraction recovery assays are shown in Table S6. The stability results are shown in Table S7, and in Figures 4 and 5, while the freeze–thaw assay results are shown in Table S8. The sample dilution impact assays are shown in Table S9.

Briefly, no carry-over and no cross-interference between the targeted compounds and internal standards were observed during the validation and the sample analysis. Except for lyso-SM (bias%: 23.6% to 57.2%, but CVs% < 15%), the three intraday assays (3 × n = 5) revealed that the CVs% and biases% were <20% for the LLOQC and <15% for the LQC, MQC, and HQC. Nevertheless, the lyso-SM interday assays showed precision results for all the levels of QCs with CV% < 15%, except for LLOQC range (CV%: 22.0%). Considering this, the lyso-SM measurements were considered semi-quantitative for this method. The LOD and LOQ were established. Lyso-GM2 and lyso-GM1 had higher limit values compared to the other biomarkers and their calibration curves were adjusted according to this result. The Pearson's correlation coefficients (r), and the coefficients of determination (R^2) were $\geq$0.995 (n = 15) for the calibration curve in the validation of all the compounds.

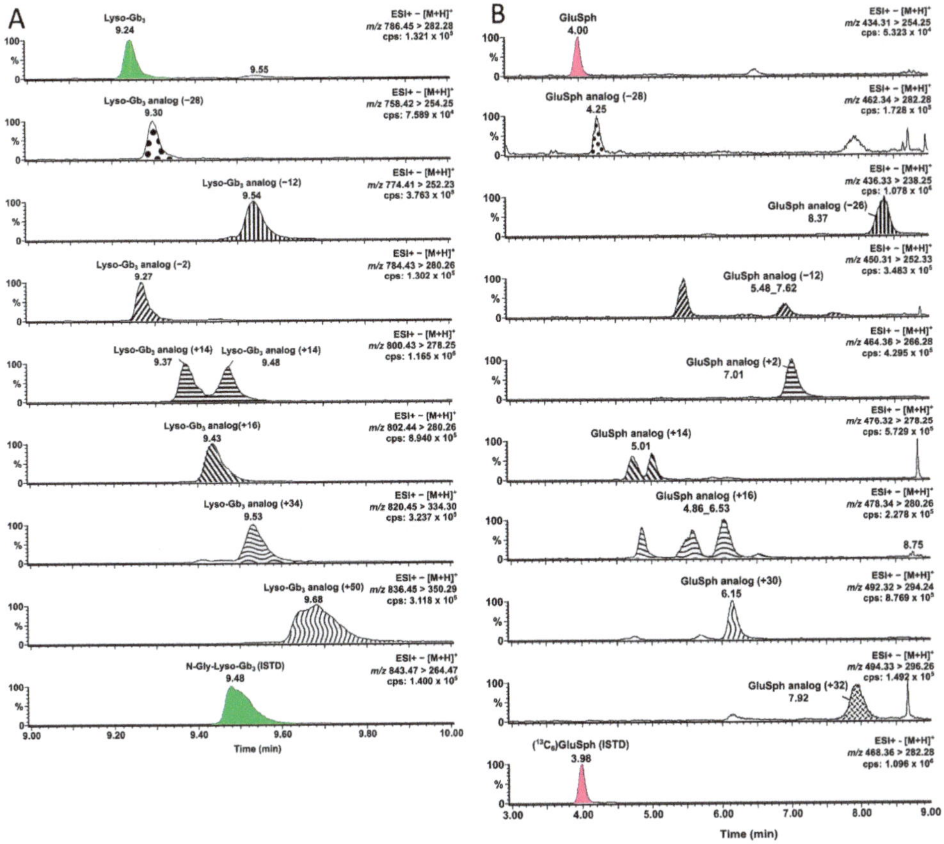

Figure 3. Chromatograms of separated (**A**) Lyso-Gb_3 and its analogs (−28, −12, −2, +14, +16, +34, and +50) using UPLC-MS/MS for Fabry disease; and (**B**) GluSph and its analogs (−28, −26, −12, +2, +14, +16, +30, and +32) by UPLC-MS/MS for Gaucher disease. Cps: counts per second; %: relative abundance; ESI+: positive electrospray ionization; *m*/*z*: mass-to-charge ratio.

The extraction recovery and matrix effect were investigated for the targeted sphingolipids and creatinine. Ideally, the internal standards selected for quantification should have similar physical and chemical properties and ionization efficiency compared to the analytes. Usually, stable isotope-labelled (SIL) standards are used [67]. SIL standards were commercially available for creatinine (d_3-creatinine), lyso-SM (d_9-lyso-SM), and GluSph ($^{13}C_6$-GluSph) during the study design. The N-Gly-lyso-Gb_3 internal standard was used for lyso-Gb_3 and its analogs, lyso-GM1, and lyso-GM2 due to their retention time proximity. The GalSph and GluSph analogs were corrected with $^{13}C_6$-GluSph, and N-Gly-lyso-Sulf was used to correct lyso-Sulf due to their structural similarities. Due to the limitation of the availability of SIL standards, the methodology was developed to limit the matrix effect and to optimize the extraction recovery. The results obtained showed that the creatinine levels do not affect the extraction recovery, and the matrix effect seems limited. The creatinine matrix spiked with high concentrations of creatinine was excluded because the MS signal of creatinine saturated the instrument detector, and the results were not reliable. However, other levels have shown similar and acceptable results for creatinine. The extraction recovery was ≥63.5% and the matrix effect varied between −12.8% and 17.0% for all the biomarkers except lyso-Gb_3, lyso-GM2, and lyso-GM1, which have higher enhancement effects, especially when the urine is more concentrated (matrix effect bias min-max range: 10.2% to 52.5%). However, within the same biomarker analysis, the matrix effect and recovery were similar independently of the creatinine and analyte levels.

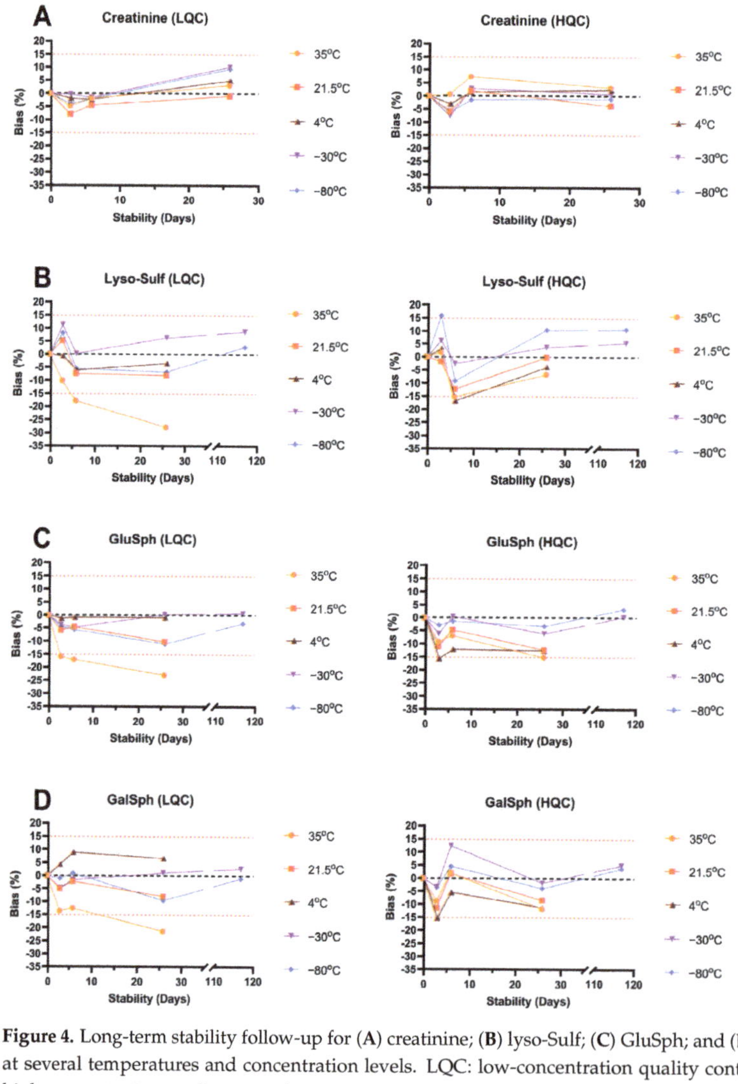

Figure 4. Long-term stability follow-up for (**A**) creatinine; (**B**) lyso-Sulf; (**C**) GluSph; and (**D**) GalSph at several temperatures and concentration levels. LQC: low-concentration quality control; HQC: high-concentration quality control; upper red dot line: maximum limit of 15% of bias compared to the nominal concentration; lower red dot line: lower limit of −15% of the bias compared to the nominal concentration.

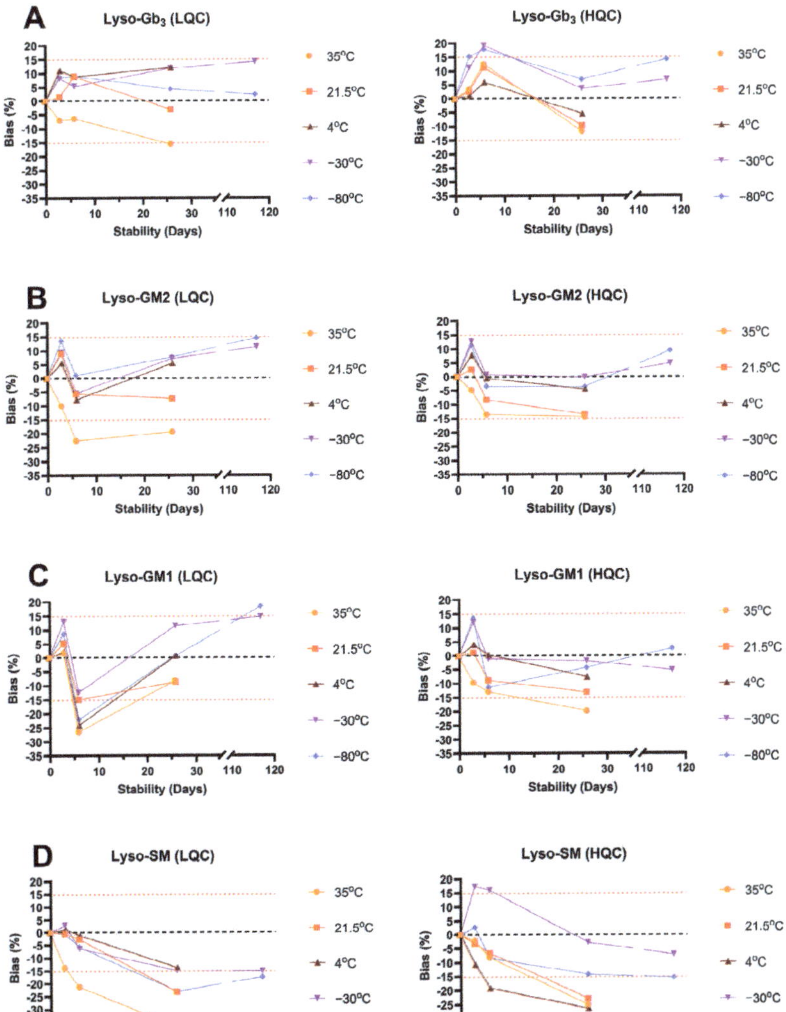

Figure 5. Long-term stability follow-up for (**A**) lyso-Gb$_3$, (**B**) lyso-GM2, (**C**) lyso-GM1 and (**D**) lyso-SM at several temperatures and concentration levels. LQC: low-concentration quality control; HQC: high-concentration quality control; upper red dot line: maximum limit of 15% of bias compared to the nominal concentration; lower red dot line: lower limit of −15% of bias compared to the nominal concentration.

The biomarker stability was assessed at 35 °C, 21.5 °C, 4 °C, −30 °C and −80 °C until 117 days for lysosphingolipids and 26 days for creatinine. These conditions were used to simulate different environmental impacts on the samples. Creatinine was stable for at least 26 days at all the concentrations and temperatures (bias between −7.6% to 10.2%). For lysosphingolipids, all the compounds were stable for at least 6 days at 35 °C but showed important degradation (>20% bias) at 26 days (nominal bias between −19.0% to −36.8%). At 21.5 °C and 4 °C, all the lysosphingolipids, except lyso-SM, were stable at least for 26 days. Lyso-SM showed higher bias (−13.7% to −26.3%) but good stability for at least

6 days. At −30 °C and −80 °C, only lysosphingolipids were evaluated over 26 days and were stable for at least 117 days. The autosampler (10 °C) stability showed nominal bias ≤7.3% for all the compounds.

The results have shown that the compounds are stable after three freeze–thaw cycles from −30 °C to 21.5 °C. Only lyso-GM1 had a bias over 20% for three and five freeze–thaw cycles, which might be caused by the large CV% for this molecule in these assays.

Finally, the results have shown that dilution factors of two and four were reliable because all the nominal biases were below 15.1%, with coefficients of variability below 9.2% between replicates. However, the MS signal of the compounds was reduced, and the results should respect the LOD and LLOQ established.

3.3. Normal Reference Values

To evaluate the normal reference values, 59 urine samples from healthy controls (24 males (<18 yrs: n = 11, ≥18 yrs: n = 13); 35 females (<18 yrs: n = 15, ≥18 yrs: n = 20)) were analyzed to establish the lysosphingolipid reference values normalized to creatinine according to the 95th percentile. Except for lyso-SM and GluSph, no lysosphingolipids were detected in the healthy controls. For the lyso-SM levels, no relation or significant differences were observed according to age and sex (p-value > 0.05). For GluSph, the levels were significantly different between male and female controls according to the Mann–Whitney U test (p-value < 0.0001). However, due to the limited Gaucher patient samples available to evaluate the normal reference values, the 95th percentile was established with the total cohort for the GluSph. The same strategy was used for all the lysosphingolipids targeted. For analogs related to lyso-Gb_3 and GluSph, the reference values were established individually and as a group. The normal reference values are summarized in Table S10. Briefly, only the GluSph (37 pmol/mmol creatinine) and lyso-SM (293 pmol/mmol) levels were quantified in the healthy controls.

3.4. Lysosphingolipid Levels in Patients and Controls

Urine samples dried on filter paper from patients diagnosed with specific sphingolipidoses (n = 119) were analyzed, and the results normalized to creatinine (pmol/mmol creatinine) are summarized in Table S11. The lyso-Sulf, GluSph, GalSph, lyso-Gb_3, lyso-GM1, lyso-GM2, and lyso-SM levels were measured in all the samples. Relative quantifications of analogs of Lyso-Gb_3 and GluSph were performed with their putative biomarker curve.

The number of patient samples available was limited for some sphingolipidoses (MLD, Krabbe, GM1, GM2, and Niemann–Pick C) due to the low prevalence of these diseases. Although it is feasible to quantify lyso-Sulf, GalSph, lyso-GM1, lyso-GM2, and lyso-SM in DUS, further validation in larger patient cohorts is required to better evaluate the sensitivity and specificity before the clinical implementation of these biomarkers in urine. However, a slight amount of lyso-SM was detected in specimens obtained from all the patients with sphingolipidoses, pathological and healthy controls, and there was no significant difference observed between the Niemann–Pick type C patient sample (53 pmol/mmol creatinine) compared to the other sphingolipidoses. Niemann–Pick type C is a less severe disease form compared to types A and B and should not affect the enzyme activity of acid sphingomyelinase [17]. A previous study showed increased plasma levels of lyso-SM in Niemann–Pick types A and B. Lyso-SM 509, a recently detected biomarker for Niemann–Pick, was increased in Niemann–Pick type C in the same study [42]. The lyso-SM level was also increased in dried blood spots of Niemann–Pick B. No study evaluating the detection of Niemann–Pick A/B using urine specimens was found [68]. Furthermore, six female Fabry patients, one male Fabry patient, four MLD patients, four MPS patients and three female healthy controls showed levels over 293 pmol/creatinine. Consequently, more samples are needed to confirm the reference value proposed in this study and to evaluate the potential of lyso-SM for the monitoring and follow-up of Niemann–Pick patients using urine dried on filter paper. Lyso-GM1 and lyso-GM2 were not detected in any groups. These biomarkers were, respectively, detected in plasma samples of GM1 patients and GM2 patients in a previous study [69]. In our current study, samples from only two GM1 treated patients were available, and no large study was found analyzing lyso-GM1 and lyso-GM2 in GM1 and GM2 patient urine samples. For lyso-sulfatide, only quarters of a 5 cm filter paper disk from diagnosed MLD patients with a concentration of creatinine near the LLOQ were available. Only one MLD patient (unknown form) had elevated lyso-Sulf, while no lyso-Sulf was detected in the three other patients.

According to the Kruskal–Wallis test (p-value < 0.0001), there were significant differences between the median levels of GluSph and its related analogs between the Gaucher patients, other patients and controls. The post hoc Dunn test analysis of the Kruskal–Wallis test revealed significant differences between groups. The results are shown in Figure 6 and in Table S12. The GluSph levels were significantly different in the Gaucher patients compared to the other groups. As expected, marked elevations of GluSph and its related analogs were found in the untreated patients compared to the treated patients. However, the differentiation was less important when the statistical analysis was performed with only the GluSph values. Interestingly, some controls and other sphingolipidoses had a slight elevation of GluSph, which may impact the median rank in the Dunn's test comparison between groups. Nevertheless, some related analogs were detected in other sphingolipidoses than Gaucher disease but were still negligible compared to levels from the Gaucher patients. The distribution of analogs −26 (21.18%), −12 (10.64%), +2 (9.71%), +14 (18.58%), +16 (17.71%), +30 (13.19%), +32 (6.70%) showed that the analog levels tend to be largely higher than GluSph (1.47%) itself (Figure 7).

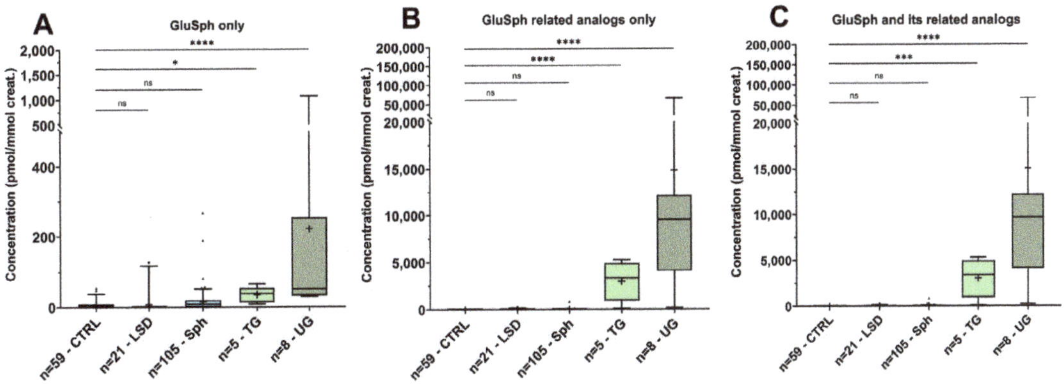

Figure 6. Box plots of the urinary levels of GluSph and the eight related analogs normalized to creatinine (pmol/mmol creatinine) in the healthy control group (CTRL, n = 59), pathological control group (LSD, n = 21), other sphingolipidoses group (Sph, n = 105), treated Gaucher patient group (TG, n = 5) and untreated Gaucher patient group (UG, n = 8) from urinary samples dried on filter paper; box plots represent the normalized urinary levels of (**A**) GluSph levels only; (**B**) GluSph related analog levels only; and (**C**) total levels of GluSph and its related analogs. The lower and upper limits shown by the box plots are the 25th and 75th percentiles, respectively. The center horizontal box line is the median. The symbol "+" is the mean. The whiskers correspond to the highest and lowest values at the 95th and 5th percentile, respectively. Values outside the 95th percentile are considered as outliers. Comparison of the CTRL group with other groups with the Kruskal–Wallis post hoc Dunn's test: ns: non-significant, *: p-value ≤ 0.05; ***: p-value ≤ 0.001; ****: p-value ≤ 0.0001.

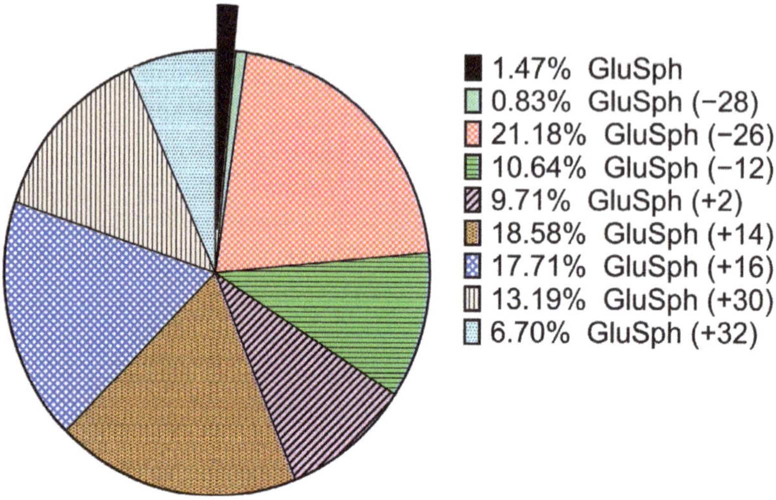

Figure 7. Mean distribution of GluSph and its eight related analogs (−28, −26, −12, +2, +14, +16, +30, +32) in the untreated Gaucher disease group (n = 5).

For lyso-Gb_3 and its related analogs, the Dunn's test results (comparison of the CTRL group to other groups for these biomarkers) and biomarker levels are shown in Figure 8 and summarized in Table S13. The Fabry patients were separated according to their sex and their predicted likely phenotype according to their mutation. Lyso-Gb_3 and its related analogs were significantly different in the Fabry disease patients compared to the pathological and healthy controls, especially for patients with classical mutations. Fabry patients with classical mutations have higher levels of lyso-Gb_3 and analog levels than patients with late-onset mutations. It was previously shown that patients with classical mutations are more likely to present an elevation of lyso-Gb_3 and related analogs compared to patients with non-classical mutations (late-onset phenotypes) [40,41,70]. Considering that only five late-onset mutations from Fabry patients were evaluated in the current study, and that these were not equally distributed in each study group, this might have affected the biomarker profiles detected. All the patients carrying the N215S mutation (5/8 UFM; 1/6 UFF; 2/2 TFM), a cardiac variant mutation, had higher analog levels compared to the patients with other late-onset mutations. This was previously observed in another cohort of Fabry patients [41,70]. As also observed in the current study, patients with intronic mutations such as c.639+919G>A (1/2 TFF; 1/6 UFF; 2/8 UFM) or c.640-801G>A (2/6 UFF; 1/8 UFM) had no urinary lyso-Gb_3 or related analogs increased. The male Fabry group tends to have higher levels of biomarkers compared to the female Fabry patient group, even if they are treated. The variation of the biomarker levels in the patient groups may partially be explained by the marked heterogeneity in the *GLA* mutations and the sex of the patients [71]. Unfortunately, the α-Gal enzyme activities for all the Fabry patients were not available and cannot be taken into consideration in our interpretation. Moreover, the X-linked inheritance of the disease, especially in heterozygous females, may lead to phenotypic variations according to the random inactivation of one of the X chromosomes in somatic cells [72]. The distributions of lyso-Gb_3 and its analogs in the untreated patients regrouped according to their sex and mutation type (classical or non-classical) are shown in Figure 9. The lyso-Gb_3-related analogs +16, +34, and +50 were the most abundant analogs in the male and female Fabry patients and were sometimes higher than lyso-Gb_3 itself. They could also be quantified in patients with late-onset mutations, even if their excretion levels were reduced, especially in the UFM group [40,70]. Considering that there are over 1000 *GLA* gene variants, including pathogenic mutations, variants of unknown significance and benign mutations [21], larger cohort studies are needed to evaluate the analog profile in the urine samples of patients with other late-onset variants.

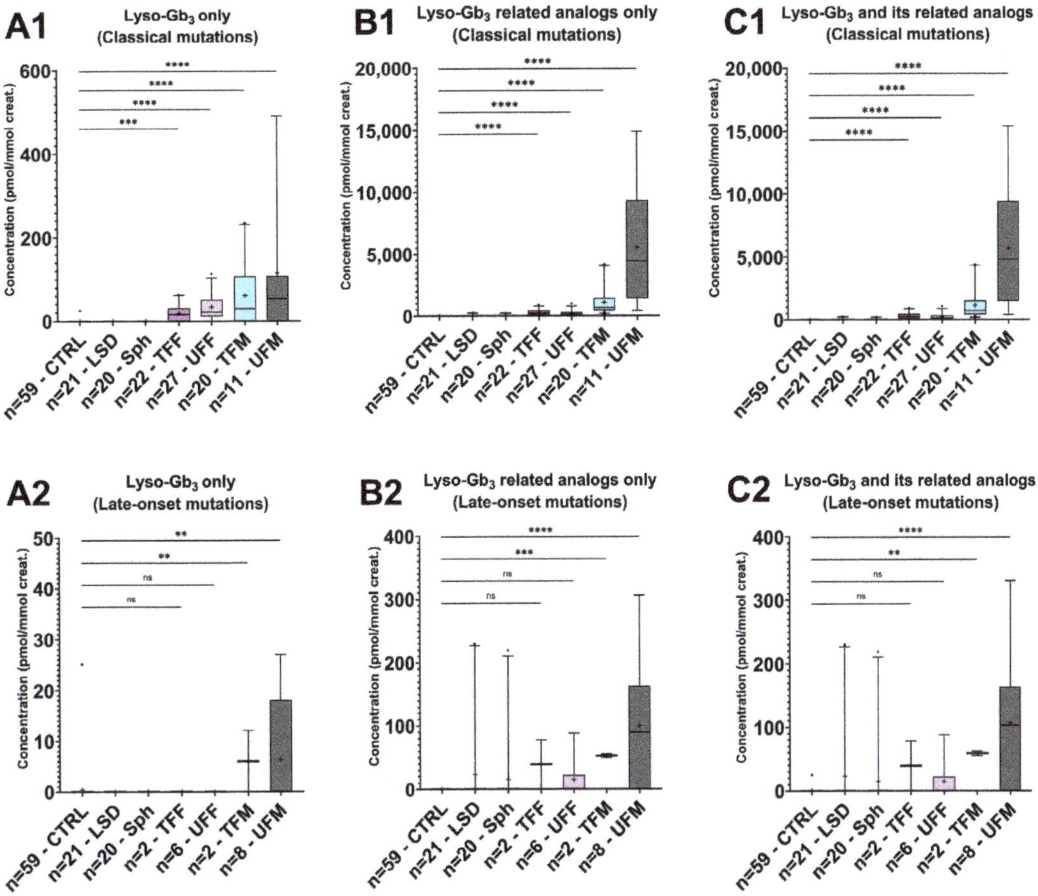

Figure 8. Box plots of the urinary levels of lyso-Gb_3 and its eight related analogs normalized to creatinine (pmol/mmol creatinine) in the healthy control group (CTRL, n = 59), pathological control group (LSD, n = 21), other sphingolipidose groups (Sph, n = 20), treated Fabry female patient group (TFF, n = 22 with classical mutations (CM) and n = 2 with late-onset mutations (LO)), untreated Fabry female group (UFF, CM: n = 27, LO: n = 6), treated Fabry male group (TFM: CM: n = 20, LO: n = 2), and untreated Fabry male group (UFM: CM: n = 11, LO: n = 8) from urinary samples dried on filter paper; box plots represent normalized urinary levels of (**A1,A2**) lyso-Gb_3 levels only in controls and patients with (1) CM and (2) LO; (**B1,B2**) lyso-Gb_3 related analog levels only in controls compared to patients with (1) CM and (2) LO; and (**C1,C2**) total levels of lyso-Gb_3 and its related analogs in controls compared to patients with (1) CM and (2) LO. The lower and upper limits shown by the box plots are the 25th and 75th percentiles, respectively. The center horizontal box line is the median. The symbol "+" is the mean. The whiskers correspond to the highest and lowest values at 95th and 5th percentile, respectively. Values outside the 95th percentile are considered as outliers. Comparison of the CTRL group with other groups with Kruskal–Wallis post hoc Dunn's test: ns: non-significant, **: p-value ≤ 0.01; ***: p-value ≤ 0.001; ****: p-value ≤ 0.0001.

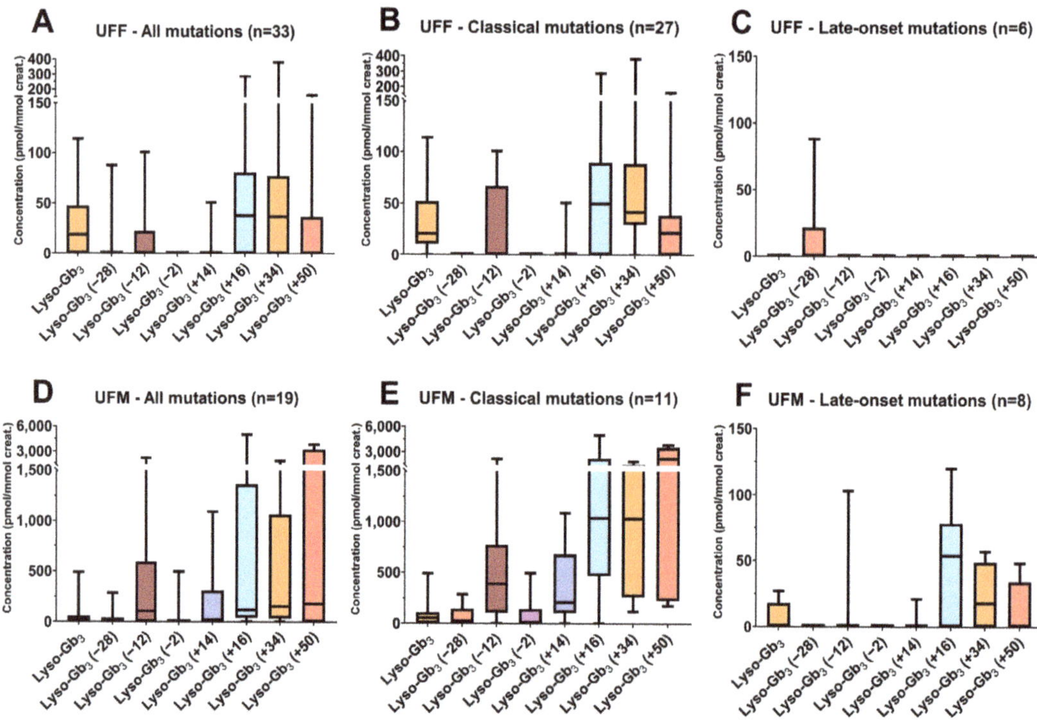

Figure 9. Box plots of the urinary levels of lyso-Gb_3 and its eight related analogs (−28, −12, −2, +14, +16, +34, +50) normalized to creatinine (pmol/mmol creatinine) in the untreated Fabry female (UFF) and untreated Fabry male (UFM) groups according to the classical and nonclassical (late-onset phenotype) mutations. (**A**) Results regrouping all the UFF; (**B**) results regrouping only UFF with classical mutations; (**C**) results regrouping only UFF with nonclassical mutations; (**D**) results regrouping all the UFM; (**E**) results regrouping only UFM with classical mutations; and (**F**) results regrouping only UFM with nonclassical mutations. The lower and upper limits shown by the box plots are the 25th and 75th percentiles, respectively. The center horizontal box line is the median. The whiskers correspond to the highest and lowest values.

3.5. Receiver Operating Characteristics (ROC Curve)

The AUCs were used to assess the efficacy of lyso-Gb_3, GluSph and their related analogs in discriminating untreated patients from healthy controls without considering their genotype. The sensitivity and specificity were evaluated at the optimal cut-off value suggested by the highest Youden index value. The AUCs, optimal cut-off levels selected, sensitivity and selectivity results are summarized in Table 3. The Youden index values are summarized in Table S14. The AUCs results were statistically significant in the analyzed subgroups (p-value < 0.0001), and over 0.75, suggesting that these urinary biomarkers are clinically useful for diagnostic reliability. For the untreated Fabry patients, the AUC results seem to be more favourable if the complete profile of lyso-Gb_3 and its related analogs was used to discriminate patients from controls (UFF: AUC = 0.877; UFM: AUC = 0.920) compared to using lyso-Gb_3 only (UFF: AUC = 0.811; UFM: AUC = 0.758). Similarly, for the untreated Gaucher patients, the AUCs results tend to be slightly favourable when using the complete profile of GluSph and its related analogs (UG: AUC = 1.000) compared to using GluSph only (UG: AUC = 0.966). Interestingly, the AUC value appears to improve in the UFM group when the analogs are included in the cut-off evaluation compared to the analysis of lyso-Gb_3 itself. According to Table S1, 11/19 UFM patients have a classical mutation and 8/19 UFM patients have a late-onset mutation. However, three patients with classical mutations and six patients with late-onset mutations have an elevation of lyso-Gb_3-related analogs, but not of lyso-Gb_3. This observation was also made in the UFF (3/33 patients) cohorts, but not in the UG group. Even if larger cohorts are needed to optimize the cut-off values for high-risk screening and to confirm these results due to the limited number of variants studied, these data suggest that analyzing a complete profile of lysosphingolipids in urine is a more comprehensive approach than analyzing the lyso-Gb_3 levels alone. For the UG patients, the analogs might be interesting in patients with low levels of GluSph, but this remains to be confirmed.

Table 3. Diagnostic reliability of lyso-Gb_3, GluSph and the respective related analog evaluation according to their ROC curve results in untreated Fabry female (UFF), untreated Fabry male (UFM) and untreated Gaucher patients (UG) compared to healthy controls.

Fabry Disease	Biomarkers Used	Lyso-Gb_3 Only	Lyso-Gb_3 Related Analogs Only	Lyso-Gb_3 and Its Related Analogs
UFF	AUCs—ROC curve 95% CI	0.811 **** 0.705 to 0.916	0.864 **** 0.769 to 0.958	0.877 **** 0.786 to 0.967
	Optimal cut-off value (pmol/mmol creatinine)	>5.36	>13.6	>27.4
	Sensitivity 95% CI	63.6 46.6% to 77.8%	72.7 55.8% to 84.9%	75.8 59.0% to 87.2%
	Specificity 95% CI	98.3 91.1% to 99.9%	100 94.0% to 100%	100 94.0% to 100%
	Highest Youden index value	0.619	0.727	0.758
UFM	AUCs—ROC curve 95% CI	0.758 *** 0.610 to 0.907	0.921 **** 0.823 to 1.00	0.920 **** 0.821 to 1.00
	Optimal cut-off value (pmol/mmol creatinine)	>12.3	>40.6	>12.6
	Sensitivity 95% CI	52.6 31.7% to 72.7%	84.2 62.4% to 94.5%	84.2 62.4% to 94.5%
	Specificity 95% CI	98.3 91.1% to 99.9%	100 94.0% to 100%	98.3 91.1% to 99.9%
	Highest Youden index value	0.509	0.842	0.825
Gaucher Disease	**Biomarkers Used**	**GluSph Only**	**GluSph Related Analogs Only**	**GluSph and its Related Analogs**
UG	AUCs—ROC curve 95% CI	0.966 **** 0.925 to 1.00	1.000 **** 1.000 to 1.000	1.000 **** 1.000 to 1.000
	Optimal cut-off value (pmol/mmol creatinine)	>29.2	>85.5	>146
	Sensitivity 95% CI	100 67.6% to 100%	100 67.6% to 100%	100 67.6% to 100%
	Specificity 95% CI	91.7 83.8% to 97.3%	100 94.0% to 100%	98.3 91.1% to 99.9%
	Highest Youden index value	0.917	1.000	0.983

UFF: untreated Fabry female; UFM: untreated Fabry male; UG: untreated Gaucher 95%. CI: confidence interval; ***: *p*-value < 0.001; ****: *p*-value < 0.0001.

4. Conclusions

The first aim of this study was to develop and validate a new, robust and sensitive UPLC-MS/MS multiplex method for the analysis of 21 creatinine-normalized lysosphingolipid biomarkers in DUS according to the FDA recommendations. The second objective was to establish normal reference values to evaluate the clinical utility of this assay for high-risk screening, monitoring, and follow-up of patients for eight sphingolipidoses. This method provides an absolute quantification of lyso-Sulf, GluSph, GalSph, lyso-Gb_3, lyso-GM1, lyso-GM2, and creatinine. The relative quantification of lyso-SM, seven related analogs of lyso-Gb_3 and eight related analogs of GluSph was also performed. The results of this study using a urine filter paper matrix show the stability of urinary lysosphingolipids at 35 °C for 3 days, at room temperature and 4 °C for at least 26 days, and at −30 and −80 °C for at least 117 days. Moreover, the use of filter paper to collect, store and transport urine specimens to laboratories instead of using "liquid urine specimens" in sample tubes allows sample shipment by regular mail (avoiding the use of dry ice), thus reducing costs. Filter paper samples also facilitate long-term storage in the freezer. Furthermore, as demonstrated in the literature, urine filter paper collection can easily be performed at home for a neonatal or pediatric population [55–63]. The current results from Fabry and Gaucher patients show that using DUS to analyze lysosphingolipids can be a feasible approach applicable to the clinical field. These results show the importance of analyzing an analog profile related to lyso-Gb_3 and GluSph to improve the biomarker sensitivity and reliability to monitor Fabry disease and Gaucher disease patients. Finally, another advantage is that the urine filter paper sample collection might facilitate patient recruitment for further collaborative studies on rare diseases.

However, this study has limitations. Except for the MLD patients, urine liquid samples were obtained from retrospective studies and were not directly collected on filter paper. The number of samples available from patients with MLD, Krabbe disease, ganglioside GM1, ganglioside GM2 and Niemann–Pick disease was small, thus limiting the interpretation of the clinical reliability of GalSph, lyso-GM1 and lyso-GM2, lyso-Sulf and lyso-SM in DUS. The interpretation of the results for lyso-Gb_3 and its analog levels might be affected by the heterogeneity of the *GLA* variants in Fabry cohorts. It was not possible to compare the plasma, cerebrospinal fluid, dried blood spots and urine levels of these patients. Clinical information, such as the patient enzymatic activity, clinical manifestations and ethnicity, was limited and it was not possible to evaluate any further correlation. Future perspectives will involve the patients' experience using filter paper urine sample collection. More clinical studies with larger cohorts are needed to measure these lysosphingolipids in urine samples from patients with various sphingolipidoses and gene variants.

Supplementary Materials: The following supporting information can be downloaded at https://www.mdpi.com/article/10.3390/biom14121612/s1, Protocol S-1. Extraction recovery and matrix effect assays; Table S1. Demographic information of participants from the sphingolipidoses group to the study where the sex, mutation, phenotype, treatment, age and biomarker levels are described; Table S2: Demographic information of participants from the healthy and pathological control groups to the study where the sex, mutation, phenotype, treatment, age and biomarker levels are described; Table S3. Concentration of working solutions of the calibrators and QCs for the evaluation of the quality criteria of the methodology; Table S4. Intra- and interday accuracy and precision assays for the targeted sphingolipids and creatinine on urine filter paper; Table S5. Limits of detection, limits of quantification and evaluation of the linearity of the calibration curve using the coefficient of determination and Pearson correlation; Table S6. Extraction recovery and matrix effect assays for the targeted sphingolipids and creatinine on urine filter paper; Table S7. Long-term stability assays at several temperatures for the targeted sphingolipids and creatinine on urine filter paper; Table S8. Freeze–thaw cycle assays (n = 3) for the targeted sphingolipids and creatinine on urine filter paper at different concentrations for 0, 3 and 5 cycles; Table S9. Dilution factor assays (n = 5) for the targeted sphingolipids and creatinine on urine filter paper using a full, half and quarter 5 cm filter paper disk containing U-MQC; Table S10. Normal reference values were established according to the 95th percentile evaluation of lysosphingolipidose levels normalized to creatinine (pmol/mmol creatinine) in healthy control samples (n = 59); Table S11. Lysosphingolipids (A) lyso-Sulf, GalSph, lyso-GM2, lyso-GM1 and lyso-SM; (B) GluSph and the eight related analogs; (C) lyso-Gb_3 and the seven related analogs; and (D) regrouped related analogs of lyso-Gb_3 and GluSph levels in urine dried on filter paper normalized to creatinine (pmol/mmol creatinine) in sphingolipidoses (Fabry disease,

Gaucher disease, MLD, GM1 and NPC) in pathological controls and healthy controls; Table S12. Kruskal–Wallis test and post hoc Dunn's test statistical results for GluSph and its analogs in the CTRL, LSD, Sph, and Gaucher disease subgroups (treated Gaucher patients (TG) and untreated Gaucher patients (UG)); Table S13. Kruskal–Wallis test and post hoc Dunn's test statistical results for GluSph and its analogs in CTRL, LSD, Sph, and Fabry disease subgroups (treated Fabry female (TFF), untreated Fabry female (UFF), treated Fabry male (TFM) and untreated Fabry male (UFM)); Table S14. Youden index evaluation of the ROC curve results obtained from (A) untreated Fabry female (UFF); (B) untreated Fabry male (UFM); and (C) untreated Gaucher (UG) patients.

Author Contributions: Conceptualization: T.M. and C.A.-B.; methodology, visualization, validation, formal analysis, investigation: T.M.; resources: C.A.-B. and B.M.; writing—original draft preparation: T.M.; writing—review and editing, T.M., B.M. and C.A.-B.; supervision: C.A.-B.; project administration: C.A.-B.; funding acquisition: C.A.-B. All authors have read and agreed to the published version of the manuscript.

Funding: This research was funded by a grant from Sanofi Canada. An educational bursary to Tristan Martineau was also provided by Sanofi Canada for this doctoral project.

Institutional Review Board Statement: This study was conducted in accordance with the Declaration of Helsinki and approved by the Research Ethics Board at the Centre intégré universitaire de santé et de services sociaux de l'Estrie—Centre hospitalier universitaire de Sherbrooke under the project ID under project ID 2023-4955 (January 2023), which was an extended study from project ID 2021-3435 (September 2020). Urine samples were obtained from consenting participants as part of the latter study.

Informed Consent Statement: As mentioned above, this project is an extended study from project ID 2021-3435 (September 2020). Extended informed consent was obtained from all the subjects involved in this study. Written informed consent has been obtained from the patient(s) to publish this paper.

Data Availability Statement: The mass spectrometry data supporting these findings are provided in the Supplementary Materials and are stored into secured data repositories at the Faculty of Medicine and Health Sciences at the Université de Sherbrooke, Sherbrooke, Quebec.

Acknowledgments: We are thankful to Sanofi Canada for providing the funds for this study. We are also grateful to Sanofi Canada for the doctoral educational bursary provided to Tristan Martineau. We are grateful to Waters Corporation for their technical expertise and support throughout this project. We would like to acknowledge the collaboration of colleagues and coordinating nurses of the Canadian Fabry Disease Initiative, Philippe Campeau (CHU Sainte-Justine, Montreal, QC), François E. Mercier (Jewish General Hospital, Montreal, QC), and Michael H. Gelb (University of Washington) for providing samples for this study. We would like to thank Pamela Lavoie for her expertise/comments in revising the manuscript.

Conflicts of Interest: T.M. received a doctoral educational bursary from Sanofi Canada. B.M. declares no conflicts of interest. C. Auray-Blais has received grants, and speaker honoraria from Takeda Pharmaceutical International AG., Sanofi Genzyme, BioMarin Pharmaceuticals Inc., Protalix Biotherapeutics, Amicus Therapeutics, and Avrobio, and traveling expenses from Waters Corp. for lectures given. She has received service agreements from Moderna Therapeutics, Sigilon Therapeutics, and 4D Molecular Therapeutics. The funders had no role in the design of the study; in the collection, analyses, or interpretation of data; in the writing of the manuscript; or in the decision to publish the results.

References

1. Platt, F.M.; d'Azzo, A.; Davidson, B.L.; Neufeld, E.F.; Tifft, C.J. Lysosomal storage diseases. *Nat. Rev. Dis. Prim.* **2018**, *4*, 27. [CrossRef]
2. Lahiri, S.; Futerman, A.H. The metabolism and function of sphingolipids and glycosphingolipids. *Cell. Mol Life Sci.* **2007**, *64*, 2270–2284. [CrossRef] [PubMed]
3. Abed Rabbo, M.; Khodour, Y.; Kaguni, L.S.; Stiban, J. Sphingolipid lysosomal storage diseases: From bench to bedside. *Lipids Health Dis.* **2021**, *20*, 44. [CrossRef]
4. Abou-Ghali, M.; Stiban, J. Regulation of ceramide channel formation and disassembly: Insights on the initiation of apoptosis. *Saudi J. Biol. Sci.* **2015**, *22*, 760–772. [CrossRef]

5. Auray-Blais, C.; Boutin, M.; Gagnon, R.; Dupont, F.O.; Lavoie, P.; Clarke, J.T. Urinary globotriaosylsphingosine-related biomarkers for Fabry disease targeted by metabolomics. *Anal. Chem.* **2012**, *84*, 2745–2753. [CrossRef]
6. Menkovic, I.; Boutin, M.; Alayoubi, A.; Mercier, F.E.; Rivard, G.E.; Auray-Blais, C. Identification of a reliable biomarker profile for the diagnosis of Gaucher disease type 1 patients using a mass spectrometry-based metabolomic approach. *Int. J. Mol. Sci.* **2020**, *21*, 7869. [CrossRef] [PubMed]
7. Menkovic, I.; Boutin, M.; Alayoubi, A.; Curado, F.; Bauer, P.; Mercier, F.E.; Auray-Blais, C. Metabolomic study using time-of-flight mass spectrometry reveals novel urinary biomarkers for Gaucher disease type 1. *J. Proteome Res.* **2022**, *21*, 1321–1329. [CrossRef] [PubMed]
8. Dupont, F.O.; Gagnon, R.; Boutin, M.; Auray-Blais, C. A metabolomic study reveals novel plasma lyso-Gb_3 analogs as Fabry disease biomarkers. *Curr. Med. Chem.* **2013**, *20*, 280–288. [CrossRef] [PubMed]
9. Ferreira, C.R.; Gahl, W.A. Lysosomal storage diseases. *Transl. Sci. Rare Dis.* **2017**, *2*, 1–71. [CrossRef] [PubMed]
10. Stirnemann, J.; Belmatoug, N.; Camou, F.; Serratrice, C.; Froissart, R.; Caillaud, C.; Levade, T.; Astudillo, L.; Serratrice, J.; Brassier, A.; et al. A review of Gaucher disease pathophysiology, clinical presentation and treatments. *Int. J. Mol. Sci.* **2017**, *18*, 441. [CrossRef]
11. Baris, H.N.; Cohen, I.J.; Mistry, P.K. Gaucher disease: The metabolic defect, pathophysiology, phenotypes and natural history. *Pediatr. Endocrinol. Rev.* **2014**, *12*, 72–81. [PubMed]
12. Won, J.S.; Kim, J.; Paintlia, M.K.; Singh, I.; Singh, A.K. Role of endogenous psychosine accumulation in oligodendrocyte differentiation and survival: Implication for Krabbe disease. *Brain Res.* **2013**, *1508*, 44–52. [CrossRef] [PubMed]
13. Bellettato, C.M.; Hubert, L.; Scarpa, M.; Wangler, M.F. Inborn errors of metabolism involving complex molecules: Lysosomal and peroxisomal storage diseases. *Pediatr. Clin. N. Am.* **2018**, *65*, 353–373. [CrossRef] [PubMed]
14. Wenger, D.A.; Luzi, P.; Rafi, M.A. Advances in the diagnosis and treatment of Krabbe disease. *Int. J. Neonatal Screen.* **2021**, *7*, 57. [CrossRef] [PubMed]
15. Ishii, S.; Taguchi, A.; Okino, N.; Ito, M.; Maruyama, H. Determination of globotriaosylceramide analogs in the organs of a mouse model of Fabry disease. *J. Biol. Chem.* **2020**, *295*, 5577–5587. [CrossRef] [PubMed]
16. Besekar, S.M.; Jogdand, S.D.; Naqvi, W.M. Fabry disease and its management: A literature analysis. *Cureus.* **2023**, *15*, e37048. [CrossRef]
17. Vanier, M.T. Chapter 176—Niemann-Pick diseases. *Handb. Clin. Neurol.* **2013**, *113*, 1717–1721. [PubMed]
18. Shaimardanova, A.A.; Chulpanova, D.S.; Solovyeva, V.V.; Mullagulova, A.I.; Kitaeva, K.V.; Allegrucci, C.; Rizvanov, A.A. Metachromatic leukodystrophy: Diagnosis, modeling, and treatment approaches. *Front. Med.* **2020**, *7*, 576221. [CrossRef]
19. Regier, D.S.; Proia, R.L.; D'Azzo, A.; Tifft, C.J. The GM1 and GM2 gangliosidoses: Natural history and progress toward therapy. *Pediatr. Endocrinol Rev.* **2016**, *13* (Suppl. S1), 663–673. [PubMed]
20. Kong, W.; Lu, C.; Ding, Y.; Meng, Y. Update of treatment for Gaucher disease. *Eur. J. Pharmacol.* **2022**, *926*, 175023. [CrossRef]
21. Umer, M.; Kalra, D.K. Treatment of Fabry disease: Established and emerging therapies. *Pharmaceuticals* **2023**, *16*, 320. [CrossRef] [PubMed]
22. Geberhiwot, T.; Moro, A.; Dardis, A.; Ramaswami, U.; Sirrs, S.; Marfa, M.P.; Vanier, M.T.; Walterfang, M.; Bolton, S.; Dawson, C.; et al. Consensus clinical management guidelines for Niemann-Pick disease type C. *Orphanet J. Rare Dis.* **2018**, *13*, 50. [CrossRef] [PubMed]
23. Khan, A.; Barber, D.L.; Huang, J.U.; Rupar, C.A.; Rip, J.W.; Auray-Blais, C.; Boutin, M.; O'Hoski, P.; Gargulak, K.; McKillop, W.M.; et al. Lentivirus-mediated gene therapy for Fabry disease. *Nat. Commun.* **2021**, *12*, 1178. [CrossRef]
24. Auray-Blais, C.; Lavoie, P.; Martineau, T.; Rupar, C.A.; Barber, D.L.; Keating, A.; Foley, R.; Khan, A.; West, M.L.; Medin, J.A. Longitudinal biomarker evaluation in Fabry disease patients receiving lentivirus-mediated gene therapy. *Rare Dis. Orphan Drugs J.* **2024**, *3*, 18. [CrossRef]
25. Adang, L.A.; Bonkowsky, J.L.; Boelens, J.J.; Mallack, E.; Ahrens-Nicklas, R.; Bernat, J.A.; Bley, A.; Burton, B.; Darling, A.; Eichler, F.; et al. Consensus guidelines for the monitoring and management of metachromatic leukodystrophy in the United States. *Cytotherapy* **2024**, *26*, 739–748. [CrossRef]
26. Yoon, I.C.; Bascou, N.A.; Poe, M.D.; Szabolcs, P.; Escolar, M.L. Long-term neurodevelopmental outcomes of hematopoietic stem cell transplantation for late-infantile Krabbe disease. *Blood* **2021**, *137*, 1719–1730. [CrossRef] [PubMed]
27. Santos, R.; Amaral, O. Advances in sphingolipidoses: CRISPR-Cas9 editing as an option for modelling and therapy. *Int. J. Mol. Sci.* **2019**, *20*, 5897. [CrossRef]
28. Gromova, M.; Vaggelas, A.; Dallmann, G.; Seimetz, D. Biomarkers: Opportunities and challenges for drug development in the current regulatory landscape. *Biomark. Insights* **2020**, *15*, 1177271920974652. [CrossRef] [PubMed]
29. Meikle, P.J.; Hopwood, J.J. Lysosomal storage disorders: Emerging therapeutic options require early diagnosis. *Eur. J. Pediatr.* **2003**, *162* (Suppl. S1), S34–S37. [CrossRef] [PubMed]
30. Mashima, R.; Okuyama, T.; Ohira, M. Biomarkers for lysosomal storage disorders with an emphasis on mass spectrometry. *Int. J. Mol. Sci.* **2020**, *21*, 2704. [CrossRef]
31. Gragnaniello, V.; Cazzorla, C.; Gueraldi, D.; Puma, A.; Loro, C.; Porcù, E.; Stornaiuolo, M.; Miglioranza, P.; Salviati, L.; Burlina, A.P.; et al. Light and shadows in newborn screening for lysosomal storage disorders: Eight years of experience in northeast Italy. *Int. J. Neonatal Screen.* **2023**, *10*, 3. [CrossRef] [PubMed]

32. Gelb, M.H.; Scott, R.C.; Turecek, F. Newborn Screening for Lysosomal Storage Diseases. *Clin. Chem.* **2015**, *61*, 335–346. [CrossRef] [PubMed]
33. Gelb, M.H.; Lukacs, Z.; Ranieri, E.; Schielen, P.C.J.I. Newborn screening for lysosomal storage disorders: Methodologies for measurement of enzymatic activities in dried blood spots. *Int. J. Neonatal Screen.* **2019**, *5*, 1. [CrossRef] [PubMed]
34. McClendon-Weary, B.; Putnick, D.L.; Robinson, S.; Yeung, E. Little to give, much to gain-what can you do with a dried blood spot? *Curr. Environ. Health Rep.* **2020**, *7*, 211–221. [CrossRef] [PubMed]
35. Gragnaniello, V.; Burlina, A.P.; Commone, A.; Gueraldi, D.; Puma, A.; Porcù, E.; Stornaiuolo, M.; Cazzorla, C.; Burlina, A.B. Newborn screening for Fabry Disease: Current status of knowledge. *Int. J. Neonatal Screen.* **2023**, *9*, 31. [CrossRef] [PubMed]
36. Gelb, M.H. Newborn screening for lysosomal storage diseases: Methodologies, screen positive rates, normalization of datasets, second-tier tests, and post-analysis tools. *Int. J. Neonatal Screen.* **2018**, *4*, 23. [CrossRef] [PubMed]
37. Burlina, A.; Brand, E.; Hughes, D.; Kantola, I.; Krämer, J.; Nowak, A.; Tøndel, C.; Wanner, C.; Spada, M. An expert consensus on the recommendations for the use of biomarkers in Fabry disease. *Mol. Genet. Metab.* **2023**, *139*, 107585. [CrossRef] [PubMed]
38. van Eijk, M.; Ferraz, M.J.; Boot, R.G.; Aerts, J.M.F.G. Lyso-glycosphingolipids: Presence and consequences. *Essays Biochem.* **2020**, *64*, 565–578.
39. Spassieva, S.; Bieberich, E. Lysosphingolipids and sphingolipidoses: Psychosine in Krabbe's disease. *J. Neurosci. Res.* **2016**, *94*, 974–981. [CrossRef]
40. Auray-Blais, C.; Lavoie, P.; Boutin, M.; Ntwari, A.; Hsu, T.R.; Huang, C.K.; Niu, D.M. Biomarkers associated with clinical manifestations in Fabry disease patients with a late-onset cardiac variant mutation. *Clin. Chim. Acta* **2017**, *466*, 185–193. [CrossRef]
41. Alharbi, F.J.; Baig, S.; Auray-Blais, C.; Boutin, M.; Ward, D.G.; Wheeldon, N.; Steed, R.; Dawson, C.; Hughes, D.; Geberhiwot, T. Globotriaosylsphingosine (Lyso-Gb_3) as a biomarker for cardiac variant (N215S) Fabry disease. *J. Inherit. Metab. Dis.* **2018**, *41*, 239–247. [CrossRef]
42. Pettazzoni, M.; Froissart, R.; Pagan, C.; Vanier, M.T.; Ruet, S.; Latour, P.; Guffon, N.; Fouilhoux, A.; Germain, D.P.; Levade, T.; et al. LC-MS/MS multiplex analysis of lysosphingolipids in plasma and amniotic fluid: A novel tool for the screening of sphingolipidoses and Niemann-Pick type C disease. *PLoS ONE* **2017**, *12*, e0181700. [CrossRef] [PubMed]
43. Polo, G.; Burlina, A.P.; Kolamunnage, T.B.; Zampieri, M.; Dionisi-Vici, C.; Strisciuglio, P.; Zaninotto, M.; Plebani, M.; Burlina, A.B. Diagnosis of sphingolipidoses: A new simultaneous measurement of lysosphingolipids by LC-MS/MS. *Clin. Chem. Lab. Med.* **2017**, *55*, 403–414. [CrossRef] [PubMed]
44. Suzuki, A.; Silsirivanit, A.; Watanabe, T.; Matsuda, J.; Inamori, K.I.; Inokuchi, J.I. Mass Spectrometry of neutral glycosphingolipids. In *Glycolipids: Methods and Protocols*; Springer: New York, NY, USA, 2023; Volume 2613, pp. 127–144.
45. Polo, G.; Burlina, A.P.; Ranieri, E.; Colucci, F.; Rubert, L.; Pascarella, A.; Duro, G.; Tummolo, A.; Padoan, A.; Plebani, M.; et al. Plasma and dried blood spot lysosphingolipids for the diagnosis of different sphingolipidoses: A comparative study. *Clin. Chem. Lab. Med.* **2019**, *57*, 1863–1874. [CrossRef] [PubMed]
46. Spiewak, J.; Doykov, I.; Papandreou, A.; Hällqvist, J.; Mills, P.; Clayton, P.T.; Gissen, P.; Mills, K.; Heywood, W.E. New perspectives in dried blood spot biomarkers for lysosomal storage diseases. *Int. J. Mol. Sci.* **2023**, *24*, 10177. [CrossRef] [PubMed]
47. Mirzaian, M.; Kramer, G.; Poorthuis, B.J. Quantification of sulfatides and lysosulfatides in tissues and body fluids by liquid chromatography-tandem mass spectrometry. *J. Lipid Res.* **2015**, *56*, 936–943. [CrossRef] [PubMed]
48. Lawrence, R.; Van Vleet, J.L.; Mangini, L.; Harris, A.; Martin, N.; Clark, W.; Chandriani, S.; LeBowitz, J.H.; Giugliani, R.; d'Azzo, A.; et al. Characterization of glycan substrates accumulating in GM1 gangliosidosis. *Mol. Genet. Metab. Rep.* **2019**, *21*, 100524. [CrossRef] [PubMed]
49. Xia, B.; Asif, G.; Arthur, L.; Pervaiz, M.A.; Li, X.; Liu, R.; Cummings, R.D.; He, M. Oligosaccharide analysis in urine by maldi-tof mass spectrometry for the diagnosis of lysosomal storage diseases. *Clin. Chem.* **2013**, *59*, 1357–1368. [CrossRef] [PubMed]
50. Iwahori, A.; Maekawa, M.; Narita, A.; Kato, A.; Sato, T.; Ogura, J.; Sato, Y.; Kikuchi, M.; Noguchi, A.; Higaki, K.; et al. Development of a diagnostic screening strategy for Niemann-Pick diseases based on simultaneous liquid chromatography-tandem mass spectrometry analyses of N-Palmitoyl-O-phosphocholine-serine and sphingosylphosphorylcholine. *Biol Pharm. Bull.* **2020**, *43*, 1398–1406. [CrossRef] [PubMed]
51. Lavoie, P.; Boutin, M.; Auray-Blais, C. Multiplex analysis of novel urinary lyso-Gb_3-related biomarkers for Fabry disease by tandem mass spectrometry. *Anal. Chem.* **2013**, *85*, 1743–1752. [CrossRef] [PubMed]
52. Menkovic, I.; Boutin, M.; Alayoubi, A.; Curado, F.; Bauer, P.; Mercier, F.E.; Auray-Blais, C. Quantitation of a urinary profile of biomarkers in Gaucher disease type 1 patients using tandem mass spectrometry. *Diagnostics* **2022**, *12*, 1414. [CrossRef] [PubMed]
53. Kelkel, M.A.; Boutin, M.; Curado, F.; Bauer, P.; Beauregard-Lacroix, É.; Mercier, F.E.; Maranda, B.; Menkovic, I.; Martineau, T.; Auray-Blais, C. Lysosphingolipid urine screening test using mass spectrometry for the early detection of lysosomal storage disorders. *Bioanalysis* **2022**, *14*, 289–306. [CrossRef]
54. Armengol, M.; Hayward, G.; Abbott, M.G.; Bird, C.; Bergmann, J.H.M.; Glogowska, M. Clinician and parent views on urine collection in precontinent children in the UK: A qualitative interview study. *BMJ Open* **2024**, *14*, e081306. [CrossRef]
55. Auray-Blais, C.; Cyr, D.; Drouin, R. Quebec neonatal mass urinary screening programme: From micromolecules to macromolecules. *J. Inherit. Metab. Dis.* **2007**, *30*, 515–521. [CrossRef] [PubMed]
56. Auray-Blais, C.; Boutin, M.; Lavoie, P.; Maranda, B. Neonatal urine screening program in the province of Quebec: Technological upgrade from thin layer chromatography to tandem mass spectrometry. *Int. J. Neonatal Screen.* **2021**, *7*, 18. [CrossRef] [PubMed]

57. Civallero, G.; Bender, F.; Gomes, A.; Marasca, G.; Guidobono, R.; De Mari, J.; Burin, M.; Giugliani, R. Reliable detection of mucopolysacchariduria in dried-urine filter paper samples. *Clin. Chim. Acta* **2013**, *415*, 334–336. [CrossRef] [PubMed]
58. de Souza, H.M.R.; Scalco, F.B.; Garrett, R.; de C Marques, F.F. Development of a kit for urine collection on filter paper as an alternative for Pompe disease screening and monitoring by LC-HRMS. *Anal. Methods* **2023**, *15*, 3932–3939. [CrossRef]
59. Auray-Blais, C.; Lavoie, P.; Abaoui, M.; Côté, A.M.; Boutin, M.; Akbari, A.; Levin, A.; Mac-Way, F.; Tr Clarke, J. High-risk screening for Fabry disease in a Canadian cohort of chronic kidney disease patients. *Clin. Chim. Acta* **2020**, *501*, 234–240. [CrossRef] [PubMed]
60. Barcenas, M.; Suhr, T.R.; Scott, C.R.; Turecek, F.; Gelb, M.H. Quantification of sulfatides in dried blood and urine spots from metachromatic leukodystrophy patients by liquid chromatography/electrospray tandem mass spectrometry. *Clin. Chim. Acta* **2014**, *433*, 39–43. [CrossRef] [PubMed]
61. Whitehead, D.; Conley, J. The next frontier of remote patient monitoring: Hospital at home. *J. Med. Internet Res.* **2023**, *25*, e42335. [CrossRef]
62. Stöth, F.; Martin Fabritius, M.; Weinmann, W.; Luginbühl, M.; Gaugler, S.; König, S. Application of dried urine spots for non-targeted quadrupole time-of-flight drug screening. *J. Anal. Toxicol.* **2023**, *47*, 332–337. [CrossRef] [PubMed]
63. Smit, P.W.; Elliott, I.; Peeling, R.W.; Mabey, D.; Newton, P.N. An overview of the clinical use of filter paper in the diagnosis of tropical diseases. *Am. J. Trop. Med. Hyg.* **2014**, *90*, 195–210. [CrossRef]
64. FDA Access Data. Elfabrio. Available online: https://www.accessdata.fda.gov/drugsatfda_docs/label/2023/761161s000lbl.pdf (accessed on 19 July 2024).
65. Matuszewski, B.K.; Constanzer, M.L.; Chavez-Eng, C.M. Strategies for the assessment of matrix effect in quantitative bioanalytical methods based on HPLC-MS/MS. *Anal. Chem.* **2003**, *75*, 3019–3030. [CrossRef] [PubMed]
66. Saville, J.T.; Smith, N.J.; Fletcher, J.M.; Fuller, M. Quantification of plasma sulfatides by mass spectrometry: Utility for metachromatic leukodystrophy. *Anal. Chim. Acta* **2017**, *955*, 79–85. [CrossRef] [PubMed]
67. Yang, S.; Mu, L.; Feng, R.; Kong, X. Selection of internal standards for quantitative matrix-assisted laser desorption/ionization mass spectrometric analysis based on correlation coefficients. *ACS Omega* **2019**, *4*, 8249–8254. [CrossRef] [PubMed]
68. Chuang, W.L.; Pacheco, J.; Cooper, S.; McGovern, M.M.; Cox, G.F.; Keutzer, J.; Zhang, X.K. Lyso-sphingomyelin is elevated in dried blood spots of Niemann-Pick B patients. *Mol. Genet. Metab.* **2014**, *111*, 209–211. [CrossRef] [PubMed]
69. Welford, R.W.D.; Farine, H.; Steiner, M.; Garzotti, M.; Dobrenis, K.; Sievers, C.; Strasser, D.S.; Amraoui, Y.; Groenen, P.M.A.; Giugliani, R.; et al. Plasma neurofilament light, glial fibrillary acidic protein and lysosphingolipid biomarkers for pharmacodynamics and disease monitoring of GM2 and GM1 gangliosidoses patients. *Mol. Genet. Metab. Rep.* **2022**, *30*, 100843. [CrossRef] [PubMed]
70. Alharbi, F.J.; Baig, S.; Rambhatla, S.B.; Vijapurapu, R.; Auray-Blais, C.; Boutin, M.; Steeds, R.; Wheeldon, N.; Dawson, C.; Geberhiwot, T. The clinical utility of total concentration of urinary globotriaosylsphingosine plus its analogues in the diagnosis of Fabry disease. *Clin. Chim. Acta* **2020**, *500*, 120–127. [CrossRef]
71. Ferreira, S.; Auray-Blais, C.; Boutin, M.; Lavoie, P.; Nunes, P.L.; Martins, E.; Garman, S.C.; Oliveira, J.P. Variations in the GLA gene correlate with globotriaosylceramide and globotriaosylsphingosine analog levels in urine and plasma. *Clin. Chim. Acta* **2015**, *447*, 96–104. [CrossRef]
72. Stamerra, C.A.; Del Pinto, R.; di Giosia, P.; Ferri, C.; Sahebkar, A. Anderson–Fabry disease: From endothelial dysfunction to emerging therapies. *Adv. Pharmacol. Pharm. Sci.* **2021**, *2021*, 5548445. [CrossRef] [PubMed]

biomolecules

MDPI

Article

Comparative Effects of GLP-1 and GLP-2 on Beta-Cell Function, Glucose Homeostasis and Appetite Regulation

Asif Ali, Dawood Khan, Vaibhav Dubey, Andrei I. Tarasov, Peter R. Flatt and Nigel Irwin *

Centre for Diabetes, School of Biomedical Sciences, Ulster University, Cromore Road, Coleraine BT52 1SA, Northern Ireland, UK; asif_ali-aa@ulster.ac.uk (A.A.); d.khan@ulster.ac.uk (D.K.); dubey-v@ulster.ac.uk (V.D.); a.tarasov@ulster.ac.uk (A.I.T.); pr.flatt@ulster.ac.uk (P.R.F.)

* Correspondence: n.irwin@ulster.ac.uk

Abstract: Glucagon-like peptide-1 (GLP-1) and glucagon-like peptide-2 (GLP-2) are related intestinal L-cell derived secretory products. GLP-1 has been extensively studied in terms of its influence on metabolism, but less attention has been devoted to GLP-2 in this regard. The current study compares the effects of these proglucagon-derived peptides on pancreatic beta-cell function, as well as on glucose tolerance and appetite. The insulin secretory effects of GLP-1 and GLP-2 (10^{-12}–10^{-6} M) were investigated in BRIN-BD11 beta-cells as well as isolated mouse islets, with the impact of test peptides (10 nM) on real-time cytosolic cAMP levels further evaluated in mouse islets. The impact of both peptides (10^{-8}–10^{-6} M) on beta-cell growth and survival was also studied in BRIN BD11 cells. Acute in vivo (peptides administered at 25 nmol/kg) glucose homeostatic and appetite suppressive actions were then examined in healthy mice. GLP-1, but not GLP-2, concentration dependently augmented insulin secretion from BRIN-BD11 cells, with similar observations made in isolated murine islets. In addition, GLP-1 substantially increased $[cAMP]_{cyt}$ in islet cells and was significantly more prominent than GLP-2 in this regard. Both GLP-1 and GLP-2 promoted beta-cell proliferation and protected against cytokine-induced apoptosis. In overnight fasted healthy mice, as well as mice trained to eat for 3 h per day, the administration of GLP-1 or GLP-2 suppressed appetite. When injected conjointly with glucose, both peptides improved glucose disposal, which was associated with enhanced glucose-stimulated insulin secretion by GLP-1, but not GLP-2. To conclude, the impact of GLP-1 and GLP-2 on insulin secretion is divergent, but the effects of beta-cell signaling and overall health are similar. Moreover, the peripheral administration of either hormone in rodents results in comparable positive effects on blood glucose levels and appetite.

Keywords: appetite; beta-cell; cAMP; incretin; insulin secretion; islet

1. Introduction

Glucagon-like peptide-1 (GLP-1) and glucagon-like peptide-2 (GLP-2) are peptide hormones derived from the same precursor gene, known as proglucagon, and co-secreted in equimolar concentrations by enteroendocrine L-cells in response to nutrient ingestion [1]. The proglucagon gene product undergoes tissue-specific post-translational processing to yield different bioactive peptides, including GLP-1 and GLP-2 [2]. Thus, in the intestine and brain, proglucagon is processed by prohormone convertase 1/3 (PC1/3) at Arg-Arg sites to yield glicentin, oxyntomodulin (OXM), intervening peptide-2 (IP-2), GLP-1 and GLP-2. Conversely, in alpha-cells of the endocrine pancreas, proglucagon is cleaved by PC2 to generate glicentin-related pancreatic polypeptide (GRPP) intervening peptide-1 (IP-1), the major proglucagon fragment (MPGF) and glucagon [3,4]. Whilst there is evidence of islet synthesis and the secretion of GLP-1 and GLP-2 under conditions of islet stress [5,6], these hormones are still largely considered as intestinal-derived peptides.

To date, there is a plethora of literature relating to the metabolic benefits of GLP-1 receptor activation, highlighted by the clinical approval of GLP-1 mimetics for both type 2 diabetes and obesity [7,8]. Thus, the positive modulation of GLP-1 receptor signaling leads to an enhancement of glucose-stimulated insulin secretion (GSIS) from pancreatic beta-cells, together with the inhibition of glucagon secretion from alpha-cells [7], the slowing of gastric emptying and the promotion of satiety [2]. On the other hand, GLP-2 receptor activation has been shown to exert benefits on intestinal growth and repair through the promotion of epithelial proliferation [9], leading to the clinical application of GLP-2 drugs for short bowel syndrome [10]. Despite the structural similarities of GLP-1 and GLP-2, alongside the fact that both hormones are secreted in response to nutrient ingestion [1], there is a lack of information about the possible metabolic benefits of GLP-2. However, GLP-2 may have the potential to protect against the dysregulation of glucose metabolism, as well as positively modulate energy homeostasis [11–13].

Therefore, the present study directly compares the impact of GLP-1 and GLP-2 on pancreatic BRIN BD11 beta-cell proliferation, survival and overall secretory function. The effects of both peptides on insulin secretion were then verified in isolated murine islets, along with their influence on islet cell cytosolic cAMP concentrations. In addition, we investigated the effects of the intraperitoneal injection of GLP-1 and GLP-2 on glucose tolerance and GSIS in mice. Finally, the influence of both peptides on appetite regulation in mice fasted overnight, as well as mice trained to eat for 3 h per day, was examined. Overall, the data further emphasize the prominent anti-diabetic and -obesity effects of GLP-1 receptor signaling. Furthermore, we also reveal, for the first time, the positive impact of GLP-2 receptor signaling on islet cell cAMP levels, as well as beta-cell turnover, meriting further investigation in terms of therapeutic strategies for diabetes.

2. Materials and Methods

2.1. Peptides

Peptides (95% purity) were obtained from a commercial source (Synpeptide, Shanghai, China) and fully characterized in our laboratory, as previously described [14].

2.2. In Vitro Insulin Secretion

The BRIN-BD11 beta-cell line [15] was used to examine the insulin secretory actions of the test peptides (n = 8; 20 min incubation; 10^{-12}–10^{-6} M) at 5.6 and 16.7 mM of glucose, as described previously [16]. Furthermore, the impact of the peptides on insulin secretion (n = 4; 60 min incubation; 10^{-8} and 10^{-6} M) was also examined in an islet isolated from 12-week-old C57BL/6 mice by collagenase digestion [16]. The subsequent acid–ethanol extraction of test islets allowed for the presentation of islet secretion data as a percentage of the islet insulin content. Samples were kept at −20 °C prior to insulin determination using an in-house radioimmunoassay [17].

2.3. Live Islet Cell Time-Lapse Imaging

C57BL/6 male mouse (12 weeks old) islets were isolated as above. To quantify the cytosolic cAMP levels, a recombinant fluorescent sensor (Upward Green cADDis, Montana Molecular, Bozeman, MT, USA) was used. The sensor was delivered to the islets via adenoviral transduction, allowing 24–48 h for gene expression. Time-lapse imaging of $[cAMP]_{cyt}$ was performed using the Green Upward cADDis sensor, as described previously [18], with image acquisition managed using μManager 2.0 software, capturing the cAMP levels in the islets every 60 s (16 mHz). For imaging, a bath perfusion solution (140 mM of NaCl, 4.6 mM of KCl, 2.6 mM of $CaCl_2$, 1.2 mM of $MgCl_2$, 1 mM of NaH_2PO_4, 5 mM of $NaHCO_3$, 10 mM of glucose, 10 mM of HEPES, pH 7.4) was used, along with GLP-1, GLP-2 (both at 10 nM) or IBMX (100 μM) as a positive control. Image sequences were analyzed with the use of the open-source FIJI software version 2.9.0 (National Institutes of Health (NIH), Bethesda, MD, USA).

2.4. Beta-Cell Proliferation and Cellular Stress Studies

The impact of GLP-1 and GLP-2 on BRIN-BD11 beta-cell proliferation (40,000 cells per well) was assessed using the Ki67 primary antibody (Ab15580, AbCam, Cambridge, UK) and the Alexa Fluor® 594 secondary antibody, as described previously in our laboratory [19]. For apoptosis studies, cellular stress was induced by the incubation of BRIN BD11 beta-cells with a cytokine cocktail (IL-1β 100 U/mL, IFN-γ 20 U/mL, TNF-α 200 U/mL), and the rate of apoptosis was monitored through TUNEL staining (Fluorescein, Roche Diagnostics, Burgess Hill, UK) [19]. The effects were visualized using a fluorescence microscope (Olympus system microscope, model BX51; Southend-on-Sea, UK) and a DP70 camera adapter system using DAPI (350 nm), TRITC (594 nm) and FITC (488 nm) filters, alongside an Olympus XM10 camera. For quantification, the cell-counter function within ImageJ Software Version 1.54 (National Institutes of Health (NIH), Bethesda, MD, USA) was employed to establish the number of positively stained Ki-67 or TUNEL cells, as appropriate; the data were then presented as a percentage of the total cells investigated.

2.5. Animal Experiments

Animal experiments were performed in male C57BL/6 (12–14 weeks of age) or NIH Swiss male mice (30 weeks of age), as appropriate; both were purchased from Harlan Ltd., Huntingdon, UK. Animals were housed individually in the Biomedical and Behavioural Research Unit (BBRU) at Ulster University for pre-clinical studies, with a standard temperature and light cycle, namely 22 ± 2 °C and a 12 h light/dark cycle, respectively. All procedures were performed in compliance with the UK Animal Scientific Procedures Act 1986.

2.6. In Vivo Experiments

The effects of GLP-1 or GLP-2 (25 nmol/kg bw, an intraperitoneal (i.p.) administration) on food intake, glucose homeostasis and insulin secretion were studied in overnight fasted C57BL/6 mice, as previously documented in our laboratory [19]. The dosing regimen employed for the test peptides was based on previous positive observations with GLP-1 and GLP-2, as well as related gut-derived peptide hormones, within the same experimental systems [19]. In a separate series, NIH male mice habituated to a daily feeding regime of 3 h/day were also used to evaluate the impact of GLP-1 and GLP-2 (25 nmol/kg bw) on the cumulative food intake, using the same protocol as described above. These mice were subject to a progressive reduction in the daily feeding period over 3 weeks at 12 weeks of age, as detailed previously [20]. Mice were maintained on this 3 h/day feeding regimen until 30 weeks of age, and experiments were then conducted.

2.7. Biochemical Analyses

Blood glucose levels were quantified using a blood glucose meter (Ascencia Contour; Bayer Healthcare, Berkshire, UK). For plasma insulin analysis, blood samples were collected into chilled fluoride/heparin glucose micro-centrifuge tubes (Sarstedt, Numbrecht, Germany), immediately centrifuged at 13,000× g for 1 min, and retained at −20 °C before insulin quantification by radioimmunoassay [17].

2.8. Statistical Analyses

Statistical analyses were conducted using GraphPad PRISM software (Version 8.0, Irvine, CA, USA). Data are presented as mean ± S.E.M. Comparative analyses between groups were performed using one-way ANOVA with a Bonferroni post hoc test or Student's unpaired t-test, as appropriate. Statistical significance was defined as $p < 0.05$.

3. Results

3.1. GLP-1, but Not GLP-2, Evokes Prominent Insulin Secretion from BRIN BD11 Beta-Cells and Isolated Islets

At 5.6 and 16.7 mM of glucose, GLP-1 (10^{-10}–10^{-6} M) significantly ($p < 0.05$–0.001) augmented insulin secretion when compared to the controls (Figure 1A,B). Conversely, GLP-2 did not impact insulin release from BRIN BD11 cells at any of the concentrations tested at either 5.6 or 16.7 mM glucose (Figure 1A,B). Equivalent observations were made at 16.7 mM of glucose in collagenase-isolated mouse islets (Figure 1C).

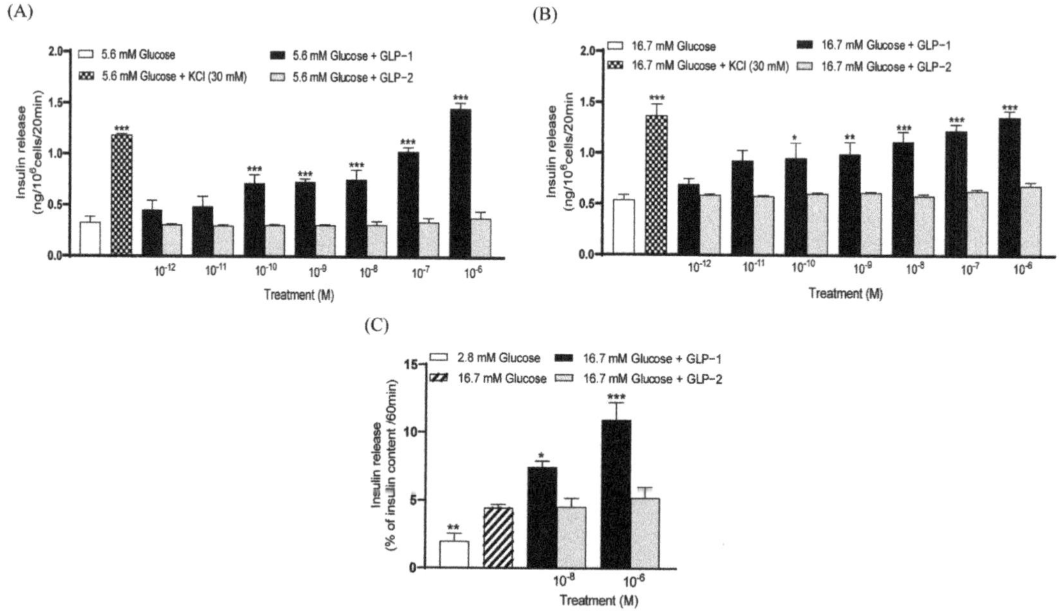

Figure 1. Effects of GLP-1 and GLP-2 on insulin secretion from (**A**,**B**) BRIN BD11 beta and (**C**) isolated mouse islets. BRIN BD11 cells were incubated for 20 min with (**A**) 5.6 or (**B**) 16.7 mM of glucose alone and in combination with test peptides (10^{-12} to 10^{-6} M), and the insulin secretion determined. (**C**) Isolated mouse islets were incubated for 60 min with peptides (10^{-8} and 10^{-6} M) at 16.7 mM of glucose (10 islets per well), and the insulin secretion was determined. Values are mean ± SEM ($n = 8$). * $p < 0.05$, ** $p < 0.01$, *** $p < 0.001$, compared to the respective glucose control.

3.2. GLP-1 and GLP-2 Elevate Cytosolic cAMP Concentrations in Isolated Mouse Islets

The incubation of murine islets with 10 mM of glucose in combination with 10 nM of GLP-1 resulted in a rapid and sustained increase ($p < 0.01$) in $[\text{cAMP}]_{cyt}$ (Figure 2A,B). GLP-2 also enhanced ($p < 0.01$) $[\text{cAMP}]_{cyt}$ in mouse islet cells, but was significantly less effective ($p < 0.01$) than GLP-1 (Figure 2A,B). As expected, the incubation of mouse islet cells with 100 μM of IBMX led to prominent increases ($p < 0.01$) in $[\text{cAMP}]_{cyt}$ (Figure 2A,B). In agreement, the impact of GLP-1 and IBMX on islet cell $[\text{cAMP}]_{cyt}$ was moderately positively correlated (r = 0.56), whereas similar correlations for GLP-2 with either IBMX (r = −0.18) or GLP-1 (r = −0.08) were not observed (Figure 2C).

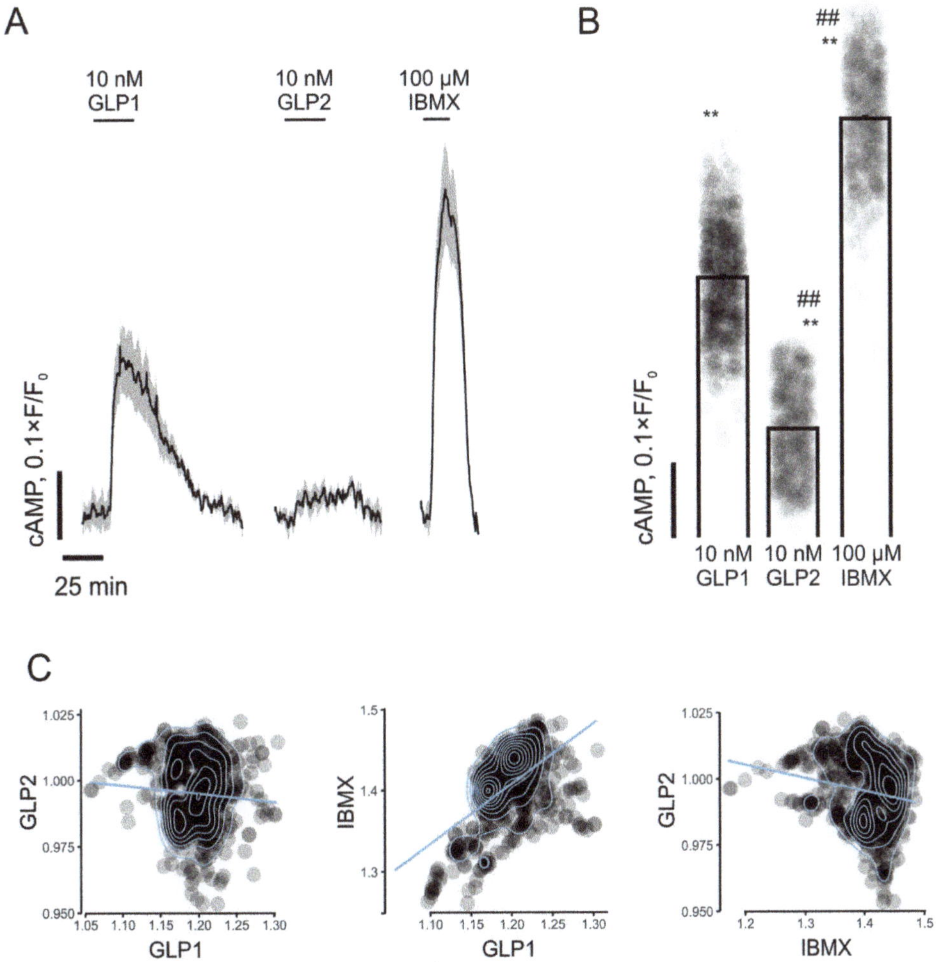

Figure 2. Effects of GLP-1 and GLP-2 on real-time cytosolic cAMP concentrations in isolated mouse islets. (**A**) Dynamics of $[cAMP]_{cyt}$ in mouse islet cells in response to incubation with GLP-1, GLP-2 (both at 10 nM) or IBMX (100 µM) in the presence of 10 mM of glucose. (**B**) Quantification of islet cell $[cAMP]_{cyt}$ responses to GLP-1, GLP-2 (both at 10 nM) or IBMX (100 µM). (**C**) Pearson correlation (per-cell) between $[cAMP]_{cyt}$ responses for GLP-1, GLP-2 and IBMX in islets. Values are mean ± SEM (n = 615 from three preparations). ** $p < 0.01$ compared to basal condition. ## $p < 0.01$ compared to 10 nM of GLP-1.

3.3. GLP-1 and GLP-2 Promote BRIN BD11 Beta-Cell Proliferation and Protect Against Cytokine-Induced Apoptosis

Both GLP-1 and GLP-2 (10^{-8} and 10^{-6} M) significantly increased ($p < 0.001$) BRIN BD11 cell proliferation following overnight culture (Figure 3A). Representative images of Ki-67-stained cells are also presented, with arrows indicating proliferating cells (Figure 3B). Similar to observations regarding proliferation, both GLP-1 and GLP-2 (10^{-8} and 10^{-6} M) significantly enhanced ($p < 0.001$) beta-cell survival and protected against cytokine-induced beta-cell apoptosis (Figure 4A). Representative images of TUNEL-stained BRIN BD11 cells are provided, with arrows indicating apoptotic cells (Figure 4B).

(A)

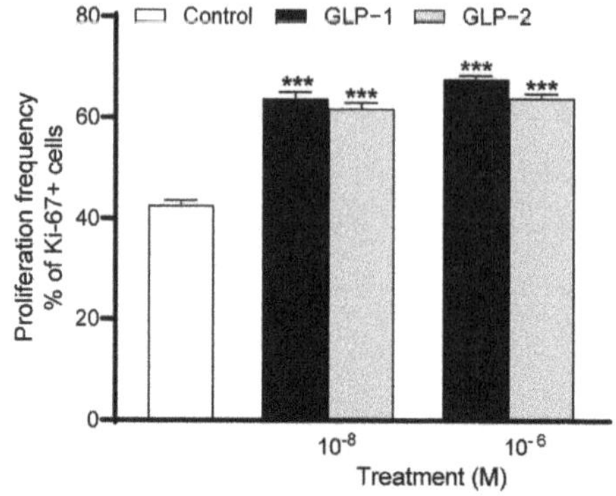

(B)

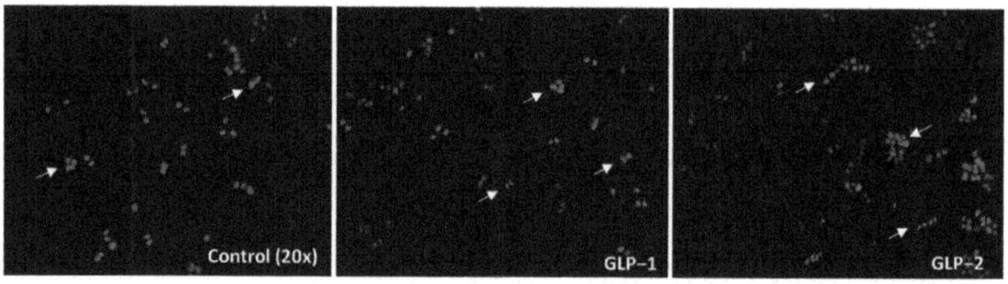

Figure 3. Effects of GLP-1 and GLP-2 on BRIN-BD11 cell proliferation. BRIN-BD11 cells were incubated with GLP-1 or GLP-2 (10^{-6} M and 10^{-8} M) for 18 h. Cells were then stained for Ki-67 (red) and DAPI (blue). The quantification of Ki-67 positive cells is shown in (**A**). Representative images (**B**) show Ki-67-positive cells indicated by arrows. Values are expressed as mean ± SEM (n = 6). *** $p < 0.001$ compared to media control.

(A)

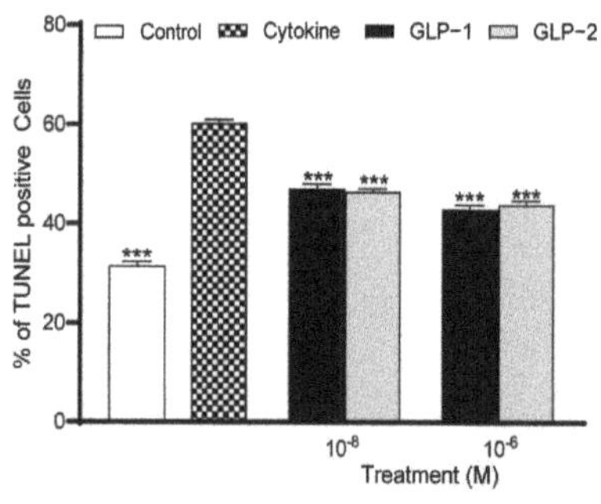

(B)

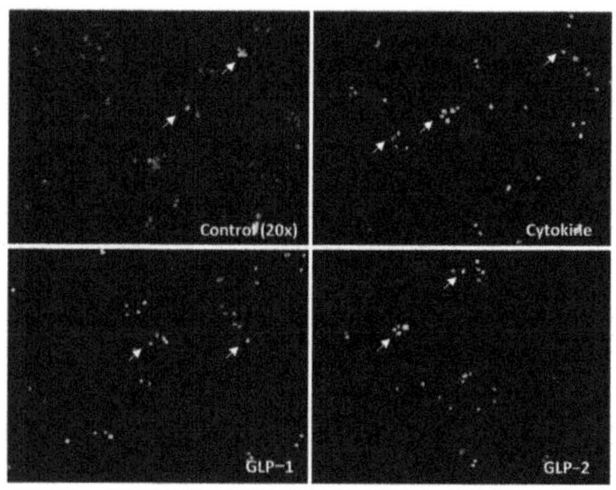

Figure 4. Effects of GLP-1 and GLP-2 on BRIN-BD11 cell apoptosis. BRIN BD11 cells were incubated in the presence of cytokines (IL-1β 100 U/mL, IFN-γ 20 U/mL, TNF-α 200 U/mL) alone or alongside GLP-1 or GLP-2 (10^{-6} M and 10^{-8} M) for 18 h before staining for TUNEL (green) or DAPI (blue). The quantification of TUNEL positive cells is shown in (**A**). Representative images (**B**) show TUNEL-positive cells indicated by arrows. Values are expressed as mean ± SEM (n = 4). *** $p < 0.001$ compared to the cytokine group.

3.4. GLP-1 and GLP-2 Improve Glucose Tolerance and Supress Appetite in Mice

The I.p. administration of GLP-1 or GLP-2 conjointly with glucose, at a dose of 25 nmol/kg, decreased individual as well as overall glucose values (Figure 5A,B). Specifically, GLP-1 reduced blood glucose levels at 60 ($p < 0.05$) and 90 ($p < 0.01$) minutes post-injection, with GLP-2 significantly reducing ($p < 0.05$) glucose only at 90 min (Figure 5A). Accordingly, the 0–90 min overall AUC glucose values were decreased by both GLP-1 ($p < 0.01$) and GLP-2 ($p < 0.05$) administration (Figure 5B). As expected, GLP-1 also augmented GSIS, corresponding to increased ($p < 0.05$) plasma insulin concentrations at 15 min post-injection, as well the overall values at 0–90 min ($p < 0.01$) (Figure 5C,D). In contrast, the effects of GLP-2 on insulin secretion failed to reach significance (Figure 5C,D). In terms of appetite suppression, both GLP-1 and GLP-2 reduced food intake ($p < 0.05$–0.001) at all observation points in the overnight fasted mice (Figure 6A). Both peptides were found to have similar significant ($p < 0.05$–0.01) appetite suppressive effects in mice trained to eat for 3 h/day, although GLP-2 lost efficacy in this model at 180 min post-injection (Figure 6B).

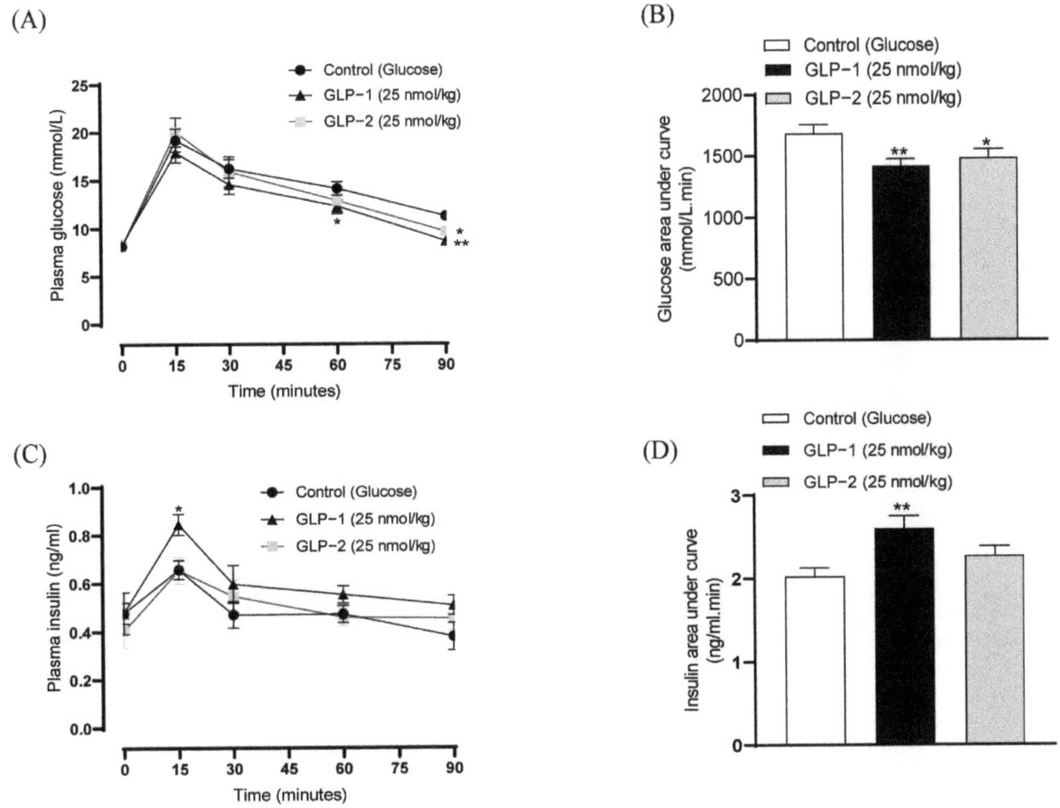

Figure 5. Effects of GLP-1 and GLP-2 on glucose tolerance and insulin secretion in mice. (**A**) Blood glucose and (**C**) plasma insulin was assessed following the administration of glucose alone (18 mmol/kg bw) or in combination with GLP-1 and GLP-2 (25 nmol/kg bw). (**B**,**D**) Overall 0–90 min AUC (**B**) glucose and (**D**) insulin values are also shown. Values are mean ± SEM ($n = 6$). * $p < 0.05$, ** $p < 0.01$, compared to the glucose control.

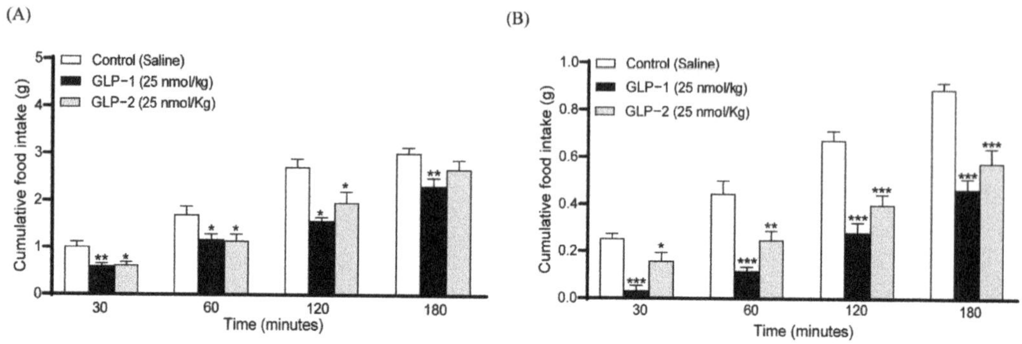

Figure 6. Effects of GLP-1 and GLP-2 on food intake in mice. Saline vehicle (0.9% NaCl), GLP-1 or GLP-2 (25 nmol/kg bw, i.p.) were administered to (**A**) overnight fasted mice or (**B**) mice trained to eat for 3 h per day, and the cumulative food intake was recorded at 30 min intervals for 3 h. Values are mean ± SEM (n = 6). * $p < 0.05$, ** $p < 0.01$, *** $p < 0.001$, compared to the respective saline control.

4. Discussion

In recent years, the products of the proglucagon gene have generated significant interest in the context of the regulation of glucose homeostasis and body composition. Among these, GLP-1 has emerged as a central player, with established therapeutic applicability for type 2 diabetes and obesity [2]. GLP-2 is a related proglucagon gene product that has structural and amino acid sequence homology with GLP-1 and has been found to promote intestinal growth and integrity, but has less well characterized effects on metabolism [21]. However, similar to GLP-1, GLP-2 receptor expression is evidenced in both human and rodent pancreatic islets, suggesting that it plays a direct role in islet function and overall metabolism [19,22]. Moreover, the GLP-2-mediated suppression of appetite has recently been observed in rodents and been demonstrated to be dependent on hypothalamic GLP-1 receptor signalling [23]. Thus, GLP-2 may have an overlapping bioactivity profile with GLP-1 that merits further consideration.

As anticipated, GLP-1 induced prominent concentration-dependent insulin secretory actions in BRIN BD11 beta-cells, as well as isolated mouse islets, that were associated with increased intracellular cAMP [24]. More intriguingly, we were also able to evidence small, but significant, GLP-2-mediated elevations in islet cell cAMP concentrations. In this regard, we have previously shown that GLP-2 does not alter the membrane potential or intracellular Ca^{2+} levels in pancreatic beta-cells [19]; thus, GLP-2 likely signals via cAMP and related downstream effectors such as protein kinase A (PKA) or exchange protein activated by cAMP (Epac) in islets, similar to GLP-1 [25]. This corresponds well with that observation that GLP-2 enhanced BRIN BD11 cell proliferation, as well as protection against cytokine-induced apoptosis, in the current study [26]. Such findings are also consistent with the well-documented proliferative actions of GLP-2 within the intestine [9]. That said, and in good agreement with others [6,19], GLP-2 did not augment GSIS within either the in vitro, ex vivo or in vivo environments. Similar differential effects of compounds on beta-cell function have been reported previously [27], as well as with agents such as Peptide YY (PYY) and Pancreatic Polypeptide (PP), which can inhibit insulin secretion but augment beta-cell growth and survival [4].

To add to the complexity, GLP-2 has been demonstrated to function as a low potent agonist of the GIP receptor [28], with GIP being well known to augment beta-cell function [29]. Thus, further investigations such as islet-cell-specific receptor knockdown studies would be required to fully delineate the effects of GLP-2 at the level of the beta-cell. This is even more relevant given the observations of local pancreatic islet GLP-2 synthesis and secretion in both rodent and human islets [6,19]. In addition, GLP-2 receptor expression is considered to be relatively low on beta-cells [6]. Nonetheless, the equivalent effects of

GLP-2 and GLP-1 on beta-cell turnover are intriguing and merit further investigation given that all types of diabetes are ultimately linked to the loss of beta-cell mass and function [30]. The anti-apoptotic role of GLP-1 in various cell types is already well documented [2], with GLP-2 previously demonstrated to protect against streptozotocin-induced DNA damage in beta-cells [19], as well as dextran sulphate-induced colitis and obesity-related neuroinflammation in mice [31,32]. Overall, it appears that GLP-2 possesses beta-cell-sparing activities that could be highly relevant for diseases such as diabetes.

Interestingly, both GLP-1 and GLP-2 improved glucose handling following combined injection with glucose in mice. For GLP-1, this can largely be attributed to GSIS [33], whereas for GLP-2, improvements in insulin action and/or insulin-independent glucose uptake would seem the most likely explanation. In this respect, there is an inverse correlation between the GLP-2 concentrations and insulin sensitivity in obese human subjects [34]. Moreover, GLP-2 receptor activation has been shown to prevent the glucose dysregulation that occurs following the induction of insulin resistance by prolonged high-fat feeding in mice [13], partly through enhancing insulin signalling [35]. In agreement with others [12,23], GLP-2 curbed food intake in mice to a similar degree as GLP-1, consistent with the knowledge that GLP-2 can directly regulate the hypothalamic neurons linked to appetite control [36]. However, unlike GLP-1 [37,38], the full translation of the appetite suppressive actions of GLP-2 is yet to be confirmed. However, our reliable observations in a well characterized rat beta-cell line and mouse model, alongside knowledge that the proglucagon gene is highly conserved across mammalian species [39], would suggest good translatability. Overall, GLP-2 exerts benefits on metabolism that are akin to those of GLP-1 [2], likely mediated through distinct pathways. It follows that a unimolecular dual GLP-1/GLP-2 receptor agonist, known as GUB09-145, suppressed caloric intake, promoted weight loss, and improved glucose tolerance in obese mice [40]. Furthermore, another recently characterized GLP-1/GLP-2 hybrid peptide, namely PG-102, was demonstrated to exert benefits on beta-cell mass and glucose control in *ob/ob* mice that were superior to either tirzepatide or retatrutide [41]. Thus, although the pancreatic beta-cell secretory actions of GLP-1 and GLP-2 are distinct, parallels in terms of their beneficial effects on beta-cell signalling and turnover, as well as glucose homeostasis and appetite regulation, provide an attractive avenue for therapeutic application. That said, there is some evidence that GLP-2 could accelerate the growth of colonic neoplasms [42]; thus, the potential intestinotrophic effects of GLP-2 would need to be appropriately moderated within this setting.

5. Conclusions

Interest in GLP-1 has increased significantly within the clinical field of diabetes and obesity. However, other related proglucagon-derived peptides, such as GLP-2, also merit attention. As such, despite GLP-2 lacking direct beta-cell secretory actions, this hormone supports overall beta-cell health by encouraging growth and preventing the destruction of these cells. In addition, we have shown, for the first time, that GLP-2 positively impacts cAMP oscillations in islet cells, giving credence to the notion that locally produced GLP-2 mediates the important crosstalk between endocrine cells [22]. When viewed alongside satiety and glucose homeostatic actions, this suggests that GLP-2 has untapped potential for the treatment of diabetes and obesity, either alone or perhaps more likely alongside GLP-1 therapies.

Author Contributions: P.R.F. and N.I. conceived/designed the study. A.A., D.K., V.D. and A.I.T. drafted the manuscript and participated in the conduct/data collection and analysis and interpretation of data. All authors have read and agreed to the published version of the manuscript.

Funding: These studies were supported by Ulster University Research Funding.

Institutional Review Board Statement: The study was conducted according to the guidelines of the Ulster Animal Welfare and Ethical Review Body (AWERB) committee and covered under a UK Home Office Animal project license (UK Home Office license certificate PPL2902 approved on 26 April 2021).

Informed Consent Statement: Not applicable.

Data Availability Statement: The authors declare that the data supporting the findings of this study are available within the article. Any additional raw data supporting the conclusions of this article will be made available by the senior author (N.I.), without undue reservation.

Conflicts of Interest: P.R.F. and N.I. are named on patents for the exploitation of peptide-based drugs for diabetes and are shareholders in Dia Beta Labs Ltd. All other authors declare no conflicts of interest.

References

1. Hein, G.J.; Baker, C.; Hsieh, J.; Farr, S.; Adeli, K. GLP-1 and GLP-2 as yin and yang of intestinal lipoprotein production: Evidence for predominance of GLP-2–stimulated postprandial lipemia in normal and insulin-resistant states. *Diabetes* **2013**, *62*, 373–381. [CrossRef] [PubMed]
2. Müller, T.D.; Finan, B.; Bloom, S.R.; D'Alessio, D.; Drucker, D.J.; Flatt, P.R.; Fritsche, A.; Gribble, F.; Grill, H.J.; Habener, J.F. Glucagon-like peptide 1 (GLP-1). *Mol. Metab.* **2019**, *30*, 72–130. [PubMed]
3. Steiner, D.F.; Smeekens, S.P.; Ohagi, S.; Chan, S.J. The new enzymology of precursor processing endoproteases. *J. Biol. Chem.* **1992**, *267*, 23435–23438. [CrossRef] [PubMed]
4. Lafferty, R.A.; O'Harte, F.P.; Irwin, N.; Gault, V.A.; Flatt, P.R. Proglucagon-derived peptides as therapeutics. *Front. Endocrinol.* **2021**, *12*, 689678. [CrossRef] [PubMed]
5. Marchetti, P.; Lupi, R.; Bugliani, M.; Kirkpatrick, C.L.; Sebastiani, G.; Grieco, F.A.; Del Guerra, S.; D'Aleo, V.; Piro, S.; Marselli, L. A local glucagon-like peptide 1 (GLP-1) system in human pancreatic islets. *Diabetologia* **2012**, *55*, 3262–3272. [CrossRef]
6. He, W.; Rebello, O.D.; Henne, A.; Nikolka, F.; Klein, T.; Maedler, K. GLP-2 is locally produced from human islets and balances inflammation through an inter-islet-immune cell crosstalk. *Front. Endocrinol.* **2021**, *12*, 697120. [CrossRef]
7. Ostawal, A.; Mocevic, E.; Kragh, N.; Xu, W. Clinical effectiveness of liraglutide in type 2 diabetes treatment in the real-world setting: A systematic literature review. *Diabetes Ther.* **2016**, *7*, 411–438. [CrossRef]
8. Wilding, J.P.; Batterham, R.L.; Calanna, S.; Davies, M.; Van Gaal, L.F.; Lingvay, I.; McGowan, B.M.; Rosenstock, J.; Tran, M.T.; Wadden, T.A. Once-weekly semaglutide in adults with overweight or obesity. *N. Engl. J. Med.* **2021**, *384*, 989–1002. [CrossRef]
9. Drucker, D.J.; Erlich, P.; Asa, S.L.; Brubaker, P.L. Induction of intestinal epithelial proliferation by glucagon-like peptide 2. *Proc. Natl. Acad. Sci. USA* **1996**, *93*, 7911–7916. [CrossRef]
10. Wallis, K.; Walters, J.R.; Gabe, S. Short bowel syndrome: The role of GLP-2 on improving outcome. *Curr. Opin. Clin. Nutr. Metab. Care* **2009**, *12*, 526–532. [CrossRef]
11. Bahrami, J.; Longuet, C.; Baggio, L.L.; Li, K.; Drucker, D.J. Glucagon-like peptide-2 receptor modulates islet adaptation to metabolic stress in the ob/ob mouse. *Gastroenterology* **2010**, *139*, 857–868. [CrossRef]
12. Baldassano, S.; Bellanca, A.L.; Serio, R.; Mul, F. Food intake in lean and obese mice after peripheral administration of glucagon-like peptide 2. *J. Endocrinol.* **2012**, *213*, 277. [CrossRef] [PubMed]
13. Baldassano, S.; Rappa, F.; Amato, A.; Cappello, F.; Mulè, F. GLP-2 as beneficial factor in the glucose homeostasis in mice fed a high fat diet. *J. Cell Physiol.* **2015**, *230*, 3029–3036. [CrossRef] [PubMed]
14. Lafferty, R.A.; Tanday, N.; McCloskey, A.; Bompada, P.; De Marinis, Y.; Flatt, P.R.; Irwin, N. Peptide YY (1–36) peptides from phylogenetically ancient fish targeting mammalian neuropeptide Y1 receptors demonstrate potent effects on pancreatic β-cell function, growth and survival. *Diabetes Obes. Metab.* **2020**, *22*, 404–416. [CrossRef] [PubMed]
15. McClenaghan, N.H.; Barnett, C.R.; Ah-Sing, E.; Abdel-Wahab, Y.H.; O'Harte, F.P.; Yoon, T.; Swanston-Flatt, S.K.; Flatt, P.R. Characterization of a novel glucose-responsive insulin-secreting cell line, BRIN-BD11, produced by electrofusion. *Diabetes* **1996**, *45*, 1132–1140. [CrossRef]
16. Pathak, V.; Vasu, S.; Gault, V.A.; Flatt, P.R.; Irwin, N. Sequential induction of beta cell rest and stimulation using stable GIP inhibitor and GLP-1 mimetic peptides improves metabolic control in C57BL/KsJ db/db mice. *Diabetologia* **2015**, *58*, 2144–2153. [CrossRef]
17. Flatt, P.R.; Bailey, C.J. Plasma glucose and insulin responses to glucagon and arginine in Aston ob/ob mice: Evidence for a selective defect in glucose-mediated insulin release. *Horm. Metab. Res.* **1982**, *14*, 127–130. [CrossRef]
18. Draper, M.; Willems, M.; Malahe, R.K.; Hamilton, A.; Tarasov, A.I. Imaging meets cytometry: Analyzing heterogeneous functional microscopic data from living cell populations. *J. Imaging* **2021**, *7*, 9. [CrossRef]
19. Khan, D.; Vasu, S.; Moffett, R.C.; Irwin, N.; Flatt, P.R. Differential expression of glucagon-like peptide-2 (GLP-2) is involved in pancreatic islet cell adaptations to stress and beta-cell survival. *Peptides* **2017**, *95*, 68–75. [CrossRef]
20. Irwin, N.; Frizelle, P.; Montgomery, I.A.; Moffett, R.C.; O'harte, F.; Flatt, P.R. Beneficial effects of the novel cholecystokinin agonist (pGlu-Gln)-CCK-8 in mouse models of obesity/diabetes. *Diabetologia* **2012**, *55*, 2747–2758. [CrossRef]
21. Drucker, D.J. Glucagon-like peptide 2. *J. Clin. Endocrinol. Metab.* **2001**, *86*, 1759–1764. [CrossRef] [PubMed]
22. Amisten, S.; Salehi, A.; Rorsman, P.; Jones, P.M.; Persaud, S.J. An atlas and functional analysis of G-protein coupled receptors in human islets of Langerhans. *Pharmacol. Ther.* **2013**, *139*, 359–391. [CrossRef] [PubMed]
23. Sun, H.; Meng, K.; Hou, L.; Shang, L.; Yan, J. GLP-2 decreases food intake in the dorsomedial hypothalamic nucleus (DMH) through Exendin (9–39) in male Sprague-Dawley (SD) rats. *Physiol. Behav.* **2021**, *229*, 113253. [CrossRef]

24. Drucker, D.J.; Philippe, J.; Mojsov, S.; Chick, W.L.; Habener, J.F. Glucagon-like peptide I stimulates insulin gene expression and increases cyclic AMP levels in a rat islet cell line. *Proc. Natl. Acad. Sci. USA* **1987**, *84*, 3434–3438. [CrossRef] [PubMed]
25. Tomas, A.; Jones, B.; Leech, C. New insights into beta-cell GLP-1 receptor and cAMP signaling. *J. Mol. Biol.* **2020**, *432*, 1347–1366. [CrossRef] [PubMed]
26. Kaihara, K.A.; Dickson, L.M.; Jacobson, D.A.; Tamarina, N.; Roe, M.W.; Philipson, L.H.; Wicksteed, B. β-Cell–specific protein kinase A activation enhances the efficiency of glucose control by increasing acute-phase insulin secretion. *Diabetes* **2013**, *62*, 1527–1536. [CrossRef]
27. Kolic, J.; Beet, L.; Overby, P.; Cen, H.H.; Panzhinskiy, E.; Ure, D.R.; Cross, J.L.; Huizinga, R.B.; Johnson, J.D. Differential effects of voclosporin and tacrolimus on insulin secretion from human islets. *Endocrinology* **2020**, *161*, bqaa162. [CrossRef]
28. Gabe, M.B.N.; Gasbjerg, L.S.; Gadgaard, S.; Lindquist, P.; Holst, J.J.; Rosenkilde, M.M. N-terminal alterations turn the gut hormone GLP-2 into an antagonist with gradual loss of GLP-2 receptor selectivity towards more GLP-1 receptor interaction. *Br. J. Pharmacol.* **2022**, *179*, 4473–4485. [CrossRef]
29. Widenmaier, S.B.; Kim, S.; Yang, G.K.; De Los Reyes, T.; Nian, C.; Asadi, A.; Seino, Y.; Kieffer, T.J.; Kwok, Y.N.; McIntosh, C.H. A GIP receptor agonist exhibits β-cell anti-apoptotic actions in rat models of diabetes resulting in improved β-cell function and glycemic control. *PLoS ONE* **2010**, *5*, e9590. [CrossRef]
30. Moon, J.H.; Choe, H.J.; Lim, S. Pancreatic beta-cell mass and function and therapeutic implications of using antidiabetic medications in type 2 diabetes. *J. Diabetes Investig.* **2024**, *15*, 669–683. [CrossRef]
31. Drucker, D.J.; Yusta, B.; Boushey, R.P.; DeForest, L.; Brubaker, P.L. Human [Gly2] GLP-2 reduces the severity of colonic injury in a murine model of experimental colitis. *Am. J. Physiol. Gastrointest. Liver Physiol.* **1999**, *276*, G79–G91. [CrossRef] [PubMed]
32. Nuzzo, D.; Baldassano, S.; Amato, A.; Picone, P.; Galizzi, G.; Caldara, G.F.; Di Carlo, M.; Mulè, F. Glucagon-like peptide-2 reduces the obesity-associated inflammation in the brain. *Neurobiol. Dis.* **2019**, *121*, 296–304. [CrossRef] [PubMed]
33. Drucker, D.J. Mechanisms of action and therapeutic application of glucagon-like peptide-1. *Cell Metab.* **2018**, *27*, 740–756. [CrossRef] [PubMed]
34. Geloneze, B.; Lima, M.M.d.O.; Pareja, J.C.; Barreto, M.R.L.; Magro, D.O. Association of insulin resistance and GLP-2 secretion in obesity: A pilot study. *Arq. Bras. De Endocrinol. Metabol.* **2013**, *57*, 632–635. [CrossRef]
35. Baldassano, S.; Amato, A.; Terzo, S.; Caldara, G.F.; Lentini, L.; Mulè, F. Glucagon-like peptide-2 analog and inflammatory state in obese mice. *Endocrine* **2020**, *68*, 695–698. [CrossRef]
36. Dalvi, P.S.; Belsham, D.D. Glucagon-like peptide-2 directly regulates hypothalamic neurons expressing neuropeptides linked to appetite control in vivo and in vitro. *Endocrinology* **2012**, *153*, 2385–2397. [CrossRef]
37. van Bloemendaal, L.; IJzerman, R.G.; Ten Kulve, J.S.; Barkhof, F.; Konrad, R.J.; Drent, M.L.; Veltman, D.J.; Diamant, M. GLP-1 receptor activation modulates appetite-and reward-related brain areas in humans. *Diabetes* **2014**, *63*, 4186–4196. [CrossRef]
38. De Silva, A.; Salem, V.; Long, C.J.; Makwana, A.; Newbould, R.D.; Rabiner, E.A.; Ghatei, M.A.; Bloom, S.R.; Matthews, P.M.; Beaver, J.D. The gut hormones PYY3-36 and GLP-17-36 amide reduce food intake and modulate brain activity in appetite centers in humans. *Cell Metab.* **2011**, *14*, 700–706. [CrossRef]
39. Irwin, D.M. Variation in the Evolution and Sequences of Proglucagon and the Receptors for Proglucagon-Derived Peptides in Mammals. *Front. Endocrinol.* **2021**, *12*, 700066. [CrossRef]
40. Madsen, M.S.A.; Holm, J.B.; Pallejà, A.; Wismann, P.; Fabricius, K.; Rigbolt, K.; Mikkelsen, M.; Sommer, M.; Jelsing, J.; Nielsen, H.B. Metabolic and gut microbiome changes following GLP-1 or dual GLP-1/GLP-2 receptor agonist treatment in diet-induced obese mice. *Sci. Rep.* **2019**, *9*, 15582. [CrossRef]
41. Won Kim, S.; Yang, S.; Sung, Y. 1874-LB: PG-102, a Bivalent GLP-1R/GLP-2R Agonist, Protects ß-Cell Mass and Enhances Glycemic Control in Obese db/db Mice, Showing Superiority over Semaglutide, Tirzepatide, and Retatrutide. *Diabetes* **2024**, *73*, 1874. [CrossRef]
42. Thulesen, J.; Hartmann, B.; Hare, K.J.; Kissow, H.; Ørskov, C.; Holst, J.J.; Poulsen, S.S. Glucagon-like peptide 2 (GLP-2) accelerates the growth of colonic neoplasms in mice. *Gut* **2004**, *53*, 1145–1150. [CrossRef]

biomolecules

MDPI

Review

A Scoping Review of Sarcoglycan Expression in Non-Muscle Organs: Beyond Muscles

Fabiana Nicita [1,2,*], Josè Freni [1,*], Antonio Centofanti [1], Angelo Favaloro [1], Davide Labellarte [1], Giuseppina Cutroneo [1], Michele Runci Anastasi [3] and Giovanna Vermiglio [1]

1 Department of Biomedical, Dental Sciences and Morphofunctional Imaging, University of Messina, 98125 Messina, Italy; antonio.centofanti@unime.it (A.C.); angelo.favaloro@unime.it (A.F.); davidelabellarte99@gmail.com (D.L.); giuseppina.cutroneo@unime.it (G.C.); gvermiglio1@unime.it (G.V.)

2 Department of Odontostomatological and Maxillofacial Sciences, "Sapienza" University of Rome, 00185 Rome, Italy

3 Department of Maxillo-Facial Surgery, "Sapienza" University of Rome, 00185 Rome, Italy; michele.runci@gmail.com

* Correspondence: fabnicita@unime.it (F.N.); jose.freni@unime.it (J.F.)

Abstract

This scoping review explores the expression patterns and molecular features of sarcoglycans (SGs) in non-muscle organs, challenging the long-standing assumption that their function is confined to skeletal and cardiac muscle. By analyzing evidence from both animal models and human studies, the review highlights the widespread presence of SG subunits in organs, including the nervous system, glands, adipose tissue, oral mucosa, retina, and other structures, with distinct regional and cell-type-specific patterns. Studies on the central nervous system demonstrate a widespread "spot-like" distribution of SG subunits in neurons and glial cells, implicating their involvement in synaptic organization and neurotransmission. Similarly, SGs maintain cellular integrity and homeostasis in glands and adipose tissue. At the same time, the altered expression of SGs is associated with pathological conditions in the gingival epithelium of the oral mucosa. These findings underscore the multifaceted roles of SGs beyond muscle, suggesting that they may contribute to cellular signaling, membrane stability, and neurovascular coupling. However, significant gaps remain regarding SG post-translational modifications and functional implications in non-muscle organs. Future research integrating molecular, cellular, and functional approaches in animal models and human tissues is essential to fully elucidate these roles and explore their potential as therapeutic targets in various diseases.

Keywords: synaptic organization; sarcoglycans; cell adhesion; non-muscle organ; tissue homeostasis; epithelial signaling

1. Introduction

Sarcoglycans (SGs) are a family of transmembrane glycoproteins essential for the maintenance of muscle cell structure and function [1–3]. These proteins, comprising six known subunits (α, β, γ, δ, ε, and ζ), assemble to form the sarcoglycan complex (SGC), a core component of the dystrophin-associated glycoprotein complex (DGC). Together with α- and β-dystroglycans (DGs), and sarcospan (SSPN), SGs establish a molecular bridge between the intracellular cytoskeleton and the extracellular matrix via laminin α2, thereby contributing to sarcolemmal integrity and enabling signal transduction during muscle contraction [4–10]. Disruption of this system leads to membrane instability and progressive muscle degeneration, as observed in several forms of limb–girdle muscular dystrophies (LGMDs) [11]. Each SG subunit displays tissue-specific expression patterns and distinct structural characteristics. α-, β-, γ-, and δ-SGs are predominantly expressed in skeletal and cardiac muscle, whereas ε-SG (which shares 44% sequence similarity with α-SG) is mainly found in smooth muscle. ζ-SG is structurally related to both γ- and δ-SG. The composition of the SGC adapts to tissue type: α-, β-, γ-, and δ-SGs are typically found in striated muscle, while ε-SG replaces α-SG in smooth muscle [12,13]. Structurally, β-, γ-, and δ-SGs are type II transmembrane proteins with intracellular N-termini, whereas α- and ε-SGs are type I proteins with extracellular N-terminal domains [13]. These differences suggest the existence of at least two distinct sarcoglycan complexes with potentially diverse functions [12,14]. Importantly, the structure and function of SGs are highly dependent on post-translational modifications and molecular interactions. Within the endoplasmic reticulum (ER), SGs undergo N-glycosylation, a sequential process involving the addition of mannose, sialic acid, and N-acetylglucosamine residues, which is essential for protein maturation, membrane targeting, and interaction with DG [3,15,16]. SGs are also localized within lipid-rich membrane microdomains (lipid rafts), composed of cholesterol and sphingolipids, which influence their clustering, localization, and signaling capacity. Moreover, SGs interact with various protein scaffolds, which support their stabilization and functional integration within cellular signaling networks.

While most research has focused on the role of SGs in muscle tissue and related pathologies, there is growing evidence of their expression in non-muscle tissues, including smooth muscle layers of internal organs (gastrointestinal and urogenital tracts, ureters, skin) [17–20], as well as in the brain, endocrine glands, and adipose tissue. Furthermore, mutations in SG-encoding genes have been linked to non-muscle phenotypes, such as the association between ε-SG mutations and myoclonus–dystonia syndrome, suggesting SG involvement in neuronal signaling and motor control [21,22]. However, the existing literature on SGs in non-muscle tissues primarily focuses on expression patterns, with limited attention to the molecular mechanisms that govern their function, specifically post-translational modifications, lipid interactions, and associations with scaffold proteins. Given that these factors are fundamental to SG structure and activity, it is likely that evaluating SGs in non-muscle contexts without considering these molecular features is incomplete or even misleading. This scoping review aims to explore and synthesize the current knowledge on SGs in non-muscle tissues, focusing primarily on their expression profiles and, secondarily, examining available data on their glycosylation patterns, interactions with lipid membranes, and associations with membrane scaffold proteins. By identifying whether and how these aspects have been investigated outside of muscle, this review seeks to fill a critical gap in the literature and support a more comprehensive understanding of SG functions across different tissue types.

2. Materials and Methods

2.1. Search Strategy

This scoping review followed the Preferred Reporting Items for Systematic Reviews and Meta-Analyses for Scoping Reviews (PRISMA-ScR) statement [23].

2.2. Eligibility Criteria

The primary question was the expression of SG proteins in various healthy non-muscle organs in both human and animal models. The secondary questions were as follows:

- Which SG subunits have been detected in non-muscle organs?
- What experimental or observational methods have been employed to assess SG expression?

- Are there reported associations between SG expression levels and specific pathological or physiological states in non-muscle organs?
- Are there molecular features, such as glycosylation, membrane lipid associations, or scaffold protein interactions, described for SGs in these organs?
- What are the main research gaps identified in the literature?

The eligibility criteria for admission in this scoping review are defined in Table 1.

Table 1. Inclusion and exclusion criteria.

Inclusion Criteria	Exclusion Criteria
Clinical and ex vivo human studies	Studies focusing exclusively on muscle
Animal studies	Studies on the expression of SGs in non-muscle organs identified in the muscular coat
Cell-based models	Studies involving sarcoglycanopathies and myoclonic dystonia
Studies that investigate the expression of one or more SG subunits (α, β, γ, δ, ε, and ζ) in non-muscle organs	Case reports, case series, reviews, meta-analyses, letters, editorials, commentaries, communications, supplements, and proceedings papers
Research articles as experimental studies, observational studies, and in vitro/in vivo analyses	Papers without full-text availability
Papers published in English	Papers published in languages other than English

2.3. Information Sources

A comprehensive literature search was conducted using the following electronic databases: PubMed/MEDLINE, Scopus, Web of Science, and ScienceDirect. In addition, Google Scholar was consulted to identify relevant grey literature. To ensure comprehensiveness, the search strategy included Boolean operators to identify articles using both MeSH (Medical Subject Headings) terms and free-text keywords: (Sarcoglycan complex OR Sarcoglycans) AND (Non-Muscle Tissue OR Epithelial tissue OR Gland OR Prostate OR Breast OR Liver OR Pancreas OR Spleen OR Nervous system OR Kidney OR Lung OR Connective tissue OR Adipose organ) AND (Immunohistochemistry OR RT-PCR OR Immunofluorescence OR mRNA). The search strategy was adapted for each database to optimize the retrieval of relevant articles (Table 2).

2.4. Screening and Selection Process

All articles identified by the search in the various databases were imported into reference management software (Mendeley Desktop, version 1.19.8), where duplicates were removed. After duplication, the titles and abstracts of the remaining studies were independently reviewed by two reviewers (FN and GV) to assess their relevance to the research questions. Studies containing the relevant search terms in the title and/or abstract were selected for full-text review. The two reviewers (FN and GV) independently reviewed the full-text articles of potentially relevant studies based on predefined eligibility criteria. Studies that met all criteria were included for data extraction, while those that did not meet the criteria were excluded, and the reasons for exclusion were recorded. Any disagreements between the two reviewers at this stage were discussed and resolved by consensus. If disagreement persisted, a third independent reviewer (JF) was consulted.

2.5. Data Extraction

The two reviewers (FN and GV) independently extracted the following data from the selected articles: authors' names, year of publication, type of study, tissue/organ types examined, SG subunits studied, and species/models. In cases where discrepancies emerged between data extractions, the third reviewer (JF) was consulted to resolve any disagreements.

Table 2. Database-specific search strategies.

Database	Search Strategy
PubMed/MEDLINE "https://pubmed.ncbi.nlm.nih.gov/ (accessed on 19 February 2025)"	(Sarcoglycan complex OR Sarcoglycans) AND (Non-Muscle Tissue OR Epithelial tissue OR Gland OR Prostate OR Breast OR Liver OR Pancreas OR Spleen OR Nervous system OR Kidney OR Lung OR Connective tissue OR Adipose organ) AND (Immunohistochemistry OR RT-PCR OR Immunofluorescence OR mRNA)
Scopus "https://www.scopus.com/home.uri (accessed on 19 February 2025)"	("Sarcoglycans" OR sarcoglycan*) AND ("Non-Muscle Tissues") AND (expression OR "immunohistochemistry" OR "immunofluorescence" OR "RT-PCR" OR mRNA)
Web of Science "https://www.webofscience.com/wos/ (accessed on 19 February 2025)"	(Sarcoglycan complex OR Sarcoglycans) AND (Non-Muscle Tissue OR Epithelial tissue OR Gland OR Prostate OR Breast OR Liver OR Pancreas OR Spleen OR Nervous system OR Kidney OR Lung OR Connective tissue OR Adipose organ) AND (Immunohistochemistry OR RT-PCR OR Immunofluorescence OR mRNA)
ScienceDirect "https://www.sciencedirect.com/ (accessed on 19 February 2025)"	(Sarcoglycan complex OR Sarcoglycans) AND "Non-muscle tissue" AND (Immunohistochemistry OR "RT-PCR" OR Immunofluorescence OR mRNA)
Google Scholar "https://scholar.google.com/ (accessed on 19 February 2025)"	(Sarcoglycan complex OR Sarcoglycans) AND "Non-muscle tissue" AND (Immunohistochemistry OR "RT-PCR" OR Immunofluorescence OR mRNA)

2.6. Data Synthesis

The extracted data from the included studies were synthesized narratively and organized by organ type. Subunits were mapped based on reported mRNA and protein expression across tissues, and evidence from human and animal studies was clearly distinguished. Consistencies and discrepancies among studies were noted, and results were summarized in descriptive tables and figures.

3. Results

3.1. Study Selection

The search process identified 438 articles. After removing duplicates, 380 articles remained for initial screening. From these, 317 papers were dismissed based on their titles and abstracts, primarily for being unrelated to SGs involvement in non-muscle organs. A comprehensive review of the remaining 63 articles revealed that 18 studies met the inclusion criteria and were included in the final analysis. Figure 1 shows the workflow of the selected studies.

3.2. Characteristics of the Included Studies

The included studies were published between 2000 and 2022, and employed various experimental approaches to investigate SG expression. Immunohistochemistry and immunofluorescence [16,24–37], fluorescence in situ hybridization (FISH) [16], and in situ hybridization (ISH) [38] were widely used to examine protein expression and localization in tissue samples. For immunohistochemistry, sections were examined using a laser scanning confocal microscope [16,24–37] or a fluorescence microscope [16,25]. For FISH, fluorophore-labeled probes were observed by a confocal microscope [16], whereas ISH employed radiolabeled cRNA probes [38]. Real-time reverse transcriptase–polymerase chain reaction (RT-PCR) [25–28,30,34,36,38–40], Western blot [25,26,28,30,33,39,40], and Northern Blot [38] were frequently applied to analyze mRNA and protein expression levels.

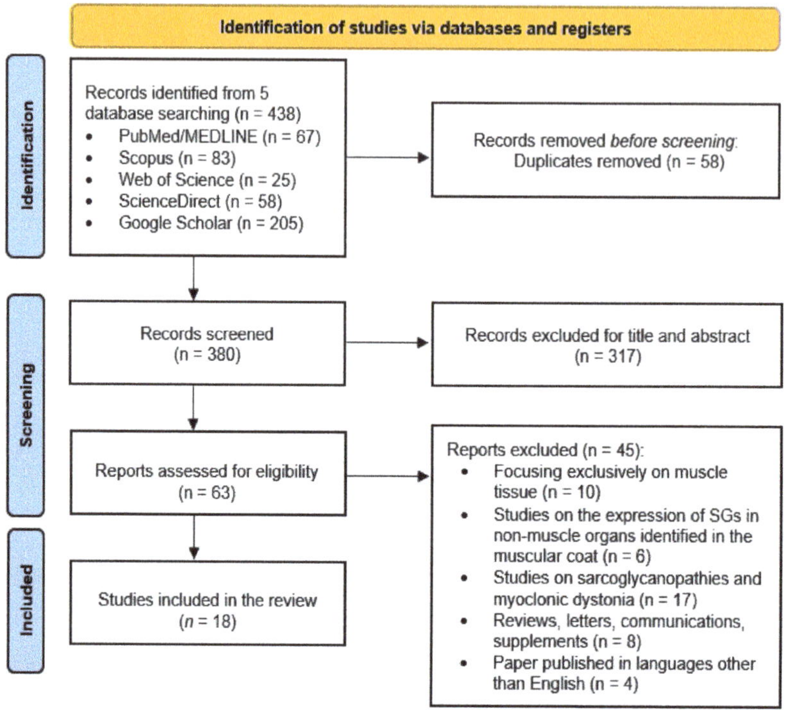

Figure 1. PRISMA flowchart summarizing the article selection process.

Studies have examined a wide range of non-muscle organ types. The nervous system, divided into central [16,24,26,28,30,31,37,38] and peripheral [25,33], and adipose [39,40] have been studied. Other organs/tissues explored included the glands (as breast [27], prostate [36], thyroid [29], pancreas [30]), and oral gingival mucosa [32,35]. The most investigated SG subunit was represented by ε [16,24–29,31–40], followed by β [24–29,31–34,36,37,39,40] and δ [24–31,34,36,37,40], γ [24,26–32,34–37], α [24,26–29,31,34–37,40], and ζ [25,26,29–31,37]. Regarding the species, the research included animal models and human tissue samples [27–29,32,35,36]. Rats [24,25,31,37–39] and mice [16,26,34,40] were frequently used. Only one study used the rabbit species [32]. Furthermore, several studies investigated human disease models, exploring conditions such as benign prostatic hyperplasia and prostate adenocarcinoma [36], fibrocystic mastopathy and breast fibroadenoma [27], Hashimoto's thyroiditis [29], and bisphosphonate-related osteonecrosis of the jaw [32,35]. The findings related to the expression of SGs, as reported in the studies, are presented in Figure 2.

3.3. Overview of SG Expression Across Non-Muscle Organs

Current evidence highlights both widespread and tissue-specific expression patterns of SG subunits across organ systems. Among them, β- and δ-SGs are the most consistently detected, being expressed in all major tissues analyzed, including the CNS [24,26,28,31,34,37], PNS [25,33], glands [27,29,36], and adipose tissue [39,40]. Their broad presence suggests a constitutive role across both neuronal and non-neuronal compartments. ε-SG also shows widespread expression but with a more variable distribution. It is abundantly expressed in the CNS [16,24,26,28,31,37,38], including in the Müller glial and ganglion cells of the retina [34], in Schwann cells of the PNS [25,33], in the prostate, breast, and thyroid [27,29,36], and in the gingival epithelium [32,35]. However, this subunit is detectable in certain organs, such as the liver, kidney, spleen, and testis, while in adipose tissue, it is only detectable during the adipogenic process [40], indicating potential tissue-specific regulation. ζ-SG displays a more restricted pattern, with predominant expression in the brain [24,30,37] and limited or absent mRNA signals in other tissues such as the pancreas, kidney, or liver [30]. By contrast, γ-SG is primarily enriched in the muscle and lung [30], with only low expression detected in the CNS (mainly in the cerebellum and brainstem) [30,31,37], and it is absent from Schwann cells and adipose tissue [25,40].

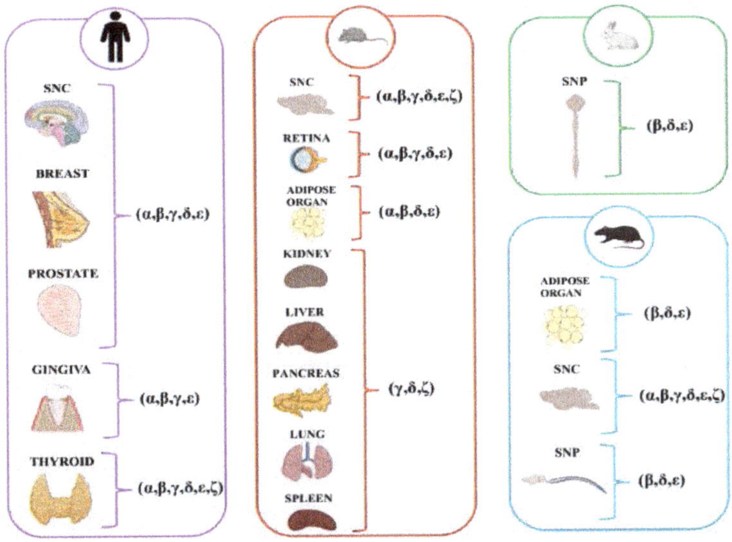

Figure 2. Expressions of SGs in non-muscle organs categorized by human and animal models (mouse, rat, and rabbit).

Other subunits (α- and γ-SGs) also exhibit more tissue-restricted profiles. α-SG is detected in cerebral cortex neurons and astrocytes [24,28,31,37], in glands [27,29,36], the oral epithelia [35], and adipose organ [40], but is consistently absent from the retina [34] and PNS [25,33]. γ-SG was primarily found in the brainstem and cerebellar regions [24,31,37], and in epithelial and oral mucosal tissues [27,29,32,35,36], but was absent in peripheral nerves and adipose tissue [25,33,39,40]. To better visualize these distribution patterns, a heatmap was generated to summarize the expression of individual SG subunits based on evidence from the included studies (Figure 3).

3.4. *Central Nervous System (CNS)*

3.4.1. Animal Studies

Xiao and LeDoux [38] detailed the molecular characteristics and expression patterns of the rat ε-SG gene. First, it was shown that this gene encodes a 437-amino-acid protein highly conserved in rats, mice, and humans, especially in its 46-amino-acid N-terminal signal sequence, which directs the protein to the cell membrane. While the rat and mouse sequences are identical, subtle variations were found in the human sequence that could affect species-specific aspects of SG function. In addition, the structural analysis showed that the extensive extracellular domain contains a conserved asparagine for glycosylation and four cysteine residues. Northern blot analysis of ε-SG mRNA expression showed moderate levels in the brain. Furthermore, quantitative RT-PCR confirmed the presence of ε-SG mRNA, revealing minimal developmental variation in the hippocampus, unlike the significant decrease in muscle tissue, where neonatal levels were over ten times higher than in adults. The in situ hybridization of brain sections demonstrated a widespread and uniform distribution of ε-SG mRNA in neurons, with the strongest signals found in areas with high neuronal density, such as the pyramidal cell layer of the hippocampus, cerebellum, and cerebral cortex, as well as in various brainstem, midbrain, thalamic, and hypothalamic nuclei. Clear hybridization signals observed in fiber pathways and white matter tracts (cerebellar peduncles and corpus callosum) suggested that glial cells also express ε-SG mRNA.

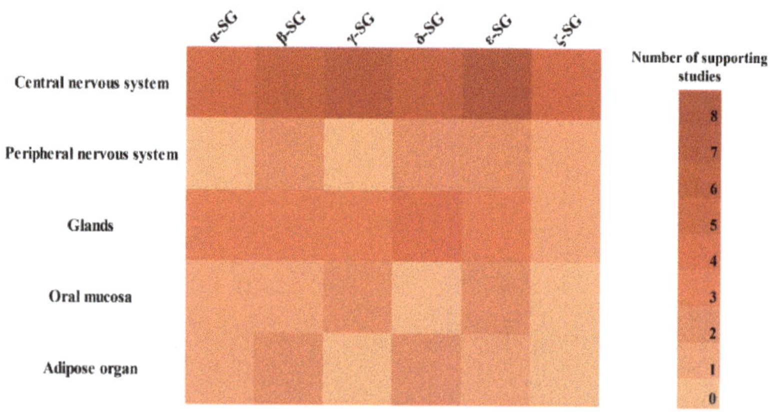

Figure 3. Heatmap summarizing the expression of SG subunits across non-muscle organs. Color intensity reflects the frequency of reported expression in the literature based on the number of studies identifying each SG subunit in a given organ.

Complementary findings by Chan et al. [16] further extended these insights by revealing that ε-SG mRNA and protein have a widespread presence across the mouse brain, as demonstrated through FISH and immunohistochemistry. High levels of ε-SG mRNA expression were particularly evident in the olfactory bulb's mitral cell layer and the cerebellum's Purkinje cell layer. Moreover, significant expression was detected in various monoaminergic cell clusters, including dopaminergic neurons within the substantia nigra pars compacta (SNc), substantia nigra pars reticulata (SNr), and ventral tegmental area (VTA), as well as serotonergic neurons in the dorsal raphe nucleus and noradrenergic neurons in the locus coeruleus. In contrast, moderate expression levels were observed in the hippocampus, multiple hypothalamic nuclei, and the amygdala, whereas lower levels were noted in regions such as the neocortex, globus pallidus, and most thalamic nuclei.

Shiga et al. [30] investigated the tissue-specific expression of ζ-SG, reporting its predominant expression in the brain. In contrast, γ-SG exhibited lower levels of expression. A more detailed regional analysis revealed a uniform distribution of ζ-SG throughout the brain, whereas γ-SG was abundant in specific areas such as the cerebellum and pons/medulla. Post-translational analysis revealed that ζ-SG, like other SGs and DGs, undergoes N-linked glycosylation, as evidenced by a shift in molecular weight after PNGase F treatment. Deglycosylation resulted in a reduction of ζ-SG's molecular weight, while maintaining a doublet migration pattern. To better understand the role of ζ-SG in SGC formation, CHO cells were used to perform the co-expression of all six known SGs with DG, which is crucial for the transportation of the SG complex to the cell plasma membrane. Similar molecular weight shifts as described above were observed for all SG and DG proteins expressed in CHO cells following PNGase F treatment. Concerning membrane interactions and scaffold proteins, ζ-SG was detected in both the plasma membrane and intracellular compartments, including the ER and Golgi apparatus. Cell surface biotinylation and immunoprecipitation experiments in CHO cells demonstrated that ζ-SG forms stable complexes with α-, β-, δ-, and ε-SG, but not with γ-SG. In particular, the simultaneous expression of all six known SGs with DG in CHO cells led to the identification of four distinct SG complexes at the membrane: (1) α- β- γ- δ, (2) α- β- ζ- δ, (3) ε- β- γ- δ, and (4) ε- β- ζ- δ. These findings indicate that ζ-SG can functionally substitute for γ-SG in SGC assembly and integrate into both α- and ε-based complexes. Moreover, SG complexes were also found to associate with DG, regardless of whether ζ-SG is present or not, suggesting that ζ-SG-containing complexes can exist as a subcomplex of the entire DGC in vivo.

Immunofluorescence studies in the cerebral and cerebellar cortices confirmed the presence of multiple SG subunits [24,37]. The research by Vermiglio et al. [24] reported that all tested SGs (α, β, γ, δ, and ε) are expressed in the rat cerebral cortex, where they exhibit a "spot-like" fluorescence pattern with spots (0.5–2 μm in diameter) around neuronal somas. All SG expressions were marked in the cerebellar cortex in the Purkinje cell layer, while the granular and molecular layers showed low or absent staining. Regarding localization, SG expression was also observed in glial cells, mainly in the central portion rather than around the soma.

Cutroneo et al. [37] further corroborated these findings by showing that all SGs (α, β, γ, δ, ε, and ζ) are expressed in rat neurons across the hippocampus, cerebral cortex, and cerebellar cortex. In these regions, the proteins exhibited a spot-like fluorescence pattern (0.5–2 μm in diameter) predominantly localized in the soma of pyramidal and granular neurons. Specifically, in the hippocampus, SGs were more expressed in pyramidal neurons of some Cornu Ammonis regions (CA1, CA2, and CA3) and dentate gyrus granular cells, which displayed more minor fluorescence spots than pyramidal neurons. In the cerebellar cortex, SG expression was evident in all layers, with a high spot-like immunostaining predominantly localized to the Purkinje cell layer. Additionally, the colocalization of each SG with the GABA_A receptor indicated colocalization at the soma level, confirming the neuronal expression of SGs.

Boulay et al. [26] investigated the expression of the SG complex in the cerebrovascular system and its regulation by astrocytic Connexin 30 (Cx] 30). RT-PCR analysis performed on purified brain vessels demonstrated the presence of transcripts encoding all known SG subunits (α, β, δ, ε, γ, and ζ) as well as SSPN, indicating that the full SG complex is transcriptionally active in the cerebrovascular system. Among these subunits, δ- and ε-SG were particularly expressed in larger-caliber vessels (diameter > 100 μm). To explore the regulatory influence of astrocytic Cx30 on vascular SG expression, the authors examined Cx30-knockout mice. Transcriptomic analysis revealed that most SG subunits, including α- and β-SG, maintained comparable expression levels between wild-type and knockout animals. However, the expression of Sgcg, the gene encoding γ-SG, was significantly upregulated in the absence of Cx30. Western blot analysis confirmed that this transcriptional upregulation of γ-SG translated into increased protein levels in brain vessels.

Rizzo et al. [31] investigated the expression of SGs in the cerebral and cerebellar cortices of rats. In the cerebral cortex, single immunofluorescence reactions revealed that all six SG subunits are expressed in a distinctive "spot-like" pattern, with spots (0.5–2 μm in diameter) primarily localized around the soma of neurons and glial cells. The statistical analysis of spot counts across different cortical regions indicated significant variability in subunit expression, particularly for α-, β-, and γ-SG, between anterior and posterior cortical areas. Conversely, in the cerebellar cortex, SG expression was consistently uniform across different regions, with the same staining pattern observed around the soma of Purkinje cells and glial cells.

In wild-type mouse retinae observed by Fort et al. [34], SGs and SSPN exhibited a distinct expression profile compared to skeletal muscle, with their localization and abundance appearing largely independent of dystrophin. Using RT-PCR and Western blot analyses in both wild-type and *mdx3cv* mice, the authors confirmed the presence of β-, δ-, γ-, and ε-SG subunits, along with SSPN, at both mRNA and protein levels. Among these, ε-SG and β-SG were the most abundantly expressed, while α-SG was undetectable in the retina of either genotype. Notably, even in *mdx3cv* mice, characterized by a severe reduction in Dp71 (a short dystrophin subunit), the expression levels and spatial distribution of SGs and SSPN remained unchanged. Immunofluorescence analyses revealed that SGs and SSPN localize predominantly at the inner limiting membrane (ILM), corresponding to the end feet of Müller glial cells (MGCs), and at the outer limiting membrane (OLM), with additional expression in the ganglion cell layer (GCL) and inner nuclear layer (INL). Double immunolabeling with markers for MGC (glutamine synthetase) and ganglion cell axons (NF68) confirmed that SGs and SSPN localize to both glial and neuronal compartments, with strong signal overlap observed particularly for ε-SG. Importantly, SGs did not colocalize with dystrophin in photoreceptor terminals or dendritic processes, and their localization was not disrupted in dystrophin-deficient mice. Based on these findings, the authors proposed the presence of two distinct SG–SSPN assemblies: one associated with Dp71 at the MGC end feet and another located in the outer retina, potentially interacting with as yet unidentified molecular partners.

3.4.2. Human Studies

Anastasi et al. [28] investigated the expression and localization of SGs in the human cerebral cortex. The study found that all SG subunits (α-, β-, γ-, δ-, ε-, and ζ-SG) are present in large neurons of the cerebral premotor cortex, with immunostaining localized predominantly along the cell surface. Three-dimensional reconstructions of full-thickness sections verified that SG immunoreactivity is distributed throughout the neuronal membranes. Furthermore, double labeling with neuronal and glial markers demonstrated that SGs are expressed in neuronal and astrocytic populations, with colocalization patterns indicating potential functional interactions at the cell surface.

Table 3 summarizes the available studies investigating SG expression in the CNS across both animal and human models.

3.5. Peripheral Nervous System (PNS)

3.5.1. Animal Studies

In the PNS, SGs are prominently expressed in the outer regions of nerve fibers, as described by Imamura et al. [33]. Western blot analysis of various mouse tissue lysates showed that the 46 kDa ε-SG is widely expressed, with the highest levels observed in peripheral nerve tissue. Immunofluorescence analysis of rabbit peripheral nerve cryosections revealed that β-SG, δ-SG, and ε-SG, together with α-DG and β-DG, are localized along the outer region of nerve fibers. In contrast, α-SG, γ-SG, and SSPN were not detected in these regions. Double staining with antibodies against ε-SG and neurofilament or laminin B1 chain indicated that ε-SG is not located on the axon but near the basal lamina in close association with Schwann cells. In addition, ε-SG colocalized with both β- and δ-SGs at the outer region of the myelin sheath and showed colocalization with DGs, Dp116, and utrophin. Wheat germ agglutinin (WGA) affinity chromatography of peripheral nerve lysates confirmed these findings. In the WGA-bound fraction from peripheral nerve, immunoblotting detected β-SG, δ-SG, and ε-SG along with DGs, Dp116, and utrophin, whereas α-SG, γ-SG, and SSPN were absent even at high detection sensitivities. Immunoprecipitation of this fraction with an anti-ε-SG antibody further confirmed that β-, δ-, and ε-SGs assemble into a complex with DGs and Dp116.

Table 3. Overview of the studies included for the CNS.

Study (Authors, Year)	Study Design	Tissue/Organ/Cell Types Examined	Types of SG Proteins Studied	Species/Models
Xiao and LeDoux (2003) [38]	Molecular study (Northern analysis, RT-PCR and ISH)	Neural (cerebellar cortex, striatum, cerebral cortex, thalamus, hippocampus)	ε	Rat [16]
Vermiglio et al. (2011) [24]	Immunofluorescence study	Cerebral and cerebellar cortex (neurons and glial cells)	α, β, γ, δ, ε	
Cutroneo et al. (2015) [37]	Immunohistochemical study	hippocampus, cerebral and cerebellar cortex	α, β, γ, δ, ε, ζ	
Rizzo et al. (2018) [31]	Immunofluorescence study	Cerebral cortex (neurons and glial cells) Cerebellar cortex (neurons and glial cells)	α, β, γ, δ, ε, ζ	
Chan et al. (2005) [16]	Immunohistochemistry and molecular (FISH) study	Olfactory bulb (mitral cell layer) Cerebellum (Purkinje cell) Hippocampal formation and neocortex Monoaminergic cell groups and brainstem nuclei	ε	Mouse
Shiga et al. (2006) [30]	Immunofluorescence and molecular (RT-PCR and immunoprecipitation) study	Brain	γ, ζ	
Fort et al. (2005) [34]	Immunohistochemistry and molecular (RT-PCR) study	Retina (Müller and ganglion cells)	α, β, γ, δ, ε	
Boulay et al. (2015) [26]	Immunofluorescence and molecular (RT-PCR, Western blot) study	Cerebrovascular system (brain vessels, cortex, and hippocampus)	α, β, γ, δ, ε, ζ	
Anastasi et al. (2012) [28]	Immunohistochemical and molecular study (RT-PCR, Western Blot)	Cerebral cortex (neurons and astrocytes)	α, β, γ, δ, ε, ζ	Human

Cai et al. [25] provided complementary evidence for the expression of SGs in Schwann cells of the PNS, where they contribute to the structural integrity of the myelin sheath. Through RT-PCR and Western blot analyses, the authors demonstrated that β-, δ-, ε-, and ζ-SGs are expressed in adult rat sciatic nerves and cultured Schwann cells, even in the absence of neuronal input. In contrast, α- and γ-SGs were not detected. Immunofluorescence localized these subunits to the abaxonal membrane of Schwann cells near the basal lamina. Notably, SG expression occurred before the onset of myelination and was upregulated by the presence of neurons, independent of myelin assembly. Co-immunoprecipitation experiments showed that β-, δ-, ε-, and ζ-SGs form a stable heteromeric complex with Dp116 and α/β-DGs. In BIO14.6 hamsters, which lack δ-SG due to a genetic deletion, the entire complex was destabilized, with reduced levels of Dp116 and α-DG in membrane fractions. Additionally, α-DG was also reduced in the supernatant. In contrast, β-DG, utrophin, and syntrophin levels remained unchanged. Electrophoretic analysis showed altered mobility of α-DG between muscle and nerve tissues. Structural analysis revealed abnormalities in the myelin sheath, including excessive folding, internal splitting, and occasional axonal compression, as well as disruption of Schmidt–Lanterman incisures. Despite normal levels of myelin proteins (MAG, MBP, P0), these structural defects were associated with mild conduction delays under thermal stress at temperatures below 20 °C.

3.5.2. Human Studies

To date, there is a paucity of human data on SG expression in the PNS.

Table 4 summarizes the available studies investigating SG expression in the PNS across both animal and human models.

Table 4. Overview of the studies included for the PNS.

Study (Authors, Year)	Study Design	Tissue/Organ/Cell Types Examined	Types of SG Proteins Studied	Species/Models
Cai et al. (2007) [25]	Immunofluorescence and molecular study (RT-PCR and Western blot)	Peripheral nervous system (sciatic nerves and Schwann cell cultures)	α, β, γ, δ, ε, ζ	Rat Hamster
Imamura (2000) [33]	Immunofluorescence and molecular (immunoprecipitation and Western blot) study	Peripheral nerves (sciatic, femoral and tibial)	α, β, γ, δ, ε	Rabbit

3.6. Glands

3.6.1. Animal Studies

Additional insights into the composition and tissue distribution of SG complexes were provided by Shiga et al. [30]. When examining mouse tissues by RT-PCR, ζ-SG mRNA levels were highest in the brain and very weak in other tissues, including those of glandular origin (e.g., pancreas). In contrast, δ-SG exhibited broad and abundant expression across all tissues examined.

3.6.2. Human Studies

Arco et al. [27] analyzed the distribution of SGs in breast glandular epithelial cells. In normal glandular breast tissue, single immunofluorescence analysis revealed a uniform staining pattern of all tested SG subunits (α-, β-, γ-, δ-, ε-SGs) in polyhedral epithelial cells. Serial sectioning combined with three-dimensional reconstruction confirmed that SG staining spanned the entire cellular thickness, with expression evident in both epithelial and myoepithelial cells, as validated by double localization with α-SMA and DAPI. In contrast, breast tissue affected by fibrocystic mastopathy and fibroadenoma exhibited a marked reduction in SG expression. In these pathological specimens, immunofluorescence showed only isolated, faint fluorescent spots, and RT-PCR analysis demonstrated significantly decreased mRNA levels of all SGs compared with normal tissue.

Cutroneo et al. [36] extended the investigation to the prostatic gland. In normal prostate, single immunofluorescence reactions demonstrated clear expression of α-, β-, γ-, δ-, ε-SGs, with a well-defined distribution localized to the apical, lateral, and basal areas of the epithelial cells arranged in a single layer. Double-localization with α-SMA and DAPI confirmed that SGs are present in epithelial and myoepithelial cells. However, these patterns were altered in prostatic diseases. Benign prostatic hyperplasia displayed only a limited SG staining pattern confined to the apical areas of epithelial cells. In contrast, in prostatic adenocarcinoma, SG expression was absent in both epithelial and myoepithelial cells. RT-PCR analysis further confirmed decreased SG transcript levels in these pathological conditions.

Favaloro et al. [29] performed an immunofluorescence study on thyroid tissue in both healthy and pathological conditions to evaluate the expression of SGs (α, β, γ, δ, ε, and ζ) and αvβ3-integrin. In healthy subjects, single-localization immunofluorescence in thyrocytes showed a uniform staining pattern for all six SG subunits and αvβ3-integrin. In contrast, thyrocytes from subjects with Hashimoto's thyroiditis exhibited a decreased staining pattern for all SGs and αvβ3-integrin, while merged images demonstrated an absence of SG signal in some of these cells. Quantitative pixel intensity analysis corroborated these observations, revealing significantly lower protein expression levels in pathological samples compared to controls.

Table 5 summarizes the available studies investigating SG expression in the glands across both animal and human models.

Table 5. Overview of the studies included for gland organs.

Study (Authors, Year)	Study Design	Tissue/Organ/Cell Types Examined	Types of SG Proteins Studied	Species/Models
Shiga et al. (2006) [30]	Immunofluorescence and molecular (RT-PCR and immunoprecipitation) study	Pancreas	δ, ζ	Mouse
Arco et al. (2012) [27]	Immunohistochemical and molecular study (RT-PCR)	Glandular breast (epithelial and myoepithelial cells)	α, β, γ, δ, ε	Human
Cutroneo et al. (2014) [36]	Immunohistochemical and molecular (RT-PCR) study	Prostate gland (epithelial and myoepithelial cells)	α, β, γ, δ, ε	
Favaloro et al. (2022) [29]	Immunofluorescence study	Thyroid	α, β, γ, δ, ε, ζ	

3.7. Oral Mucosa

3.7.1. Animal Studies

There is little to no animal data on SG expression in the oral mucosa for the studies reviewed.

3.7.2. Human Studies

Nastro-Siniscalchi et al. [32] investigated the effects of bisphosphonate treatment on oral mucosa, explicitly focusing on the expression of proteins critical for cell viability and signaling, including SGs. In control samples, the immunofluorescence analysis showed a clear and consistent expression of three SG subunits (β, γ, and ε) within the basal lamina. In samples from bisphosphonate-treated subjects without osteonecrosis, the expression of these SGs was markedly reduced. Basal lamina exhibited an almost complete absence of detectable staining for these SGs, which indicated a disruption of normal protein expression patterns. Conversely, in bisphosphonate-treated samples from patients with osteonecrosis, increased SG staining was observed on the basal lamina, with a concomitant qualitative and quantitative rise in vascular-associated fluorescence.

Further supporting these observations, the study by De Ponte et al. [35] on oral mucosa biopsies from untreated subjects revealed a regular and continuous expression of α-, ε-, and γ-SGs from the basal keratinocyte layer to the superficial layers. In biopsies from patients treated with zoledronate for 24 months, a general decrease in SG fluorescence pattern was noted, while in specimens from patients undergoing bisphosphonate therapy for 36 months, SG expression was nearly undetectable. Quantitative analysis using fluorescence intensity profiles confirmed these findings, with control samples displaying distinct peaks of SG fluorescence that were markedly diminished or absent in long-term bisphosphonate-treated specimens.

Table 6 summarizes the available studies investigating SG expression in the oral mucosa across both animal and human models.

Table 6. Overview of the studies included for the oral mucosa.

Study (Authors, Year)	Study Design	Tissue/Organ/Cell Types Examined	Types of SG Proteins Studied	Species/Models
Nastro Siniscalchi et al. (2010) [32]	Immunohistochemical study	Gingival epithelium	β, γ, ε	Human
De Ponte et al. (2013) [35]	Immunohistochemical study	Gingival epithelium	α, γ, ε	

3.8. Adipose

3.8.1. Animal Studies

Groh et al. [40] investigated the expression and functional importance of SGs in white adipose tissue. RT-PCR and protein analyses in freshly isolated white adipocytes demonstrated the presence of α-, β-, and δ-SGs, along with SSPN and DG. In contrast, γ- and ε-SG are not expressed in these cells. Immunofluorescence confirmed membrane localization of these SGs in adipocytes, and expression levels were consistent with transcript abundance. Functional experiments using mouse models deficient in individual DGC components revealed that the adipocyte SG complex functions as a tightly interdependent unit: loss of either β- or δ-SG led to destabilization of the entire complex, including SSPN. This also resulted in a significant reduction in the glycosylated form of α-DG, thereby compromising its ability to bind laminin despite the continued presence of the core α-DG protein. In contrast, the deletion of α-SG did not affect the expression of the adipocyte SG complex.

Complementing these observations, Romo-Yáñez et al. [39] investigated the expression and regulation of SGs and associated DGC components in white adipose tissue, with a focus on adipocyte differentiation. Their study demonstrated that β-, δ-, and ε-SGs are consistently expressed at the mRNA level throughout the adipogenic process in both visceral and subcutaneous fat depots of adult rats. These transcripts were detected by RT-PCR and remained present throughout all stages of adipocyte differentiation in vitro, including preconfluent preadipocytes, quiescent preadipocytes, committed adipocytes, and mature adipocytes. In contrast, no expression of α- or γ-SG was reported. Additionally, SGs were co-expressed with key DGC components, including β-dystroglycan and utrophin. Both proteins were present at all differentiation stages, implying that a functional DAPC is assembled in adipose cells. In addition, multiple subunits of syntrophins (α, βI, βII, and cII) and dystrobrevins (α and β) were expressed, forming a cytoplasmic interface for signaling and cytoskeletal anchorage. Among these, α-dystrobrevin was selectively expressed only after adipogenic commitment. Conversely, βI-syntrophin displayed depot-specific regulation, with stable expression during visceral adipogenesis and a progressive decrease in subcutaneous adipocytes.

3.8.2. Human Studies

Currently, direct studies on SG expression in human adipose tissue are limited.

Table 7 summarizes the available studies investigating SG expression in the adipose across both animal and human models.

Table 7. Overview of the studies included for the adipose organ.

Study (Authors, Year)	Study Design	Tissue/Organ/Cell Types Examined	Types of SG Proteins Studied	Species/Models
Groh et al. (2009) [40]	Molecular study (RT-PCR and Western blot)	White adipocytes	α, β, δ	Mouse
Romo-Yáñez et al. (2011) [39]	Molecular study (RT-PCR and Western blot)	Adipose tissue (differentiation adipocyte)	β, δ, ε	Rat

3.9. Other Organs

3.9.1. Animal Studies

Xiao and LeDoux. [38] investigated the expression of ε-SG mRNA in non-neural tissues. Using Northern analysis, ε-SG transcript was present in all rat tissues examined, with the highest levels detected in the kidney, moderate in the lung, and lower levels in the liver, spleen, and testis. Shiga et al. [30] also focused on ζ-SG expression in other organs. The RT-PCR analysis of mouse tissues revealed that ζ-SG mRNA levels are faintly detectable in the liver, lung, and spleen. In contrast, γ-SG exhibited high expression in organs such as the lung, while δ-SG showing broad expression in all examined tissues.

3.9.2. Human Studies

There is limited direct evidence from human studies.

Table 8 summarizes the available studies investigating SG expression in the other organs across both animal and human models.

Table 8. Overview of the studies included for other organs.

Study (Authors, Year)	Study Design	Tissue/Organ/Cell Types Examined	Types of SG Proteins Studied	Species/Models
Xiao and LeDoux (2003) [38]	Molecular study (Northern analysis, RT-PCR and ISH)	Liver Kidney Lung Spleen Testis	ε	Rat
Shiga et al. (2006) [30]	Immunofluorescence and molecular (RT-PCR and immunoprecipitation) study	Kidney Liver Lung Spleen	δ, γ, ζ	Mouse

4. Discussion

This scoping review highlights the widespread and heterogeneous expression of SGs across various non-muscle organs, including the central and peripheral nervous systems, glands, oral mucosa, and adipose tissue. Traditionally studied in the context of muscle, SGs are increasingly recognized as structural and possibly signaling components in multiple cellular environments. The data reviewed here reveal not only the presence of SGs in diverse organ types but also substantial variability in subunit composition, regional distribution, and cellular localization, features that suggest functional specialization beyond their canonical role in the DGC.

4.1. Expression and Functional Roles of SGs in Non-Muscle Organs

4.1.1. CNS

The expression of SGs within the CNS is both widespread and subunit-specific, suggesting distinct and possibly specialized roles beyond their classical functions in muscle. Multiple studies have consistently reported strong expression of ε- and ζ-SG in neurons and glial cells across various brain regions, including the cortex, hippocampus, cerebellum, thalamus, and monoaminergic nuclei [16,30,37,38]. In particular, ζ-SG appears to be predominantly brain-specific, suggesting possible neural specificity [30]. Immunohistochemical studies reveal that SGs are not only widely expressed but also follow a distinct "spot-like" distribution around neuronal somas, particularly in pyramidal and granular neurons, as well as in glial cells [24,31,37]. This spot-like pattern, uniform in some regions (e.g., cerebellum) and variable in others (e.g., cortex), could represent membrane microdomains implicated in synaptic stability or signal transduction. Despite these morphological observations, the exact molecular composition and function of these SG-positive spots remain unknown. In humans, Anastasi et al. [28] confirmed the presence of all six SG subunits in the cerebral premotor cortex, expressed in both neurons and astrocytes. This evidence not only validates the rodent findings but also suggests that SGs could play conserved and functionally relevant roles in the human brain. The membrane-associated distribution observed in full-thickness reconstructions reinforces the idea that SGs may contribute to membrane specialization or intercellular signaling at neuron–glia interfaces.

An interesting case is presented by Shiga et al. [30], who examined both the tissue distribution and biochemical assembly of ζ-SG. While ζ-SG showed strong expression in the brain, it was weakly detectable in skeletal muscle and peripheral tissues. In contrast, γ-SG displayed the opposite expression pattern, suggesting that these two subunits are deployed in a mutually exclusive, tissue-specific manner. In addition, in CHO cell models, ζ-SG was able to assemble into stable DGC complexes when co-expressed with other SG subunits and DG. These complexes mimicked those containing γ-SG, but ζ- and γ-SG did not co-assemble, implying a level of structural redundancy as well as functional exclusivity. These results seem to reflect a divergence between biochemical competence and physiological deployment. The ability of ζ-SG to form complexes in vitro does not imply its incorporation into muscle DGCs in vivo, likely due to transcriptional or trafficking constraints. Instead, ζ-SG could fulfill unique roles in neural-specific DGC variants, particularly in regions where γ-SG is absent. These findings underscore the modularity of the SG complex and highlights the need for in vivo validation to determine whether ζ-SG–based complexes support distinct signaling or structural roles in the CNS.

A key functional insight comes from the observed co-localization of SGs with GABA_A receptors, particularly in hippocampal pyramidal neurons and cerebellar Purkinje cells [37]. This spatial association suggests a critical role for SGs, particularly ε-SG, as scaffolding or regulatory elements in inhibitory synaptic domains. Supporting this hypothesis, the study by Cazurro-Gutiérrez et al. [41] provides compelling evidence that ε-SG directly contributes to the stabilization of GABAergic synapses, influencing receptor clustering and tonic inhibitory transmission. In ε-SG-deficient mice, these functions are impaired, leading to altered GABAergic neurotransmission, disrupted synaptic inhibition, and motor disturbances resembling myoclonus-dystonia (MD), a movement disorder associated with SGCE (ε-SG gene) mutations. The evidence suggests that SGCE plays a crucial role in synaptic function, and its dysfunction in MD may contribute to an imbalance in neurotransmission, particularly a deficit in GABAergic inhibitory signaling.

The CNS-centered view of SGs contrasts with their expression and likely function in the retina. Fort et al. [29] demonstrated that in both wild-type and Dp71-deficient mouse retinas, multiple SG subunits, notably β-, δ-, γ-, and ε-SG, are expressed at the inner and outer limiting membranes, where they predominantly colocalize with Müller glial cell end feet. This distribution occurs independently of dystrophin, indicating that SGs could participate in alternative membrane complexes in the retina, distinct from canonical DGC assemblies. This distinctive pattern suggests that retinal SGs could support non-synaptic functions, such as maintaining glial polarity, epithelial adhesion, or barrier function, rather than participating directly in neurotransmission. Their persistent expression in dystrophin-deficient conditions further implies the existence of alternative membrane complexes or molecular partners yet to be identified. Therefore, the divergence between the brain and retina underscores the context-dependent versatility of SGs. Unlike in the brain, where ε-SG loss affects GABAergic synapses and contributes to motor dysfunction associated with MD, the functional implications of retinal SGs remain speculative but could involve non-synaptic scaffolding roles.

Adding further complexity, Boulay et al. [26] conducted one of the few focused investigations into the expression of SGs in the cerebrovascular system, revealing important but still largely unexplored aspects. The study identified all SG transcripts, as well as SSPN, in purified mouse brain vessels. Notably, δ- and ε-SG were enriched in larger vessels, suggesting spatially regulated expression. These findings raise the possibility that SGs contribute to the molecular architecture of the neurovascular unit, perhaps supporting blood–brain barrier integrity or mechanotransduction. However, no cellular localization or functional data were provided to distinguish expression in endothelial or astroglial components. Of particular interest is the upregulation of γ-SG in Cx30-deficient mice, suggesting that astrocyte-endothelial communication mediated by gap junctions could influence the molecular composition of the vascular DGC. Nevertheless, the biological relevance of this upregulation remains unclear in the absence of mechanistic data.

4.1.2. PNS

Compared to the CNS, the expression and function of SGs in the PNS remain less explored but nonetheless reveal important tissue-specific features. The available data indicate that SGs are expressed in peripheral nerves with a distinct subunit composition and spatial localization, suggesting a specialized role in Schwann cell function and myelin architecture [25,33]. Immunohistochemical and biochemical studies show that β-, δ-, and ε-SG are the predominant subunits present in peripheral nerve fibers, where they localize to the outer region of the myelin sheath, particularly the abaxonal membrane of Schwann cells. In contrast, α-, γ-SG, and SSPN appear to be absent or below detection thresholds in this compartment [33]. These SGs co-localize with known DGC components such as α- and β-DG, Dp116, and utrophin, suggesting the existence of a Schwann cell-specific variant of the DGC. Importantly, the functional role of SGs in the PNS extends beyond structural presence. Co-immunoprecipitation studies in tissues and cultured Schwann cells demonstrate that β-, δ-, ε-, and ζ-SG form stable complexes with Dp116 and DGs even in the absence of neuronal input [25]. The formation of these complexes appears to be developmentally regulated and responsive to axonal signals, as SG expression increases in the presence of neurons even before active myelination occurs. This suggests that SGs may be involved in Schwann cell differentiation or in preparing the membrane environment for myelin formation.

Evidence from δ-SG–deficient BIO14.6 hamsters further supports a functional role for SGs in myelin integrity. In these animals, the absence of δ-SG leads to destabilization of the DGC complex, reductions in Dp116 and α-DG levels, and ultrastructural abnormalities of the myelin sheath, including excessive folding and disrupted Schmidt–Lanterman incisures [25]. These defects occur despite preserved expression of major myelin proteins such as MBP, MAG, and P0, indicating that SGs contribute to myelin organization independently of canonical myelin gene expression. While functional deficits were subtle under normal conditions, thermal stress exacerbated nerve conduction delays, implying that SG-containing complexes may confer mechanical or homeostatic resilience to the peripheral myelin sheath.

These observations suggest that, in contrast to their synaptic roles in the CNS, SGs in the PNS may function primarily as scaffolds for cytoskeletal or ECM interactions that stabilize Schwann cell morphology or regulate myelin architecture under stress. The absence of α- and γ-SG, which are essential in muscle tissue, also reinforces the idea of tissue-specific DGC configurations tailored to the mechanical and molecular environment of peripheral nerves. Notably, human data on SG expression in the PNS remain lacking, leaving open questions about the translational relevance of these findings. Future investigations should focus on the dynamics of SG complex assembly during development and regeneration, as well as its interaction with axonal cues and potential role in human peripheral neuropathies.

4.1.3. Glands and Oral Mucosa

Beyond the nervous system, SGs are also expressed in various epithelial tissues, where their roles remain incompletely understood. Evidence from both human and animal studies consistently indicates that SG expression is spatially organized and sensitive to pathological changes, hinting at potential structural or regulatory roles.

In the pancreas, Shiga et al. [30] observed the slight expression of ζ-SG mRNA, in contrast to its strong and widespread expression in the brain. This minimal transcriptional activity could suggest that ζ-SG does not play a major role in pancreatic physiology, or that its expression is limited to a specialized subpopulation of cells, possibly within islets or ductal epithelium. While δ-SG was more broadly expressed across multiple tissues in the same study, no detailed regional or cellular localization in the pancreas was provided. This limits functional interpretation but raises questions about whether different SGs contribute to distinct cellular compartments, such as endocrine versus exocrine regions. The possibility that SGs participate in maintaining the structural integrity of pancreatic acini or islets remains open but unexplored. Given the critical role of cell polarity, adhesion, and membrane organization in secretory function, even low-level SG expression might serve stabilizing or scaffolding roles. However, without protein-level confirmation or immunolocalization data, such roles remain hypothetical.

In contrast, human studies in glandular organs such as the breast, prostate, and thyroid have provided more extensive data. Under physiological conditions, all SG subunits (α-, β-, γ-, δ-, ε-, and ζ-SG) are expressed with a uniform distribution throughout the apical, lateral, and basal surfaces of epithelial cells [27,29,36]. These patterns, confirmed by 3D reconstructions and colocalization with myoepithelial markers, suggest that SGs are integral components of epithelial cell architecture and may contribute to cell polarity, adhesion, or barrier function. Importantly, SG expression is sensitive to disease state. In healthy and pathological conditions of the breast and prostate, SG expression is markedly reduced or even lost [27,36]. This downregulation is not limited to protein localization but also involves decreased mRNA levels, as shown by RT-PCR. Similarly, in Hashimoto's thyroiditis, SG expression in thyrocytes is significantly diminished and often completely absent in affected cells [29]. These observations raise the possibility that SG loss may not be merely a consequence of cellular dysfunction but could also contribute to epithelial disorganization, impaired intercellular signaling, or altered immune interactions. The mechanisms behind this downregulation remain uncertain. It is unclear whether SG expression in epithelial tissues is directly modulated by inflammatory signals, altered ECM composition, or hormonal control. Moreover, the functional implications of SG loss, whether it leads to impaired epithelial barrier integrity, altered mechanotransduction, or increased susceptibility to cell transformation, have yet to be addressed experimentally.

The oral mucosa provides another example of SG sensitivity to environmental and pharmacological stress. Studies in human tissue reveal that SGs are expressed in the basal and suprabasal layers of the oral epithelium, but their expression is progressively reduced in patients undergoing long-term bisphosphonate therapy, especially those with medication-related osteonecrosis of the jaw [32,35]. Interestingly, while SG expression decreases in tissue without osteonecrosis, it appears paradoxically upregulated in samples from patients with established necrosis [32], possibly reflecting a reactive or compensatory remodeling response.

This evidence points to a broader, context-dependent role for SGs in epithelial biology. Their expression appears tightly regulated under physiological conditions and dynamically modulated in response to pathological conditions. Downregulation of these proteins in pathological states suggests functional relevance, but a mechanistic understanding is lacking. Without functional studies, it remains uncertain whether SGs in these tissues form classical DGC-like complexes or participate in alternative membrane scaffolds. The co-expression of SGs with DGC components such as DG or SSPN in these tissues has not been examined. Future studies should explore whether loss of epithelial SGs plays a causal role in disease progression at the level of glandular and mucosal epithelia.

4.1.4. Adipose

Emerging data suggest the relevance of SGs in adipose biology. Studies in rodents have identified a selective pattern of SG expression in white adipose tissue, particularly implicating β-, δ-, and ε-SG, alongside core components of the DGC, such as β-DG, utrophin, and SSPN [39,40]. These results suggest that adipocytes may form a non-canonical and tissue-specific DGC. Romo-Yáñez et al. [39] showed that SG transcripts are present throughout adipocyte differentiation, from preadipocytes to mature cells, in both visceral and subcutaneous depots. Notably, α- and γ-SG are consistently absent, reinforcing the idea that SG assembly is highly context-dependent and adapted to local structural and signaling needs. The persistence of β-, δ-, and ε-SG throughout all stages of adipogenesis further suggests a role in membrane stabilization or intracellular signaling, rather than a transient developmental function. In addition, Groh et al. [40] demonstrated that deletion of either β- or δ-SG leads to destabilization of the entire adipocyte SG complex, including a reduction in glycosylated α-DG, which compromises laminin binding despite the presence of the core DG protein. This disruption suggests that SGs in adipose tissue are not redundant but necessary to maintain proper DGC assembly and their interaction with the extracellular matrix. The fact that α-SG deletion does not destabilize the complex highlights a divergence from muscle biology, where α-SG is indispensable for complex integrity. These observations suggest that the adipocyte SG complex may play a role in anchoring the cytoskeleton to the extracellular matrix, regulating cell shape, or maintaining membrane stability under metabolic or mechanical stress. Additionally, the co-expression of syntrophins and dystrobrevins in differentiating adipocytes suggests possible signaling roles, although the exact pathways involved remain undefined [39].

Despite this growing body of evidence, direct functional studies on SGs in adipose physiology are lacking. It is unclear whether SG deficiencies influence adipocyte metabolism, insulin signaling, or inflammation, factors critically relevant to obesity and metabolic syndrome. Moreover, human data are absent, limiting translational interpretation.

4.1.5. Other Organs

Compared to previously described tissues, SGs expression in visceral organs such as the kidney, liver, lung, spleen, and testis has been less extensively explored and remains poorly understood. Nevertheless, early transcriptomic data from rodent models indicate that certain SG subunits, particularly ε-SG and δ-SG, are expressed at varying levels across these tissues [30,38].

In particular, ε-SG mRNA was found at moderate to high levels in the kidney and lung, and at lower levels in the liver, spleen, and testis [38], suggesting a broad but non-uniform distribution. Shiga et al. [30] further reported that ζ-SG mRNA is faintly detectable in several non-neural tissues, including the liver and kidney, in contrast to its strong expression in the brain. These data suggest that while SGs are not exclusive to excitable tissues, their expression is likely tissue-specific and finely regulated. However, despite these transcript-level findings, no functional analyses or protein-level localization studies have been reported for most of these organs. The cell types on which to assess SG expression, whether epithelial, endothelial, stromal, or immune-derived, remain unidentified. Furthermore, it is unclear whether these SGs participate in classical DGC assemblies or form alternative complexes with distinct partners, as proposed for retina and adipose tissue.

This lack of data represents a significant knowledge gap. The possibility that SGs could contribute to organ-specific membrane stability, epithelial polarity, or mechanotransduction remains speculative. Moreover, whether SG expression is altered in diseases affecting these organs—such as nephropathies, hepatic fibrosis, or pulmonary hypertension—has not been explored. Given the diversity of mechanical and signaling demands across these tissues, it is plausible that SGs may perform supportive or regulatory roles that differ markedly from those observed in muscle and brain. Nonetheless, without cellular resolution, protein localization, or disease-model data, these hypotheses remain untested.

4.2. Focus of the Post-Translational Modifications (PTMs) of SGs: Glycosylation

Glycosylation is the most consistently reported post-translational modification of SGs across non-muscle organs. Evidence from the CNS clearly demonstrates that ζ-SG, like other SGs, undergoes N-linked glycosylation [30]. This was confirmed by molecular weight shifts following PNGase F treatment, observed not only in native brain tissue but also in cellular systems (CHO cells) where SGs were co-expressed with DGs, confirming a conserved biochemical nature. These results suggest that glycosylation is integral to the formation and stability of the SG complex. However, the functional consequences of this modification in the brain remain completely unexplored. In the retina, Fort et al. [34] confirmed the presence of glycosylated SG subunits, including β-, δ-, and ε-SG, even in the absence of dystrophin, indicating glycosylation as an independent feature from DGC integrity. However, no glycosylation-specific experiments were conducted, leaving open the question of whether SGs contribute to the formation of unique glycoprotein complexes in Müller glia or photoreceptors. In the PNS, Imamura et al. [33] indirectly confirmed the glycosylated state of SG-associated DGC proteins in Schwann cells via WGA chromatography. On the other hand, the loss of δ-SG in BIO14.6 hamsters caused destabilization of glycosylated α-DG [25]. These data suggest that SG-dependent glycosylation may modulate DGC composition and function, particularly in glial and Schwann cells.

Similarly, in adipose tissue, Groh et al. [40] showed that the absence of β- or δ-SG destabilized the glycosylated form of α-DG, impairing laminin binding. This observation suggests that SGs could indirectly control the glycosylation status of associated DGC components. However, SGs themselves were not directly analyzed for glycosylation status in this context. Despite the clear expression of SGs in epithelial and glandular tissues, no data are currently available on the glycosylation patterns of SGs. Given the role of glycosylation in adhesion and cell–matrix interactions, this represents a significant gap, especially in tissues where polarity and tight junctions are crucial.

5. Limitations and Gaps in the Literature

Despite growing interest in the role of SGs beyond skeletal and cardiac muscle, our understanding of their biological significance across non-muscle organs remains incompletely characterized. Much of the current knowledge stems from animal models and relies heavily on descriptive immunolocalization studies. Although these investigations have revealed consistent SG expression in various tissues, including the central and peripheral nervous systems, glands, adipose tissue, and the oral mucosa, they often lack the molecular depth and functional integration necessary to establish the physiological or pathological relevance of these findings.

One of the most pervasive limitations is the technical heterogeneity across studies. Differences in experimental protocols, including antibody specificity, tissue fixation methods, staining conditions, and imaging analysis, undermine the comparability of results. For example, the widely reported "spot-like" distribution of SGs in neurons, glia, and epithelial cells has been variably interpreted, but without standardized criteria for subcellular annotation or sufficient functional validation. The specificity of SG staining is further complicated by inconsistencies in antibody performance and the lack of rigorous controls in some studies, which raises questions about data reproducibility and the interpretation of results.

The distribution of studies across organs is highly uneven. The CNS remains the most extensively investigated, but the bulk of available evidence derives from animal models, with limited validation in human tissues. Notably, SG expression in the eye is supported by only a single study focused on the retina, underscoring a significant gap in ocular research. Glandular organs have been studied in a handful of human investigations; however, these studies are largely descriptive and limited to comparisons between normal and pathological states. Functional implications remain speculative and cross-species validation is lacking. The oral mucosa, although represented by two human studies, remains poorly understood from a mechanistic perspective. Similarly, in adipose tissue, SG expression has been documented in rodents, but the functional relevance and regulation of these complexes, especially in humans, remain unexplored. Furthermore, the molecular characterization of SGs in non-muscle contexts remains incomplete and uneven across organs. While certain PTMs, most notably N-linked glycosylation, have been confirmed in the CNS and adipose tissue, other important PTMs remain largely unclear. However, no study to date has directly tested whether SGs undergo these modifications, leaving a critical gap in understanding how membrane retention and spatial compartmentalization are regulated in diverse cellular environments.

Adding to this complexity, the functional consequences of SG expression in non-muscle organs remain largely unexplored. While data from the nervous system suggest roles in GABAergic signaling, myelin architecture, and neurovascular interactions, these remain exceptional cases. In contrast, tissues such as the thyroid, prostate, mammary gland, and oral mucosa exhibit marked SG downregulation under pathological conditions (e.g., Hashimoto's thyroiditis, adenocarcinoma, or drug-induced osteonecrosis), yet the biological significance of these changes is unknown. Whether SG loss reflects disrupted cell–matrix adhesion, altered membrane polarity, or impaired signaling remains speculative in the absence of mechanistic or functional studies.

Finally, the lack of standardized methodological frameworks and multi-modal analyses hinders the development of a coherent model for SG function outside the muscle. Most studies fail to integrate transcriptomic, proteomic, and functional data within the same tissue, preventing meaningful cross-organ comparisons. The use of different animal models, developmental stages, and tissue processing protocols further compounds this issue, especially given the tissue-specific diversity in SG subunit composition and PTM landscapes. Future research should aim to bridge these gaps through multi-modal approaches that integrate molecular, cellular, and physiological analyses. Priorities include validating SG expression and localization in human tissues, characterizing their post-translational modifications and binding partners across organ systems, and investigating their roles in maintaining tissue architecture and function under both physiological and pathological conditions. Such comprehensive studies are essential for clarifying the roles of SGs in non-muscle organs and for evaluating their potential as biomarkers or therapeutic targets beyond the muscular system.

6. Conclusions

This scoping review emphasizes that SGs, classified as muscle-specific proteins, are widely expressed in non-muscle organs and tissues with heterogeneous expression patterns and subcellular localizations, suggesting novel roles in synaptic organization, cell adhesion, and neurovascular integrity. Nevertheless, considerable gaps exist concerning the molecular composition and functional implications of SG complexes in these tissues. Although evidence of conserved molecular features, such as N-linked glycosylation and interactions with scaffold proteins, is emerging, the composition, regulation, and mechanistic roles of SG complexes in non-muscle contexts remain incompletely defined. Future research should aim to elucidate these aspects using integrated molecular, cellular, and functional approaches. Advanced techniques such as proteomics and gene editing may reveal novel interactors and tissue-specific regulatory mechanisms. To support a broader conceptual framework for these emerging functions, we tentatively propose the functional term *Anastasi Linkage Proteins* (ALPs) to describe SG-containing complexes in non-muscle contexts. This terminology, honoring Professor Giuseppe Pio Anastasi's contributions, is not intended as a formal genetic reclassification, but rather to support a conceptual shift toward their recognition as widely distributed, functionally diverse membrane complexes.

Author Contributions: Conceptualization, F.N. and G.V.; methodology, F.N.; formal analysis, A.F. and D.L.; investigation, J.F.; data curation, J.F.; writing—original draft preparation, F.N.; writing—review and editing, F.N. and G.V.; visualization, A.C., G.C. and M.R.A.; supervision, G.V. All authors have read and agreed to the published version of the manuscript.

Funding: This research received no external funding.

Institutional Review Board Statement: Not applicable.

Informed Consent Statement: Not applicable.

Data Availability Statement: All data are contained within the article.

Conflicts of Interest: The authors declare no conflicts of interest.

References

1. Durbeej, M.; Campbell, K.P. Biochemical Characterization of the Epithelial Dystroglycan Complex. *J. Biol. Chem.* **1999**, *274*, 26609–26616. [CrossRef] [PubMed]
2. Matsumura, K.; Saito, F.; Yamada, H.; Hase, A.; Sunada, Y.; Shimizu, T. Sarcoglycan Complex: A Muscular Supporter of Dystroglycan-Dystrophin Interplay? *Cell. Mol. Biol.* **1999**, *45*, 751–762. [PubMed]
3. Noguchi, S.; Wakabayashi, E.; Imamura, M.; Yoshida, M.; Ozawa, E. Formation of Sarcoglycan Complex with Differentiation in Cultured Myocytes. *Eur. J. Biochem.* **2000**, *267*, 640–648. [CrossRef]
4. Younus, M.; Ahmad, F.; Malik, E.; Bilal, M.; Kausar, M.; Abbas, S.; Shaheen, S.; Kakar, M.U.; Alfadhel, M.; Umair, M. SGCD Homozygous Nonsense Mutation (p.Arg97∗) Causing Limb-Girdle Muscular Dystrophy Type 2F (LGMD2F) in a Consanguineous Family, a Case Report. *Front. Genet.* **2019**, *9*, 727. [CrossRef]
5. Xiao, H.F.; Gong, S.Z.; Zhao, Y.; LeDoux, M.S. Developmental Expression of Rat TorsinA Transcript and Protein. *Dev. Brain Res.* **2004**, *152*, 47–60. [CrossRef]
6. Wolburg, H.; Wolburg-Buchholz, K.; Fallier-Becker, P.; Noell, S.; Mack, A.F. Structure and Functions of Aquaporin-4-Based Orthogonal Arrays of Particles. In *International Review of Cell and Molecular Biology*; Kwang, W.J., Ed.; Academic Press: Cambridge, MA, USA, 2011; Volume 287, pp. 1–41.
7. Cirak, S.; Arechavala-Gomeza, V.; Guglieri, M.; Feng, L.; Torelli, S.; Anthony, K.; Abbs, S.; Garralda, M.E.; Bourke, J.; Wells, D.J.; et al. Exon Skipping and Dystrophin Restoration in Patients with Duchenne Muscular Dystrophy after Systemic Phosphorodiamidate Morpholino Oligomer Treatment: An Open-Label, Phase 2, Dose-Escalation Study. *Lancet* **2011**, *378*, 595–605. [CrossRef]
8. Chawla, G.; Lin, C.-H.; Han, A.; Shiue, L.; Ares, M.J.; Black, D.L. Sam68 Regulates a Set of Alternatively Spliced Exons during Neurogenesis. *Mol. Cell. Biol.* **2009**, *29*, 201–213. [CrossRef]
9. Henry, M.D.; Campbell, K.P. Dystroglycan: An Extracellular Matrix Receptor Linked to the Cytoskeleton. *Curr. Opin. Cell Biol.* **1996**, *8*, 625–631. [CrossRef]
10. Straub, V.; Campbell, K.P. Muscular Dystrophies and the Dystrophin-Glycoprotein Complex. *Curr. Opin. Neurol.* **1997**, *10*, 168–175. [CrossRef]
11. Ceccarini, M.; Grasso, M.; Veroni, C.; Gambara, G.; Artegiani, B.; Macchia, G.; Ramoni, C.; Torreri, P.; Mallozzi, C.; Petrucci, T.C.; et al. Association of Dystrobrevin and Regulatory Subunit of Protein Kinase A: A New Role for Dystrobrevin as a Scaffold for Signaling Proteins. *J. Mol. Biol.* **2007**, *371*, 1174–1187. [CrossRef]

12. Straub, V.; Ettinger, A.J.; Durbeej, M.; Venzke, D.P.; Cutshall, S.; Sanes, J.R.; Campbell, K.P. Epsilon-Sarcoglycan Replaces Alpha-Sarcoglycan in Smooth Muscle to Form a Unique Dystrophin-Glycoprotein Complex. *J. Biol. Chem.* **1999**, *274*, 27989–27996. [CrossRef] [PubMed]
13. Wheeler, M.T.; McNally, E.M. Sarcoglycans in Vascular Smooth and Striated Muscle. *Trends Cardiovasc. Med.* **2003**, *13*, 238–243. [CrossRef] [PubMed]
14. Wheeler, M.T.; Zarnegar, S.; McNally, E.M. Zeta-Sarcoglycan, a Novel Component of the Sarcoglycan Complex, Is Reduced in Muscular Dystrophy. *Hum. Mol. Genet.* **2002**, *11*, 2147–2154. [CrossRef] [PubMed]
15. Shi, W.; Chen, Z.; Schottenfeld, J.; Stahl, R.C.; Kunkel, L.M.; Chan, Y.-M. Specific Assembly Pathway of Sarcoglycans Is Dependent on Beta- and Delta-Sarcoglycan. *Muscle Nerve* **2004**, *29*, 409–419. [CrossRef]
16. Chan, P.; Gonzalez-Maeso, J.; Ruf, F.; Bishop, D.F.; Hof, P.R.; Sealfon, S.C. Epsilon-Sarcoglycan Immunoreactivity and MRNA Expression in Mouse Brain. *J. Comp. Neurol.* **2005**, *482*, 50–73. [CrossRef]
17. Chakrabarty, B.; Sharma, M.C.; Gulati, S.; Kabra, M.; Pandey, R.M.; Sarkar, C. Skin Biopsy: A New Tool to Diagnose Sarcoglycanopathy. *J. Child Neurol.* **2014**, *29*, NP5–NP8. [CrossRef]
18. Arena, S.; Favaloro, A.; Cutroneo, G.; Consolo, A.; Arena, F.; Anastasi, G.; Di Benedetto, V. Sarcoglycan Subcomplex Expression in Refluxing Ureteral Endings. *J. Urol.* **2008**, *179*, 1980–1986. [CrossRef]
19. Arena, S.; Cutroneo, G.; Favaloro, A.; Sinatra, M.T.; Trimarchi, F.; Scarvaglieri, S.; Mallamace, A.; Arena, F.; Anastasi, G.; Di Benedetto, V. Abnormal Distribution of Sarcoglycan Subcomplex in Colonic Smooth Muscle Cells of Aganglionic Bowel. *Int. J. Mol. Med.* **2010**, *25*, 353–359. [CrossRef]
20. Anastasi, G.; Cutroneo, G.; Sidoti, A.; Rinaldi, C.; Bruschetta, D.; Rizzo, G.; D'Angelo, R.; Tarone, G.; Amato, A.; Favaloro, A. Sarcoglycan Subcomplex Expression in Normal Human Smooth Muscle. *J. Histochem. Cytochem.* **2007**, *55*, 831–843. [CrossRef]
21. Muni-Lofra, R.; Juanola-Mayos, E.; Schiava, M.; Moat, D.; Elseed, M.A.; Michel-Sodhi, J.; Harris, E.; McCallum, M.; Moore, U.; Richardson, M.T.; et al. Longitudinal Analysis of Respiratory Function of Different Types of Limb Girdle Muscular Dystrophies Reveals Independent Trajectories. *Neurol. Genet.* **2023**, *9*, e200084. [CrossRef]
22. Li, J.; Liu, Y.; Li, Q.; Huang, X.; Zhou, D.; Xu, H.; Zhao, F.; Mi, X.; Wang, R.; Jia, F.; et al. Mutation in E-Sarcoglycan Induces a Myoclonus-Dystonia Syndrome-Like Movement Disorder in Mice. *Neurosci. Bull.* **2020**, *37*, 311–322. [CrossRef] [PubMed]
23. Tricco, A.C.; Lillie, E.; Zarin, W.; O'Brien, K.K.; Colquhoun, H.; Levac, D.; Moher, D.; Peters, M.D.J.; Horsley, T.; Weeks, L. PRISMA Extension for Scoping Reviews (PRISMA-ScR): Checklist and Explanation. *Ann. Intern. Med.* **2018**, *169*, 467–473. [CrossRef] [PubMed]
24. Vermiglio, G.; Runci, M.; Scibilia, A.; Biasini, F.; Cutroneo, G. Preliminary Study on Sarcoglycan Sub-Complex in Rat Cerebral and Cerebellar Cortex. *Ital. J. Anat. Embryol.* **2012**, *117*, 54–64.
25. Cai, H.; Erdman, R.A.; Zweier, L.; Chen, J.; Shaw, J.H.; Baylor, K.A.; Stecker, M.M.; Carey, D.J.; Chan, Y.M. The Sarcoglycan Complex in Schwann Cells and Its Role in Myelin Stability. *Exp. Neurol.* **2007**, *205*, 257–269. [CrossRef]
26. Boulay, A.; Saubaméa, B.; Cisternino, S.; Mignon, V.; Mazeraud, A.; Jourdren, L.; Blugeon, C.; Cohen-Salmon, M. The Sarcoglycan Complex Is Expressed in the Cerebrovascular System and Is Specifically Regulated by Astroglial Cx30 Channels. *Front. Cell. Neurosci.* **2015**, *9*, 9. [CrossRef]
27. Arco, A.; Favaloro, A.; Gioffrè, M.; Santoro, G. Sarcoglycans in the Normal and Pathological Breast Tissue of Humans: An Immunohistochemical and Molecular Study. *Cells Tissues Organs* **2012**, *195*, 550–562. [CrossRef]
28. Anastasi, G.; Tomasello, F.; Di Mauro, D.; Cutroneo, G.; Favaloro, A.; Conti, A.; Ruggeri, A.; Rinaldi, C.; Trimarchi, F. Expression of Sarcoglycans in the Human Cerebral Cortex: An Immunohistochemical and Molecular Study. *Cells Tissues Organs* **2012**, *196*, 470–480. [CrossRef]
29. Favaloro, A.; Rizzo, G.; Santoro, G.; Pergolizzi, S.; Furci, A.; Centofanti, A.; Cutroneo, G. Sarcoglycans and Integrins in Human Thyrocytes: An Immunofluorescence Study. *Ital. J. Anat. Embryol.* **2022**, *126*, 125–138. [CrossRef]
30. Shiga, K.; Yoshioka, H.; Matsumiya, T.; Kimura, I.; Takeda, S.; Imamura, M. Zeta-Sarcoglycan Is a Functional Homologue of Gamma-Sarcoglycan in the Formation of the Sarcoglycan Complex. *Exp. Cell Res.* **2006**, *312*, 2083–2092. [CrossRef]
31. Rizzo, G.; Di Mauro, D.; Cutroneo, G.; Schembri-Wismayer, P.; Brunetto, D.; Spoto, C.; Vermiglio, G.; Centofanti, A.; Favaloro, A. An Immunofluorescence Study About Staining Pattern Variability of Sarcoglycans in Rat's Cerebral and Cerebellar Cortex. *Eur. J. Exp. Biol.* **2018**, *8*, 7. [CrossRef]
32. Nastro Siniscalchi, E.; Cutroneo, G.; Catalfamo, L.; Santoro, G.; Allegra, A.; Oteri, G.; Cicciù, D.; Alonci, A.; Penna, G.; Musolino, C.; et al. Immunohistochemical Evaluation of Sarcoglycans and Integrins in Gingival Epithelium of Multiple Myeloma Patients with Bisphosphonate-Induced Osteonecrosis of the Jaw. *Oncol. Rep.* **2010**, *24*, 129–134. [CrossRef] [PubMed]
33. Imamura, M.; Araishi, K.; Noguchi, S.; Ozawa, E. A Sarcoglycan-Dystroglycan Complex Anchors Dp116 and Utrophin in the Peripheral Nervous System. *Hum. Mol. Genet.* **2000**, *9*, 3091–3100. [CrossRef] [PubMed]
34. Fort, P.; Estrada, F.-J.; Bordais, A.; Mornet, D.; Sahel, J.-A.; Picaud, S.; Vargas, H.R.; Coral-Vázquez, R.M.; Rendon, A. The Sarcoglycan-Sarcospan Complex Localization in Mouse Retina Is Independent from Dystrophins. *Neurosci. Res.* **2005**, *53*, 25–33. [CrossRef] [PubMed]

35. De Ponte, F.S.; Favaloro, A.; Siniscalchi, E.N.; Centofanti, A.; Runci, M.; Cutroneo, G.; Catalfamo, L. Sarcoglycans and Integrins in Bisphosphonate Treatment: Immunohistochemical and Scanning Electron Microscopy Study. *Oncol. Rep.* **2013**, *30*, 2639–2646. [CrossRef]
36. Cutroneo, G.; Bramanti, P.; Favaloro, A.; Anastasi, G.; Trimarchi, F.; Di Mauro, D.; Rinaldi, C.; Speciale, F.; Inferrera, A.; Santoro, G.; et al. Sarcoglycan Complex in Human Normal and Pathological Prostatic Tissue: An Immunohistochemical and RT-PCR Study. *Anat. Rec.* **2014**, *297*, 327–336. [CrossRef]
37. Cutroneo, G.; Bramanti, P.; Anastasi, G.; Bruschetta, D.; Favaloro, A.; Vermiglio, G.; Trimarchi, F.; Di Mauro, D.; Rizzo, G. Sarcoglycans and Gaba(a) Receptors in Rat Central Nervous System: An Immunohistochemical Study. *Ital. J. Anat. Embryol.* **2015**, *120*, 105–116.
38. Xiao, J.; LeDoux, M.S. Cloning, Developmental Regulation and Neural Localization of Rat Epsilon-Sarcoglycan. *Brain Res. Mol. Brain Res.* **2003**, *119*, 132–143. [CrossRef]
39. Romo-Yáñez, J.; Montañez, C.; Salazar-Olivo, L.A. Dystrophins and DAPs Are Expressed in Adipose Tissue and Are Regulated by Adipogenesis and Extracellular Matrix. *Biochem. Biophys. Res. Commun.* **2011**, *404*, 717–722. [CrossRef]
40. Groh, S.; Zong, H.; Goddeeris, M.M.; Lebakken, C.S.; Venzke, D.; Pessin, J.E.; Campbell, K.P. Sarcoglycan Complex: Implications for Metabolic Defects in Muscular Dystrophies. *J. Biol. Chem.* **2009**, *284*, 19178–19182. [CrossRef]
41. Cazurro-Gutiérrez, A.; Marcé-Grau, A.; Correa-Vela, M.; Salazar, A.; Vanegas, M.I.; Macaya, A.; Bayés, À.; Pérez-Dueñas, B. ε-Sarcoglycan: Unraveling the Myoclonus-Dystonia Gene. *Mol. Neurobiol.* **2021**, *58*, 3938–3952. [CrossRef]

biomolecules

MDPI

Review

Biomolecule-Based Coacervation: Mechanisms, Applications, and Future Perspectives in Biomedical and Biotechnological Fields

Dong Hyun Kim [1], Mi-Ran Ki [2], Da Yeon Chung [1] and Seung Pil Pack [1,*]

[1] Department of Biotechnology and Bioinformatics, Korea University, Sejong 30019, Republic of Korea; jklehdgus@korea.ac.kr (D.H.K.); anthic83@korea.ac.kr (D.Y.C.)

[2] Institute of Industrial Technology, Korea University, Sejong 30019, Republic of Korea; allheart@korea.ac.kr

* Correspondence: spack@korea.ac.kr

Abstract: Coacervate is a form of liquid–liquid phase separation (LLPS) in which a solution containing one or more charged components spontaneously separates into two immiscible liquid phases. Due to their ability to mimic membraneless cellular environments and their high biocompatibility, coacervates have found broad applications across various fields of life sciences. This review provides a comprehensive overview of recent advances in biomolecule-based coacervation for biotechnological and biomedical applications. Encapsulation via biomolecule-based coacervation enables high encapsulation efficiency, enhanced stability, and the sustained release of cargos. In the field of tissue engineering, coacervates not only support cell adhesion and proliferation but also serve as printable bioinks with tunable rheological properties for 3D bioprinting. Moreover, biomolecule-based coacervates have been utilized to mimic membraneless organelles, serving as experimental models to understand the origin of life or investigate the mechanisms of biochemical compartmentalization. This review discusses the mechanisms of coacervation induced by various types of biomolecules, evaluates their respective advantages and limitations in applied contexts, and outlines future research directions. Given their modularity and biocompatibility, biomolecule-based coacervates are expected to play a pivotal role in next-generation therapeutic development and the construction of controlled tissue microenvironments, especially when integrated with emerging technologies.

Keywords: liquid–liquid phase separations; coacervates; membraneless organelles; complex coacervation; simple coacervation

1. Introduction

Coacervation is liquid–liquid phase separation (LLPS) that occurs in a solution containing one or more polymeric colloids that spontaneously separate into two immiscible liquid phases [1]. In this process, the phase with a higher concentration of colloidal components is referred to as the coacervate phase, which exists in equilibrium with a lower-density liquid phase [2]. The coacervate phase typically appears as amorphous liquid droplets and, over time, these droplets can fuse together, forming a more stable structure that can be obtained in large quantities [1,3]. Due to these physicochemical properties, coacervation has found various applications in the biological and biomedical fields, where it serves as a crucial mechanism for molecular organization and regulation under physiological conditions [4–6].

In biological systems, the organization and compartmentalization of biomolecules play an essential role in maintaining cellular function [7]. Traditionally, intracellular compartmentalization has been attributed mainly to the membrane organelles [8,9]. However, recent studies have highlighted the significance of LLPS-driven compartmentalization as an alternative mechanism for organizing cellular components [3]. LLPS enables the selective enrichment and segregation of biomolecules, leading to the formation of coacervate phases [10]. As a result, biomolecule-based coacervation has attracted increasing attention in life sciences and bioengineering, particularly for its role in various biological processes.

Biomolecule-based coacervation is driven by interactions among biomolecules such as proteins, nucleic acids, and polysaccharides. These interactions involve electrostatic attractions, hydrogen bonding, and hydrophobic forces [10–13]. Compared to synthetic polymer-driven systems, biomolecule-based coacervates offer superior biocompatibility, greater responsiveness to physiological stimuli, and intrinsic functional significance for biological systems [8,14]. For example, protein-based coacervates can create microenvironments that regulate enzyme activity, while nucleic acid-based coacervates play crucial roles in gene expression and RNA metabolism [15–17].

Although synthetic polymer-based coacervation systems have been widely explored, they often exhibit low biocompatibility and pose challenges related to polymer residue and purification [1,11]. In contrast, biomolecule-based coacervates provide a more biomimetic and biologically relevant platform, exhibiting features like stimuli-responsive behavior and enhanced integration with cellular systems [18–20].

Given these advantages, biomolecule-based coacervation is emerging as a key research focus in fields such as synthetic biology, biotechnology, and biomedicine. Despite the growing interest and recent progress in this system, a comprehensive understanding of its biotechnological and biomedical applications remains limited. As shown in Figure 1, this review summarizes the mechanisms, types, and formation factors of coacervation, as well as recent advances in this area, focusing on the biological relevance and application potential of coacervation systems derived from natural biomolecules.

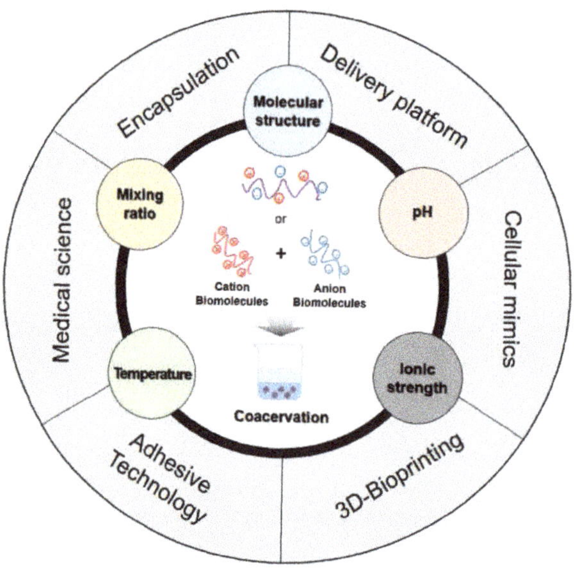

Figure 1. Illustration of factors and applications of coacervation based on cation biomolecules and anion biomolecules.

2. Mechanism of Coacervation

Various theories have been proposed to understand the complex mechanism of coacervation. Bungenberg de Jong and Kruyt were pioneers in the study of coacervation, first describing the phenomenon in the late 1920s [2]. Their research revealed that polymeric coacervates consist of four non-water species: a polyanion, a polycation, a cation, and an anion. Their work provided essential insights into the interactions between charged polymers and the conditions necessary for coacervation to occur [2]. This early research facilitated the establishment of an empirical basis for explaining the coacervate systems. The first successful theory of complex coacervation was published by Voorn and Overbeek [21]. This model introduced a theoretical framework to describe coacervate systems, highlighting a two-phase coexistence region where phase separation occurs. It described the phenomenon as a spontaneous polymer charge neutralization process driven by electrostatic interactions between the polyelectrolytes used in the experiment. It also explained the conditions for coacervation in terms of polymer charge density, molecular weight, temperature, and dielectric constant [2,21]. Furthermore, this theory described the condition of coacervation based on the Flory–Huggins theory [22,23] for entropy contributions and the Debye–Hückel theory [24] for electrostatic interactions, where σ and r represent the charge density and polymer molecular weight, respectively [21,25].

$$\sigma^3 r \geq 0.53$$

Subsequently, Veis and Aranyi refined the Voorn–Overbeek theory by modifying the Flory–Huggins interaction parameter and replacing the electrostatic term with entropy gain to achieve a more precise characterization of coacervation, leading to the development of the following model where A^+ and B^- represent the polycation and polyanion, respectively [26,27].

$$[A^+] + [B^-] \rightarrow [A^+B^-]_{agg.}$$
$$[A^+B^-]_{agg.} \rightarrow [A^+B^-]_{agg.}^{Dilute\ pahse} + [A^+ + B^-]_{Random.C_c}^{Conc.\ pahse}$$

This theory states that a neutralized aggregate ($[A^+B^-]_{agg.}$) is formed through electrostatic interactions, resulting in low configurational entropy. An increase in configurational entropy leads to LLPS that forms randomly mixed concentrated phases and diluted aggregate phases. However, this theory has a limitation in that it is only applicable to polymeric systems with a low charge density.

In 1980, Tainaka advanced the Veis–Aranyi theory by proposing that phase separation in polyelectrolyte systems is primarily driven by attractive forces between aggregates [28]. The Tainaka model expands the applicability beyond low-charge-density systems to include high-charge-density polyelectrolytes, suggesting that coacervation is possible even in systems without specific ionic bonds. Experimental observations indicate that deviating from critical parameters leads to different physical outcomes. When the charge density is excessively high, flocculation occurs instead of coacervation, while insufficient charge density prevents the phase separation necessary for coacervation. However, this theory fails to explain why coacervation is suppressed under low ionic strength conditions and cannot fully describe the complex coacervation processes in multi-polymer systems [2,28].

A variety of contemporary theoretical frameworks have developed to deal with the intricacies of coacervation. These include Random Phase Approximation, Field Theoretic approaches, Counterion Release models, Molecular Simulation techniques, and the Polymer Reference Interaction Site model [29]. These advancements indicate an increasing comprehension of the physical characteristics of polymeric coacervates and the persistent challenges within this field. Although a multitude of theories elucidates various aspects of complex coacervation, no singular theory can thoroughly cover all the characteristics evident in these systems [29]. The variety of approaches signifies continuous investigation and the necessity for an integrated theory capable of forecasting coacervate characteristics across various parameters and scales.

3. The Type of Coacervation

Coacervation is classified into simple coacervation and complex coacervation based on its formation process. Simple coacervation involves a single macromolecular species and can be induced by dehydration or desolvation through the addition of salts, alcohols, or other additives, as well as by changes in temperature [18]. This process reduces the interactions between the macromolecule and the solvent while promoting intermolecular interactions among the macromolecules [9,30].

In contrast, complex coacervation is primarily driven by electrostatic interactions between oppositely charged molecules, with charge neutralization being modulated by factors such as ionic strength and pH [31,32]. However, in addition to electrostatic interactions, other intra- and intermolecular forces—such as π–π interactions, cation–π interactions, hydrogen bonding, dipole–dipole interactions, and their combinations—can also influence the coacervation process [33,34].

3.1. Simple Coacervation

Simple coacervation, also known as self-coacervation, is a process that occurs when the condition of an aqueous solution is meticulously modulated to specific parameters. These parameters may include the pH level, the ionic strength, or the temperature of the solution. In such instances, the macromolecules within the solution interact intrinsically, resulting in phase separation into a coacervate phase and a dilute phase [18,35].

Cai et al. provided a comprehensive analysis of the molecular mechanisms underlying the simple coacervation of histidine-rich beak protein 1 (HBP-1) and histidine-rich beak protein 2 (HBP-2). Their findings revealed that the process of simple coacervation of HBPs was predominantly driven by hydrophobic interactions, which occurred after charge neutralization through salt screening and pH regulation. An increase in ionic strength was observed to result in enhanced protein–protein interactions, thereby facilitating the process of simple coacervation. This study demonstrated that the sequence modularity of the C-terminus of HBPs played a significant role in their ability to form simple coacervates. This finding suggests that the specific arrangement of amino acids influences their behavior in a solution [36].

Research is also being conducted on simple coacervation utilizing chitosan, a material extensively employed in drug delivery. Jing et al. explored the simple coacervation process of carboxymethyl chitosan (CMCS) [37]. This process was achieved by adjusting the pH of the aqueous solution close to its isoelectric point. The simple coacervation of CMCS occurred across a wide pH range of 3.0 to 6.0, which involved electrostatic interactions, hydrogen bonding, and hydrophobic interactions [37].

Kaushik et al. investigated the simple coacervation of elastin, focusing in particular on the effects of temperature and ionic strength on the coacervation process [38]. The simple coacervation in elastin was explored near its isoelectric point (pI = 4.7). The result of the experiment was that the coacervation temperature was found to decrease from 38 °C to 33 °C as the ionic strength of the solution increased from 0 to 40 mM NaCl. The results indicated that hydrophobic interactions were significant forces in simple coacervation in elastin [38].

3.2. Complex Coacervation—Binary and Ternary

Binary coacervation is the most common type of coacervation that is formed through the mixing of two oppositely charged molecules. The stoichiometric properties of two polyelectrolytes, salt, and pH have been demonstrated to exert a significant influence on the processes of their formation, stability, and viscoelastic behavior [1].

Gulão et al. [39] conducted a study on the subject of binary coacervation between polypeptide-leucine and gum arabic (GA), with a view to regulating GA and salt concentration at different pH values ranging from 1.0 to 12.0. The formation of insoluble complex coacervates was most prevalent at a pH of 4.0. The formation of coacervate was indicated by higher turbidity and larger particle sizes in samples containing 0.2% polypeptide-leucine and 0.03% GA without salt. The precipitate was dissociated at a pH of 2.0, which is the pKa of the GA [39].

Ternary coacervation refers to the phase separation process that involves the addition of a third component to the binary coacervation composed of two components. The physical properties and responses of ternary coacervation are largely governed by this third component, which renders its selection critical for maintaining the characteristic features of binary coacervation [18]. The incorporation of a third material has been shown to result in ternary coacervation, thereby providing enhanced protection against environmental factors such as salt concentration and pH. This process enables the formation of the coacervate over a broader compositional range compared to that observed in binary coacervation [18].

Black et al. [40] proposed a method for the stable encapsulation of proteins using complex coacervation driven by electrostatic interactions between charged polypeptides. In this study, PLys was used as the cationic component, while PGlu was employed as the anion counterpart. Bovine serum albumin (BSA) was selected as the third component to induce coacervation. The process of ternary coacervation was initiated by the formation of an intermediate complex between the positively charged PLys and the negatively charged BSA under physiological pH (7.4) through electrostatic interaction. Subsequently, the addition of PGlu to this complex induced ternary coacervation [40].

4. Factors Influencing the Coacervation Process

The process of coacervation is initiated by a delicate balance among the electrostatic interaction, hydrophobic interaction, hydrogen bonding, and van der Waals forces between distinct molecular species [41]. A disruption of this balance can lead to the disassembly of coacervation into a homogeneous single-phase solution, or alternatively, induce phase separation, leading to precipitation. Consequently, the physicochemical parameters of the system—including the molecular structure, ionic strength, pH, and molecular ratio of the components—play a significant role in the formation and stability of the coacervate phases [1]. Figure 2 shows the coacervation formation of the mechanisms and regulatory factors.

4.1. Molecular Structure

Molecular characteristics are a key factor influencing coacervation. The molecular structure is essential in determining the degree and properties of coacervation [1]. The arrangement, location, and density of the charges within the molecule can fluctuate based on its structure, thereby affecting the intermolecular interactions [42]. Variations in molecular size and topology (e.g., linear, branched, or secondary structures) alter the ion distribution, potentially influencing coacervate formation [43–46]. In addition, the chirality of a molecule can induce steric hindrance, resulting in unique aggregation behaviors that may cause precipitation instead of coacervation [47].

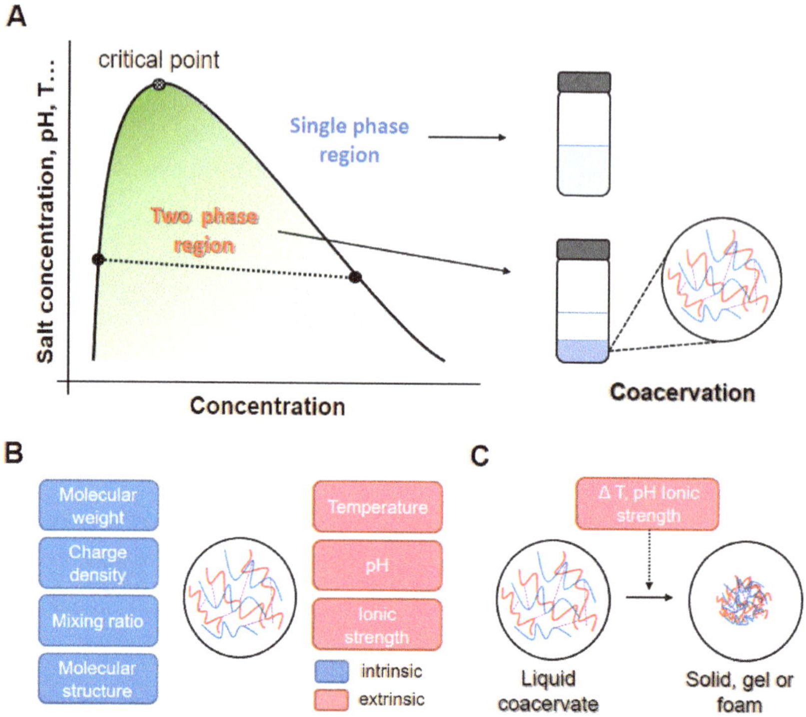

Figure 2. Illustration of (**A**) the range of LLPS of coacervation, (**B**) the intrinsic and extrinsic factors influencing the coacervation process, and (**C**) the phase transition of coacervates depending on temperature, pH, or ionic strength.

Perry et al. investigated the complex coacervation of PLys with PGlu by varying the polymer chirality [47]. The phase separation between oppositely charged PLys and PGlu was established by electrostatic interactions and hydrogen bonding. Coacervation occurred when at least one of the polypeptides was racemic, which disrupted the hydrogen bonding among the chains. In contrast, when both polypeptides possessed pure chirality, they formed solid precipitates with β-sheet structures due to steric hindrance [47].

Pramanik et al. demonstrated that histidine-rich peptides, specifically $(GHGXY)_4$ (where X = Leu/Val/Pro), have a significant impact on coacervate formation [48]. The hydrophobicity of the amino acid residues exerted a substantial influence on the coacervation process. Leucine demonstrated enhanced coacervate stability and increased droplet sizes in comparison to those containing less hydrophobic residues such as valine. However, the impact of chirality (D- and L-amino acid variants) on coacervate characteristics did not show significant differences [48].

Vieregg et al. examined the formation of coacervates based on the structural characteristics of oligonucleotides [49]. Single-stranded DNA underwent coacervation with PLys due to its lower charge density. However, double-stranded DNA with its higher charge density led to the formation of solid precipitate [49].

4.2. *pH*

In the process of coacervation, pH serves as a key parameter by modulating the charge state of the molecules, thereby influencing electrostatic interactions and ultimately determining the occurrence and nature of phase separation. This is particularly significant for molecules containing weakly acidic or basic functional groups, as pH adjustments can effectively tune the degree of ionization. An appropriate pH environment enables the molecules to possess opposite charges, which is essential for the formation of coacervates [50,51].

Kayitmazer et al. investigated the formation of hyaluronic acid (HA) and chitosan (CS)-based coacervates as a function of pH [52]. Coacervation occurred within a pH range between the pKa values of HA and CS, where the ionic strength of both molecules increases. As the pH increased, the HA/CS system underwent a transition from a single-phase solution to coacervation, eventually resulting in a precipitated state [52].

Doshi et al. reported on the occurrence of pH-induced coacervation in legume protein mixtures [53]. The proteins remained soluble in the higher pH range, while the precipitation of protein particles occurred within the lower pH range. Within the intermediate pH range, coacervates were formed by the simple coacervation of proteins [53].

Jamshidian et al. investigated the complex coacervation of wheat germ protein and high methoxy pectin [54]. The pH of the solution was found to have a significant effect on the process of coacervation, with alterations to the surface charges of biomolecules. Changes in particle size and zeta potential were observed with varying pH levels. At pH_{max}, the result exhibited the highest turbidity and the largest particle size [54].

4.3. Temperature

In contrast to specific systems that exhibit significant temperature sensitivity, the majority of coacervation systems demonstrate relatively weak temperature dependence. This thermo-responsive behavior is primarily governed by the intrinsic physicochemical properties of the individual macromolecule involved, such as its hydrophilicity, polyelectrolyte characteristics, and molecular conformation [1]. Macromolecules that exhibit thermo-responsive solubility can be classified based on two critical solution temperature behaviors: the Lower Critical Solution Temperature (LCST) and the Upper Critical Solution Temperature (UCST) [55,56]. In LCST systems, macromolecules are fully soluble at lower temperatures; however, upon heating, phase separation occurs. This behavior is predominantly driven by entropy. At low temperatures, macromolecule–solvent interactions are dominant. Elevated temperature enhances macromolecule–macromolecule interactions while diminishing macromolecule–solvent interactions. This transition facilitates phase separation. Conversely, UCST systems display the opposite trend. These systems remain a homogeneous mixture at elevated temperatures but undergo phase separation upon cooling. In this case, this behavior is predominantly driven by enthalpy [57,58]. Complex coacervation often exhibits phase behavior similar to that of LCST systems. Strong macromolecule–macromolecule interactions are present at higher temperatures, which weaken as the temperature decreases. The reduction in intramolecular and intermolecular interactions upon cooling leads to phase separation, given that the process is generally governed by entropic contributions. Coacervation under UCST conditions is rare, but has been reported in systems with strong enthalpic driving forces [59].

Fu et al. induced coacervation using five types of N-substituted polypeptoids (NNCAs) with different alkyl chain lengths: allyl, propargyl, butyl, hexyl, and octyl [60]. The polypeptoids exhibited both LCST and UCST systems. The phase transition temperatures were tunable within a range of 29–55 °C, depending on the chemical composition and the length of the alkyl side chains. This study demonstrated that the cloud point of the polypeptoids decreased with an increase in alkyl chain length. The cloud point decreased from 55.0 °C to 39.0 °C as the alkyl chain length increased from butyl to octyl [60].

The research conducted by Nie et al. presented the coacervation of a globular protein known as lipoate-protein ligase A (LplA) from *E. coli* [61]. LplA exhibited temperature-sensitive, reversible LCST coacervation. The coacervation began at approximately 14 °C, with significant structural changes occurring as the temperature increased. The emergence of larger particles between 12 °C and 16 °C in the protein solution specifically indicated the initial stages of the transition from monomeric LplA to oligomeric structures. This study also demonstrated that the coacervation of LplA could be selectively induced and reversed through the dissolution of the condensates by LplA's natural substrate, lipoic acid, and its analogue, lipoamide [61].

4.4. Mixing Ratio

A pivotal factor in coacervation, especially in complex coacervation, is the charge stoichiometry between the positively and negatively charged biomolecules. Coacervation typically necessitates an electrically neutral state [1]. When synthetic polymers comprise solely one type of charged group, the calculation of charge stoichiometry to determine the mixing ratio is relatively straightforward [62]. However, the prediction of the mixing ratio is much more challenging for biomolecules like proteins that contain both positive and negative charges. The driving force for phase separation may arise not from the overall net charge but rather from the specific charge distributions on the molecular surface [63].

Wei et al. investigated the complex coacervation between theabrownin (TB) and whey protein isolate (WPI) depending on their mixing ratio. It was found that an intermediate pH induced complex coacervation with the strongest electrostatic interactions. The findings further demonstrated that the most robust electrostatic interaction was observed at a ratio of 10:1 of TB and WPI. The process of coacervation was mainly driven by electrostatic interactions, in conjunction with hydrogen bonding and hydrophobic interactions [64].

Cui et al. examined the coacervation of Antarctic krill protein isolate (AKPI) and GA. The coacervation process was conducted at a ratio of AKPI to GA of approximately 3:1 [65]. The electrostatic interaction and hydrogen bonding were identified as the main driving forces of coacervation. It was demonstrated that the mixing ratio not only influences the formation of coacervate but also has the potential to affect the ratio of wall to core [65].

Li et al. investigated the coacervation of chia seed gum (CSG) and WPI under WPI/CSG mass ratios (16:1–1:1, *w*/*w*). The stronger formation occurred at comparatively elevated mass ratios of 16:1, 8:1, and 4:1. With the rising proportion of WPI, there was a corresponding increase in the zeta potential value. The most pronounced electrostatic interaction was observed at the 4:1 ratio [66].

4.5. Ionic Strength

Ionic strength is determined by the concentration of ionic species present in the solution [67]. It regulates the electrostatic interactions among the charged components, thereby influencing the formation of coacervation [66,68]. Moderate ionic strength may facilitate coacervation; however, an excessive concentration of salt can inhibit this process. This concept is identified as the critical salt concentration [69,70]. Elevated ionic strength diminishes the electrostatic attractions between oppositely charged molecules, thereby significantly reducing their ability to undergo coacervation. However, coacervation exhibiting significant resistance to salt can still manifest LLPS and maintain stability under increased ionic strength [71,72].

Kayitmazer et al. demonstrated that HA and CS form a coacervate due to strong electrostatic interactions between them [52]. In their study, the coacervation was markedly suppressed at an ionic strength $\geq$ 1.5 M NaCl. At higher ionic strengths, the presence of additional ions in the solution screens the electrostatic interactions between the charged groups of HA and CS. This screening effect reduces the attractive forces that drive coacervation, making it more challenging for the polymers to come together and form coacervates. The study also demonstrated that, as ionic strength increases, there is a transition from coacervation to flocculation. This means that instead of forming a stable coacervate phase, the system may lead to the aggregation of particles (flocculation) due to the reduced electrostatic repulsion between them [52]. It is noteworthy that non-stoichiometric coacervation was observed when the zeta potential values were higher than zero, particularly at charge ratios less than 0.46, indicating that coacervation can happen even when the system is not electrically neutral, which is contrary to typical expectations [52].

Onuchic et al. investigated the effects of divalent cations (Mg^{2+}, Ca^{2+}, and Sr^{2+}) on the formation of coacervates in a system consisting of an arginine-rich peptide (RP3) and polyU (RP3–polyU) [73]. They found that an elevated concentration of Mg^{2+} reduced the formation of coacervates. This result was consistent with the effects observed for monovalent cations (Na^{+}), whereas divalent cations demonstrated a more pronounced influence. A comparable phenomenon was observed for other divalent cations, each demonstrating the reduced concentration thresholds necessary to initiate coacervation [73].

5. Biomolecule-Based Coacervation

5.1. Proteins

Proteins are well-known materials commonly used in coacervation processes. In recent years, protein-based coacervation has been increasingly studied not only in the field of life sciences but also in diverse applications such as food and cosmetics [66]. Protein-based coacervation primarily utilizes animal-derived charged proteins such as heparin, elastin, gelatin, tau, and BSA. More recently, protein extracts obtained from plants, such as soy, pea, and canola, have also been used in coacervation. This type of coacervation is mainly induced through electrostatic interactions between proteins and other proteins or polysaccharides [55,63,66,74–78]. Compared to synthetic polymers, protein-based coacervates offer advantages such as enhanced stability, biocompatibility, controlled release, permeability, and the retention of biological function.

Allahyartorkaman et al. conducted an investigation into the phosphorylation-induced simple coacervation of tau protein, as well as RNA-assisted complex coacervation [79]. The phosphorylation of tau protein triggered a simple coacervation upon a temperature transition from 4 °C to room temperature. The investigation revealed that, at concentrations below 2 μM, neither tau nor phosphorylated tau (p-tau) demonstrated coacervation. However, a binodal pattern of nucleation was observed when the concentration of p-tau was elevated to a range of 2 to 100 μM. The p-tau, in contrast to its non-phosphorylated counterpart, exhibited unique phase separation characteristics when subjected to different pH conditions. This indicates that phosphorylation is fundamental to the response of tau proteins to variations in pH, especially regarding their capacity for the formation of coacervates [79].

As shown in Figure 3, Liu et al. investigated the complex coacervation between ovalbumin (OVA) and dextran sulfate (DS) and their properties [80]. The research demonstrated that complex coacervation is significantly influenced by the ratio of OVA to DS. With an increase in the OVA/DS ratio, the critical pH values for coacervation were observed to shift towards higher values. The salt concentration also influenced coacervation. At low salt concentrations ($\leq$100 mM), the solubility of OVA and DS was enhanced by reducing electrostatic repulsion. However, at elevated concentrations ($\geq$100 mM), complex coacervation was inhibited. An increase in the concentration of salt ion within the solution resulted in a decrease in the critical pH, thereby reducing the electrostatic interaction. The thermodynamic parameters were employed to examine the stable coacervate formation between OVA and DS. The enthalpy changes suggested that the interaction between OVA and DS was exothermic, predominantly influenced by electrostatic attractions. The observed entropy change indicated that hydrophobic interactions play a role in coacervate formation, serving as an additional driving force for this process [80].

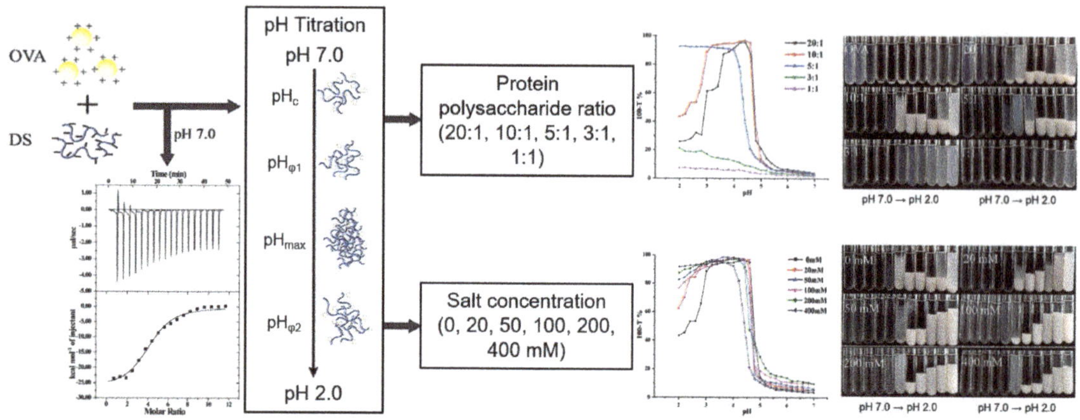

Figure 3. Schematic of the coacervation between ovalbumin and dextran sulfate under biopolymer ratio and salt concentration. The coacervation with OVA and dextran sulfate was induced by the protein/polysaccharide ration and salt concentration. (Adjusted from ref. [80]. Copyright © 2021 Elsevier Ltd.).

Archut et al. explored the interactions between various fractions of pea protein and pectin [81]. Soluble pea protein (SPP) was isolated from pea protein (PP) concentrate using isoelectric precipitation. Subsequently, the turbidity of SPP/pectin and PP/pectin coacervation was assessed to examine the interaction behavior. The study revealed that pea globulins played a substantial role in the increase in turbidity as a result of the protein's simple coacervation. In contrast, the SPP did not affect the turbidity, indicating that these proteins primarily formed complexes with pectin without simple coacervation. A molecular weight distribution analysis further demonstrated that pea albumin and low-molecular-weight fractions below 20 kDA were essential for forming stable coacervates with pectin [81].

5.2. Nucleacids

Nucleic acids, such as negatively charged ssDNA or RNA [82], can induce coacervation through electrostatic interactions with oppositely charged cationic peptides or proteins. Nucleic acid-based coacervation is widely used in studies that mimic membraneless organelles such as stress granules and nucleoli [10,83]. Furthermore, it finds applications in targeted therapy and gene editing through the stable encapsulation of functional RNAs including mRNA and siRNA. Due to the sequence specificity of DNA/RNA, nucleic acid-based coacervation enables selective molecular recognition and binding. This process provides biocompatibility and allows for the precise regulation of biological responses [84–86].

Vieregg et al. investigated the phenomenon of coacervation involving oligonucleotides, specifically single-stranded and double-stranded DNA, in conjunction with cationic peptides [49]. Their findings revealed that coacervates were successfully formed between the peptides and single-stranded DNA, whereas no such interactions were observed with double-stranded DNA. At salt concentrations of 500–700 mM NaCl, single-stranded DNA and peptide complexes transitioned from precipitates to coacervates, while, at concentrations above 1 M, the formed coacervates dissociated. Under 300 mM NaCl conditions, 10 bp dsDNA exhibited a phase transition from precipitation to coacervation at 50 °C. Although most single-stranded sequences formed coacervates, certain purine-rich sequences induced precipitation even in the absence of hybridization [49].

Morita et al. showed DNA-based coacervation from branched DNA nanostructures [87]. The results demonstrated that the physical properties of these structures can be controlled by manipulating factors such as cooling rates and DNA concentrations (Figure 4). The process of coacervation of three single-stranded DNAs (ssDNA1, ssDNA2, and ssDNA3), which are designed to form Y-motifs, was examined under varying temperature conditions. When the solution temperature T exceeds TmY (T_{mY} = melting temperature of the Y-motif stem), the ssDNAs are completely dissociated. Under conditions where TmY > T > TL (T_L: temperature for the formation of liquid-like DNA coacervates), dispersed DNA Y-motifs are formed. In the range of TL > T > TG (T_G: temperature for the formation of gel-like DNA coacervates), the Y-motifs dynamically bind and unbind via their sticky ends, leading to the formation of liquid-like DNA coacervates. When $T < T_G$, the Y-motifs form gel-like DNA coacervates through static binding via their sticky ends. Under the experimental conditions, T_{mY} was 75 °C and the measured T_L and T_G were approximately 64 °C and 35 °C, respectively [87].

In a study, van Haren et al. investigated how coacervates (biomolecular condensates) are formed through the enzymatic deacetylation of peptides in the presence of DNA [88]. They found that coacervation did not occur when only DNA and peptides were present. However, when deacetylation was induced using the enzyme SIRT3, the coacervation between DNA and peptides was successfully triggered under the same conditions. At low concentrations of the enzyme, the weak interactions between DNA and peptides led to the formation of gel-like coacervates. In contrast, at higher enzyme concentrations, stronger interactions were promoted, resulting in a transition to liquid-like coacervates [88].

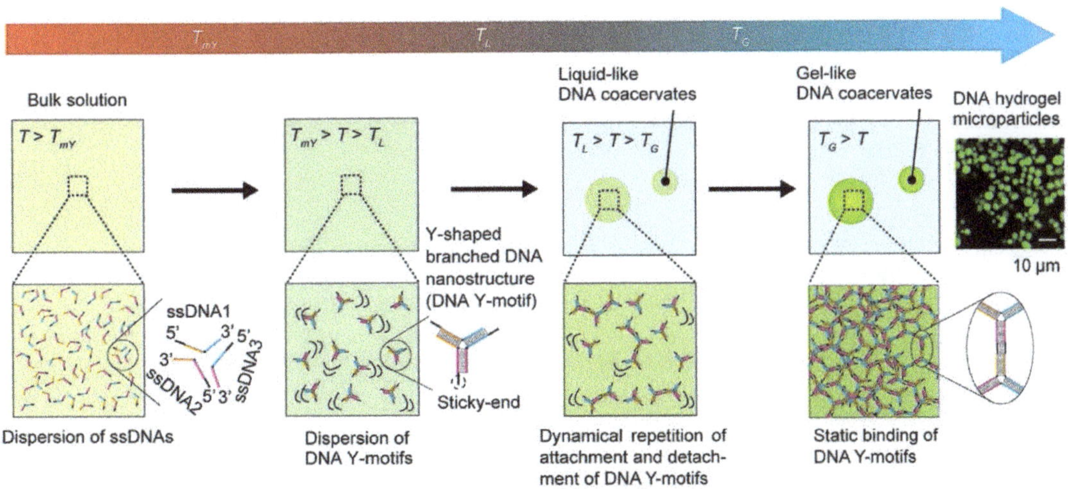

Figure 4. Schematic illustration of the formation of liquid- and gel-like DNA coacervates. When the solution temperature $T > T_{mY}$, three kinds of single-stranded DNAs (ssDNA1, ssDNA2, and ssDNA3) are dispersed in the bulk solutions; when $T_{mY} > T > T_L$, the three ssDNAs form a DNA Y-motif with self-complementary sticky ends; when $T_L > T > T_G$, liquid-like coacervates are formed by the dynamical repetition of attachment and detachment of the sticky ends of the DNA Y-motifs; when $T_G > T$, gel-like DNA coacervates are formed by the static binding of the DNA Y-motifs via their sticky ends. T_{mY} is the melting temperature of the Y-motif stem. T_L is the formation temperature of liquid-like DNA coacervates. T_G is the formation temperature of gel-like DNA coacervates. (Adjusted from ref. [87]. Copyright 2024, the Authors, published open access by *Advanced Materials Interfaces* under the terms of the Creative Commons CC BY License).

5.3. Peptides

Peptides possess a simple backbone structure and various functional groups on their side chains, contributing to molecular self-assembly [89,90]. Hydrophobic interactions and hydrogen bonding within the peptide chains promote the formation of special secondary structures such as α-helices and β-sheets, which can further organize into unique architectures like fibers, nanotubes, and nanowires [91–93]. Based on these structural advantages, peptides are highly responsive to external stimuli. In particular, due to their specific sequences, peptides can not only undergo self-assembly but also interact with other molecules such as nucleic acids and polysaccharides to induce coacervation [94,95].

Gulão et al. investigated the complex coacervation based on polypeptide leucine with GA. The optimal pH for coacervation was found to be below 4.0 with increased turbidity and particle size at this pH range. When the concentration of polypeptide leucine and GA was 0.2% and 0.03% and there was a no salt condition, the coacervation had higher turbidity and the highest particle size. The elastic behavior of coacervates also increased under the same conditions [39].

The coacervation based on elastin-like polypeptides (ELPs) was investigated by Fisher et al. [96]. The ELPs were engineered to include either positively charged residues (e.g., K30, KQ30) or negatively charged residues (D/E30, D/EQ30) to enable their use in complex coacervation. The charge balance was found to be a critical factor for effective coacervation, with the most pronounced turbidity and material yield observed when the mixture contained a slight excess of positively charged ELPs. Coacervation was highly sensitive to salt concentration; for instance, the coacervates composed of K30 and DE30 completely disappeared upon the addition of 50 mM NaCl. Similarly, for ELPs containing Q and fewer charged residues, coacervation was suppressed even at 30 mM NaCl. The formation of coacervates was analyzed through turbidity measurements and optical microscopy observations [96].

Joshi et al. investigated the complex coacervation with polypeptides to encapsulate nonenveloped viruses, specifically porcine parvovirus (PPV) and human rhinovirus (HRV) [97] (Figure 5). The peptide properties, such as peptide chemistry, chain length, charge patterning, and hydrophobicity, effected the virus incorporation. Virus encapsulation is primarily driven by electrostatic interactions, with the optimal charge fraction (i.e., the ratio of cationic monomers) typically falling between 0.4 and 0.6. When analyzing coacervation according to polypeptide chain length—short (N = 48), medium (N = 400), and long (N = 800)—the long chains (N = 800) shifted the optimal coacervation condition toward a lower cationic charge fraction. In the case of polypeptides with alternating charged and neutral residues (e.g., (K2G2)12), charge clustering was shown to influence encapsulation efficiency. Regarding the charge density, the polypeptides with a high charge density demonstrated the highest encapsulation efficiency, while introducing neutral residues to reduce the charge density led to decreased efficiency. Increasing the hydrophobicity of the peptides (e.g., K2G2→K2A2→K2L2) affected PPV encapsulation moderately, whereas HRV encapsulation was significantly enhanced when leucine was incorporated into the peptide sequence [97].

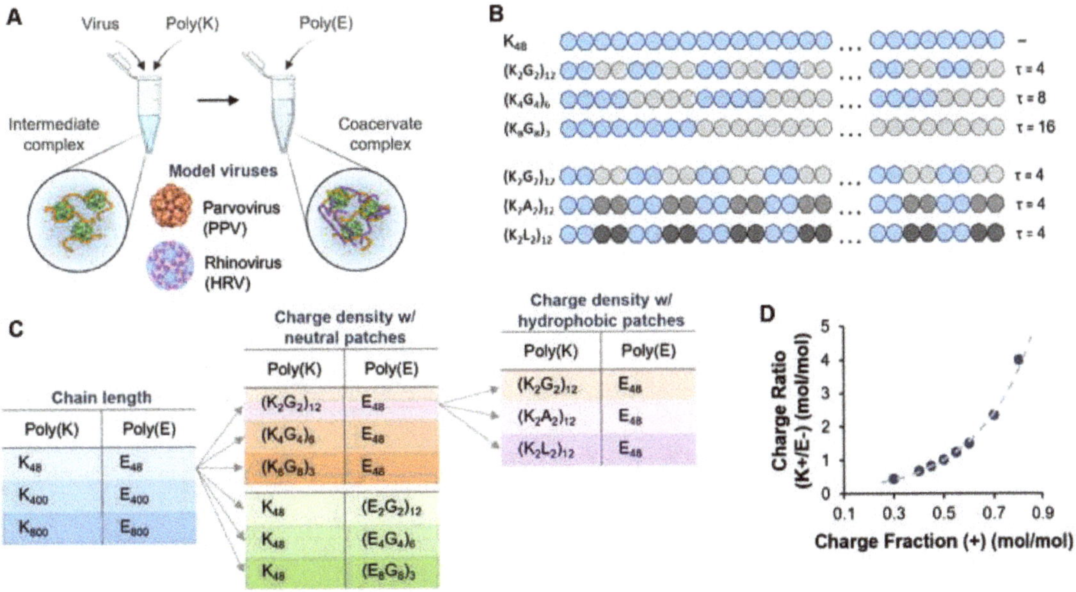

Figure 5. Coacervate design using sequence-defined block-*co*-polypeptides. (**A**) Schematic depiction of virus-containing coacervate formulation. (**B**) Example depiction of the variations in polypeptide charge density and hydrophobicity of the cationic lysine (K)-containing polymers. Charge blockiness is defined by the parameter τ, while hydrophobicity is indicated by the color of the gray blocks as the neutral amino acid spacers go from glycine to alanine to leucine. (**C**) Experimental design matrix to study the effect of polypeptide characteristics such as charge patterning and hydrophobicity on virus encapsulation. (**D**) Plot of the charge ratio (K+/E−) as a function of the total cationic charge fraction from the polypeptides present in the system. (Reused from ref. [97]. Copyright 2023, the authors published open access by *Biomacromolecules* under the terms of the CC-BY-NC-ND 4.0).

5.4. Polysaccharides

Polysaccharides are natural polymers with high biocompatibility, biodegradability, and functional groups, making them widely studied for coacervation in various biological fields. Commonly used polysaccharides include alginate, HA, GA, pectin, and dextran. Due to their chemical properties, polysaccharides typically carry a negative charge and form coacervates through electrostatic interactions with positively charged molecules such as proteins and peptides [32,63,98,99]. Being naturally derived, polysaccharides exhibit low toxicity and excellent biocompatibility, and they can stably protect encapsulated biomolecules from external environments. In addition, they allow for reversible phase separation in response to changes in pH or ionic strength [100].

Li et al. investigated the interaction involved in coacervation between different types of polysaccharides (CS and carboxymethyl cellulose) and soy protein isolate under various conditions. The coacervate with carboxymethyl cellulose (CMC) and CS and soy protein isolate (SPI) exhibited gel-type rheological behaviors (elastic modulus (G′) values = 1000 Pa). The addition of salt ions (50 mM NaCl) decreased the elastic and viscous modulus (G′ and G″) of the SPI/CS and SPI/CMC coacervates from 1000 Pa to 100 Pa. The peak viscosity of SPI simple coacervates occurred at a pH near its isoelectric point of 4.5. However, the coacervate of SPI with CS and CMC demonstrated higher viscosity than the coacervates of SPI alone [101].

Yuan et al. investigated complex coacervation with SPI and CS for the microencapsulation of algal oil in a separate study on coacervation utilizing CS [102]. The optimal pH and CS/SPI ratio were 6.0 and 0.125 g/g, respectively. The interaction between CS and SPI showed a strong affinity (Ka = $4.16 \times 10^5 \pm 0.09 \times 10^5$ M^{-1}). The CS/SPI coacervates exhibited higher encapsulation efficiency (97.36 ± 1.16) and improved oxidative stability compared to the simple coacervation of SPI [102].

As shown in Figure 6, Wang et al. developed a blend film of soluble soybean polysaccharide (SSPS) and fish gelatin (FG) through complex coacervation [103]. The coacervates were fabricated using an SSPS/FG mass ratio of 1:3. The electrostatic interaction between SSPS and FG at a pH of 5.0 promoted complex coacervation in the film-forming solution. The coacervation with SSPS and FG exhibited increased surface roughness, enhanced thermal stability, and improved water vapor barrier properties. The SSPS/FG coacervate delayed the release of curcumin from the films into food simulants during a release test [103].

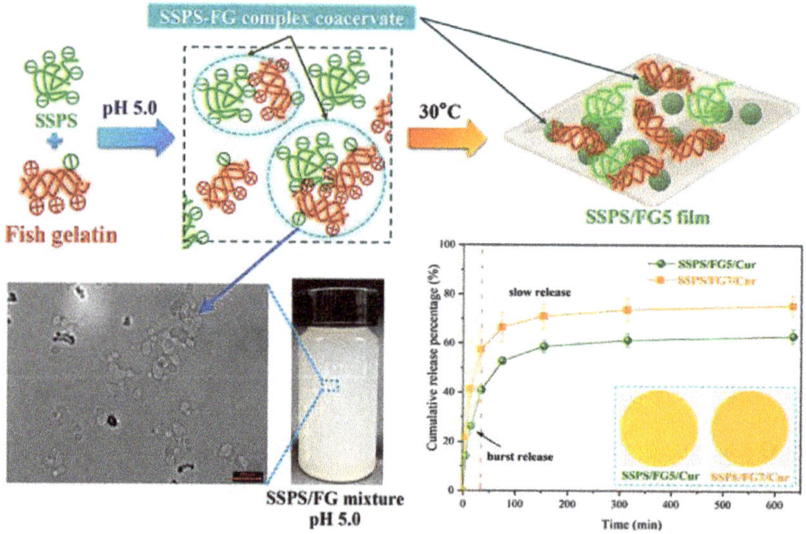

Figure 6. Polysaccharide-based coacervation with soy soluble soybean polysaccharide (SSPS) and fish gelatin. The coacervates were induced at a pH of 5.0 and coated for film. The curcumin-encapsulated coacervate film exhibited both burst release and slow release. (Reused from ref. [103]. Copyright © 2024 Elsevier Ltd.).

6. Application of Biomolecule-Based Coacervation

Coacervation is a liquid–liquid phase separation process that enables the formation of concentrated biomolecular droplets. In contrast to coacervation using synthetic polymers with defined charges, coacervation with natural biomacromolecules such as proteins, polysaccharides, nucleic acids, and polypeptides provides superior biocompatibility and biodegradability [104–106]. Biomolecule-based coacervates are being actively explored for their multifunctional roles in biomedical and industrial applications [107–109]. They have been utilized in the encapsulation of bioactive compounds, the stabilization of vaccines, and the formulation of functional foods [110–112]. In particular, their responsiveness to environmental stimuli such as pH and temperature makes them ideal for site-specific release [109,113]. Recently, numerous studies have explored the application of coacervation in underwater adhesion and tissue engineering [114,115]. Furthermore, coacervates can mimic intracellular compartments, contributing to research on the origin of life and the development of artificial cells [116]. Table 1 presents the recent trends in coacervation for the biotechnology and biomedical fields.

Table 1. Recent study of coacervation for biotechnology and biomedical applications.

Bioapplication	Biomolecule Component	Cargo	Study Objective	Ref.
Encapsulation	SA/CS	Walnut oil	Improved higher loading capacity and oxidation stability	[117]
Encapsulation	GA/CS or GA/trehalose/CS	*Lacticaseibacillus rhamnosus*	Promoted the stability of probiotic bacteria	[118]
Encapsulation	Gelatin/chia mucilage	Oregano essential oil	Developed encapsulation system for spray drying and improved encapsulation efficiency	[119]
Encapsulation	Zein/CS	Resveratrol	Improved encapsulation efficiency and dispersion stability via atmospheric cold plasma	[120]
Encapsulation	Plum seed protein isolate/polysaccharides	Essential oils	Enhanced stability, storage, emulsification, and encapsulation	[121]
Encapsulation	Zein–gallic acid/CS	Gallic acid	Induced structural modifications of encapsulation and enhanced thermal stability	[122]
Encapsulation	SPI/CS	Deer oil	Enhanced the stability of encapsulation against oxidative stress and encapsulation efficiency.	[123]
Encapsulation	β-conglycinin/lysozyme	Curcumin	Improved encapsulation efficiency, loading capacity, and stability against light and heat treatment	[124]
Encapsulation	GA/Krill protein isolate	Antarctic Krill oil	Developed stable and biocompatible encapsulation for oil	[65]
Encapsulation	WPI/GA	Tuna oil	Enhanced oxidative stability and made encapsulation more thermosensitively suitable	[125]
Encapsulation	WPI/flaxseed gum/monodiglyceride fatty acids	Resveratrol	Improved stability, encapsulation efficiency, and sustained antioxidant ability	[126]
Delivery platform	Cholesterol-modified DNA/histone	Virus particle, mRNA, cytokines, peptides	Enhanced the stability of the delivery vehicle biological agents	[127]
Delivery platform	Fungal CS/GA	α-tocopherol	Promoted stable and easy-to-prepare encapsulation materials for harsh conditions	[128]
Delivery platform	LMWG or HMWG/SA	miRNA-497	Developed a biocompatible delivery system to enhance cellular uptake and stability	[129]
Delivery platform	Heparin GAG/tyrosine- and arginine-based peptide	Tannic acid	Developed a stable and permeable delivery system that released drugs in response to biological triggers	[130]
Delivery platform	Dextran graft copolymer	DNA	Enhanced release capabilities and transfection	[131]
Delivery platform	Ellagic acid/casein	Ellagic acid	Enhanced oral absorption and improved solubility	[132]
Delivery platform	Peptide	pDNA, mRNA/sgRNA, RNP	Developed a redox-triggered delivery vehicle for CRISRP-Cas9 genome editing	[133]
Delivery platform	Peptide	siRNA, pDNA, mRNA	Developed a pH-responsive delivery nanocarrier for cancer therapy	[134]

Table 1. *Cont.*

Bioapplication	Biomolecule Component	Cargo	Study Objective	Ref.
Delivery platform	Single-stranded DNA/PLL	Emamectin benzoates	Improved loading capacities and stability against photodegradation	[135]
Delivery platform	PEAD/heparin	BMP-2	Developed a protein delivery platform with enhanced colloidal stability	[136]
Tissue engineering	CMC/gelatin	-	Developed 3D extrusion printing hydrogel with higher printing fidelity and without any discontinuities during the printing process	[137]
Tissue engineering	Gleatin/GA/CMC	β-carotene	Improved thermal, pH, and ionic strength stability and evaluated its potential applications in surimi	[138]
Tissue engineering	HA/CS	Rat BMSCs	Developed chondro-inductivity scaffold for encapsulating BMSCs	[139]
Tissue engineering	Gelatin/QHECE	Glucose oxidase	Developed glucose-responsive microneedle loaded with glucose oxidase and enhanced drug release	[140]
Tissue engineering	LMWC/HA or HMWC/HA	-	Developed biocompatible hydrogel with shape adaptability and enhanced wet adhesion	[141]
Tissue engineering	SPI/chelator-soluble pectin	β-conglycinin, glycinin	Developed food inks for 3D printing with enhanced particle distribution and mechanical properties	[142]
Tissue engineering	Theabrownin/whey protein isolate	-	Developed coacervate for modulating energy metabolism and mitochondrial apoptosis to strengthen muscle cells	[64]
Tissue engineering	ApoEVs/GelMA or curcumin/CMCS/GelMA	Apoptotic extracellular vesicles, curcumin	Developed multifunctional 3D-printed scaffold for enhancing skin regeneration and promoting antibacterial activity and ROS scavenging activity	[143]
Tissue engineering	PEAD/heparin	Cargo IGF-1	Enhanced bioactivity of cargo IGF-1 and sustained release to embed in cartilage regeneration hydrogel	[144]
Tissue engineering	Egg yolk/CMC	β-carotene	Enhanced stability of interfacial layer and structural strength	[145]
Adhesive technology	Methacrylated LMWC/HA	-	Developed coacervate with enhanced wet tissue adhesion and tunable properties	[146]
Adhesive technology	CS/HA	-	Enhanced underwater adhesion strength against salt switch conditions and promoted antibacterial properties	[147]
Adhesive technology	Peptide/polyoxometalate	-	Developed injectable, self-solidifying underwater adhesion and enhanced its properties	[148]
Adhesive technology	Tyramine-conjugated alginate/RGD peptide	Calcium phosphate	Developed a photo-mineralized hydrogel and enhanced adhesiveness and bioactivity of bones	[149]

Table 1. *Cont.*

Bioapplication	Biomolecule Component	Cargo	Study Objective	Ref.
Cellular mimicking	PDDA/ATP	Dextran	Developed a demembranization system that reconfigures in response to biological signals with enhanced permeability	[150]
Cellular mimicking	RNA/peptide	RNA	Developed fuel-dependent RNA-containing coacervation that mimics membraneless organelles	[83]
Cellular mimicking	Dextran/polyaspartic acid	DNA, enzymes	Developed a biomimetic platform capable of biomacromolecule segregation, reaction control, and morphological reconfiguration	[151]
Cellular mimicking	PEG/dextran	DNA	Developed compartmentalized artificial cell structures to mimic and investigate spatiotemporal control mechanisms	[152]

6.1. Encapsulation

Encapsulation enables the isolation and protection of sensitive molecules by surrounding them with a phase-separated environment. The ability of coacervation to accommodate so-called client molecules is one of its most intriguing features [111,112,153]. Client molecules, such as proteins and other small compounds, can be encapsulated during the coacervate phase to study the biological processes in a more precise and controlled environment [154,155]. In particular, the presence of coacervates within an aqueous solution has been observed to exhibit remarkably low interfacial energy sufficiency, thereby enabling the absorption of a wide range of substances [156,157]. Encapsulation can be achieved through two distinct mechanisms. The first mechanism involves incorporating the cargo as a component of the coacervate system. The second mechanism occurs through specific interactions and preferential partitioning [158–161]. Cargo encapsulation via coacervation has been demonstrated to stabilize and protect the cargo within an aqueous environment, while also allowing for controlled release in response to external stimuli or environmental changes through the dissociation of the coacervation [162,163]. Coacervation-based encapsulation offers two key advantages: (1) the ability to perform encapsulation entirely in an aqueous environment, and (2) the potential to significantly concentrate the target molecule in the macromolecule-rich coacervate phase compared to the initial solution [40,164].

Barajas-Álvarez et al. demonstrated coacervation in GA/CS or GA/trehalose/CS crosslinking with tripolyphosphate to encapsulate *Lacticaseibacillus rhamnosus* probiotic [118]. This study evaluated the encapsulation efficiency, physicochemical properties, and probiotic survival under storage conditions and simulated gastrointestinal fluids. Furthermore, crosslinking with tripolyphosphate improved the encapsulation efficiency after the drying process. The probiotics encapsulated within coacervate showed a higher level of short-chain fatty acid production in comparison to the non-encapsulated probiotics [118].

As shown in Figure 7, Wang et al. demonstrated the encapsulation of tuna oil (TO) through the process of coacervation with WPI and various reducing sugars, including glucose, fructose, maltose, and lactose [125]. The coacervates fabricated with WPI, GA, and maltose demonstrated an enhanced storage stability of TO. The encapsulation efficiency of WPI/maltose/GA was found to be 87.41%, which is higher than the encapsulation efficiency of WPI self-coacervation. Microencapsulated-protected TO had a 35.78% lower peroxide value than free TO after 16 days of accelerated oxidation at 55 °C [125].

Research has also been conducted on the modification of coacervation components to enhance their functionality. Xue et al. [121] modified plum seed protein isolate (PSPI) by enzymatic hydrolysis using alcalase, pepsin, and flavourzyme to conjugate with polyphenols such as atechin, curcumin, and proanthocyanidin. The modified PSPI was then used to form coacervates with various polysaccharides, and their structural and functional properties were evaluated. As a result, the modified PSPI/polysaccharide coacervates effectively maintained the stability of essential oils and exhibited improved emulsifying and encapsulating properties [121].

Despite the broad application of coacervation in the encapsulation of various hydrophobic and hydrophilic cargos, there is still a lack of systematic studies examining how the molecular characteristics of small molecules influence the encapsulation mechanisms, phase behavior, and material properties [165]. The work of Zhao et al. demonstrated that the partitioning of small molecule dyes into various polymer-based coacervates is primarily determined by their charge and hydrophobicity [166,167].

On the other hand, while coacervation enables the stable encapsulation of proteins, not all target proteins possess strong charges, which is an ongoing challenge in the field. To address this, Obermeyer et al. proposed a solution to the problem of binary protein–polyelectrolyte coacervates. Using conjugation chemistry, they artificially enhanced the native charge of proteins by modifying them with succinic anhydride, thereby generating supercharged proteins. Their study further demonstrated that a relatively low degree of supercharging—with a negative-to-positive charge ratio of approximately 1.1 to 1.4—was sufficient to induce coacervation with cationic polymers [168].

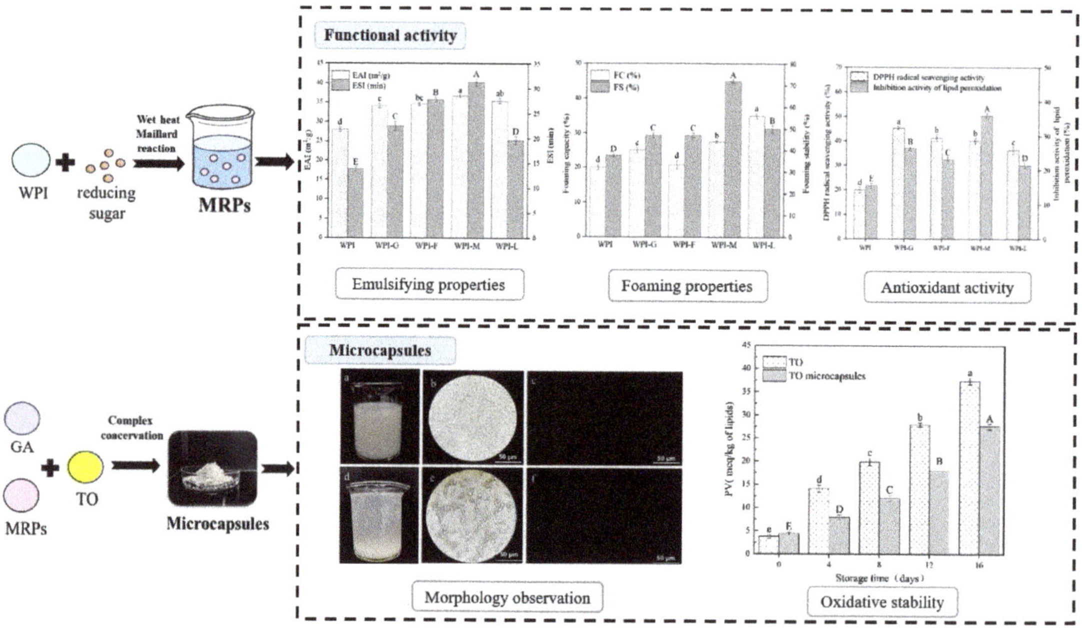

Figure 7. Schematic illustration of coacervation with whey protein isolate (WPI) and reducing sugar for encapsulating based on Maillard reaction. Functional activity (emulsifying properties, foaming properties, and antioxidant activity) of coacervates between WPI and different kinds of polysaccharides were analyzed. Tuna oil (TO) was encapsulated by complex coacervation with gum arabic (GA) and Maillard reaction products (MRPs). Different lowercase and uppercase letters

indicate statistically significant differences among groups. Lowercase letters represent significant differences within each analytical parameter across different times or conditions, while uppercase letters indicate significant differences among different analytical parameters under the same condition ($p < 0.05$). (Reused from ref. [125]. Copyright © 2025 Elsevier Ltd.).

6.2. Delivery Platform

The use of coacervation for biomedical applications is based on the advantages of complex coacervation for encapsulation. Recently, coacervation systems using naturally derived polymers such as heparin, CS, and alginate have been investigated as delivery platforms [169,170]. For example, coacervates composed of CS and nucleic acids have been developed to deliver DNA and RNA to mammalian cells [171,172]. In contrast, recent research has demonstrated that peptide-based coacervation can be employed to deliver functional nucleic acids for gene therapy or cancer therapy [133,134].

Delaporte et al. investigated animal-free coacervation. The study used fungal CS to induce coacervation with GA. While the electrostatic interactions of fungal CS/GA coacervation decreased in different solvents (isohexadecance, ethylhexyl stearate, and ethanol), the wetting properties of the coacervates were improved. The encapsulation efficiency of α-tocopherol by coacervates was 82.6% at a 1:1 material-to-active mass ratio [128].

Sun et al. used peptide coacervates for CRISPR/Cas9 genome editing (Figure 8). Peptide coacervation was induced in 20 mM phosphate buffer (pH 6.5, ionic strength 100 mM) and successfully encapsulated pDNA, mRNA, and Cas9 nuclease. The delivered cargo resulted in improved transfection and gene editing efficiency compared to conventional reagents [133].

Wang et al. investigated pH-responsive coacervates with peptides for nucleic acid delivery that respond to the different physiological pH conditions of various microenvironments such as tumors, lysosomes, and the bloodstream. The peptide coacervates delivered siRNA, pDNA, and mRNA into cancer cells [134].

In addition, Li et al. reported the application of coacervation with zein and CS for the sustained release of curcumin [173]. Lee et al. loaded injectable hydrogel coacervates with bortezomib, an anticancer drug [174]. Huei et al. applied iron (Fe)-crosslinked CMC-based coacervates for the controlled release of ibuprofen [175]. The strategy of using coacervates as a delivery platform has been shown to be effective in various studies. This approach utilizes the flexible and modular nature of charge-driven coacervate formation to address the challenges of (1) the protection and/or isolation of the drug, (2) the targeted distribution and absorption into specific cells or tissues, and (3) the controlled release of the drug over time.

6.3. Tissue Engineering

Tissue engineering aims to restore, maintain, improve, or replace tissues and organs using a combination of cells, scaffolds, and growth factors. These scaffolds are designed to mimic the extracellular matrix (ECM) and provide physical support for cell adhesion and proliferation [176,177]. Scaffold fabrication has traditionally relied on the gelation of precursor materials through chemical crosslinking methods such as photo-polymerization, enzymatic reactions, and click chemistry. However, such scaffolds often form simple, homogeneous gel structures, making it difficult to implement microscale patterning. Because conventional scaffold design focuses primarily on external cellular support, it does not adequately mimic the intracellular environment. As a strategy to enhance the functional properties of conventional scaffolds, recent studies have begun to incorporate coacervation [178]. Scaffolds produced by coacervation can form membraneless compartments that resemble intracellular organelles, thereby enabling the mimicry of intracellular microenvironments. Such coacervate-based scaffolds allow for the selective concentration of bioactive molecules, such as functional RNAs or growth factors, and help to maintain their bioactivity. Through electrostatic interactions, these encapsulated functional biomolecules can be released in a stimulus-responsive manner (e.g., in response to pH or ATP).

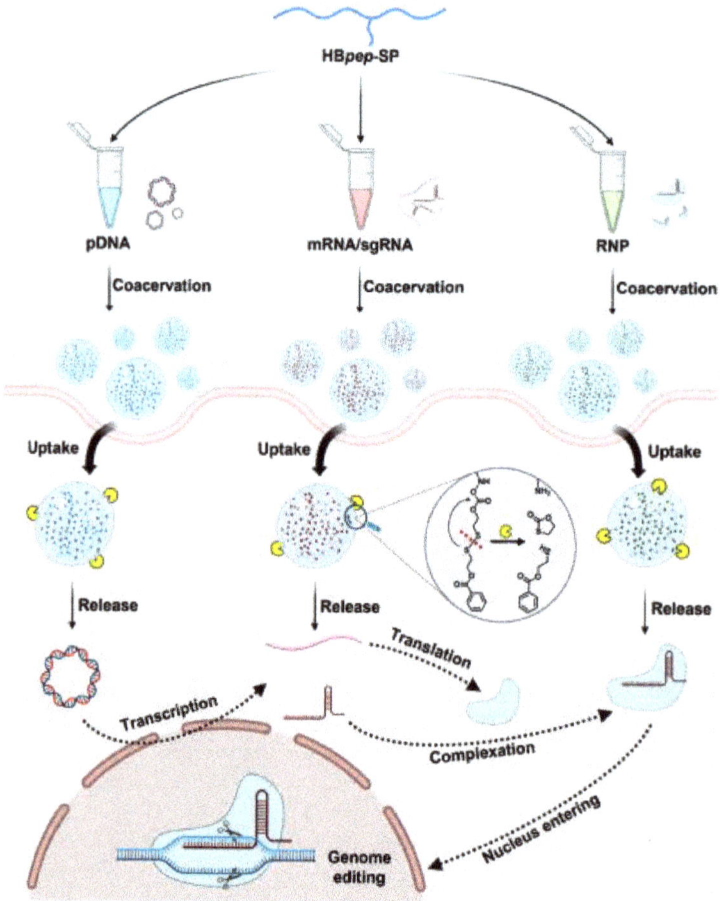

Figure 8. Schematic illustration of the universal delivery of CRISPR/Cas9 genome editing machineries mediated by HB*pep*-SP coacervates. Three types of CRISPR/Cas9 genome editing machineries including pDNA, mRNA/sgRNA, and Cas9 ribonucleoprotein (RNP) can be readily recruited during the LLPS of the HB*pep*-SP peptide. The cargo-loaded coacervates are internalized by the cell and then reduced by glutathione (the yellow spheres) in the cytosol, triggering cleavage of the side chain modification and disassembly of the cargo-loaded coacervates. The Cas9 RNP, which is directly released from coacervates or produced by the transcription and translation of pDNA and mRNA, enters the nucleus and induces double-strand breaks on the genomic DNA. (Reused from ref. [133]. Copyright © 2023 American Chemical Society.).

For example, Karabıyık Acar et al. developed a coacervate-based scaffold using HA and CS for cartilage repair. The HA/CS coacervates effectively encapsulated bone marrow mesenchymal stem cells (BMSCs) and demonstrated their efficacy in vitro. Chondrogenic markers such as ACAN, COL2A1, and SOX9 were upregulated following chondrogenic induction. Notably, even in the absence of chondrogenic stimuli, the encapsulated cells within the coacervates exhibited increased expression of cartilage-related markers, indicating their intrinsic chondro-inductive potential [137].

In another line of research, advances in tissue engineering have also led to progress in 3D printing-based scaffold fabrication for precise microenvironmental control. Gharanjig et al. proposed an extrudable hydrogel based on CMC/gelatin complex coacervation. Their study analyzed the rheological properties required for 3D printing and demonstrated that a CMC-to-gelatin ratio of 1:15 resulted in the highest print fidelity, with no observed discontinuities during extrusion [137].

Jiang et al. developed a unique 3D-printed bilayer multifunctional scaffold incorporating regenerative apoptotic extracellular vesicles (ApoEVs) and antibacterial coacervates that showed strong potential for full thickness wound healing and revealed the underlying healing mechanisms (Figure 9). The upper layer, mimicking the epidermis, formed a dense structure to protect the wound from mechanical impact and pressure. The lower layer, mimicking the dermis, contained larger pores to promote cell migration and proliferation. Furthermore, the top layer included pH-responsive curcumin coacervates that exhibited antibacterial and reactive oxygen species (ROS) scavenging properties. The lower layer supported the sustained release of ApoEVs and enhanced fibroblast proliferation and migration, as well as angiogenesis in the endothelial cells. In vivo experiments confirmed that the scaffold accelerated wound healing and reduced scar tissue formation. Moreover, the RNA sequencing analysis revealed the molecular mechanisms by which the bilayer scaffold contributed to wound repair [143].

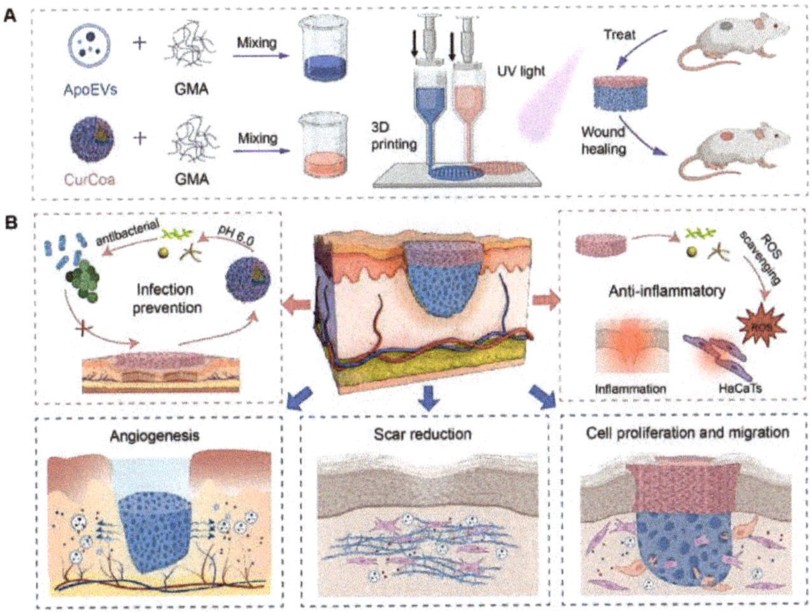

Figure 9. (**A**) Schematic illustration of ApoEVs/GelMA and CurCoa/GelMA coacervation for 3D-printed tissue engineering. (**B**) Scaffold based on ApoEVs/GelMA and CurCoa/GelMA coacervates showed various properties such as infection prevention, anti-inflammatory, angiogenesis, scar reduction, and cell proliferation and migration. (Reused from ref. [143]. Copyright © 2025 Elsevier Ltd.).

6.4. Adhesive Technology

The use of coacervation as a strategy for the development of surgical adhesives has recently attracted growing research interest [115,179]. This process has been observed in natural biological mechanisms, where various species use underwater adhesives to attach to different surfaces [180]. Representative examples include the sandcastle worm and mussels, in which coacervate-based adhesive systems were first identified [115,181,182]. In these organisms, coacervation has been found to play a critical role in the formation of adhesive materials under wet conditions [183]. Coacervation based on biomacromolecules has shown great potential as a wet adhesive, exhibiting excellent performance even in hydrated environments [184].

Recently, research has continued to focus on the development of underwater adhesives based on the immiscibility of coacervation in water and the strong adhesive properties of mussel adhesive proteins (MAPs) [185]. Lim et al. developed a MAP-based encapsulated coacervate system and applied it as a smart tissue adhesive capable of drug delivery. In this study, cationic recombinant hybrid MAPs (fp-131 or fp-151) and anionic HA were complexed by coacervation to form an adhesive. On aluminum surfaces, the bulk adhesive strength of this coacervate was more than twice as strong as that of the protein alone [186].

Furthermore, in a study by Yun et al. [149], an injectable adhesive hydrogel was prepared by coacervation using tyramine-conjugated alginate and RGD peptide-fused mussel adhesive proteins. Based on this coacervate hydrogel platform, a photoreactive agent, calcium ions, and phosphonodiol were incorporated to provide dual functionality: photocrosslinking and the formation of amorphous calcium phosphate, both activated by visible light irradiation. In in vivo experiments using a rat femoral tunnel defect model, the developed adhesive hydrogel demonstrated easy applicability to irregular defect sites, rapid bone regeneration without the need for bone grafting, and excellent integration with the surrounding tissue [149]

Several additional studies have been conducted on mussel adhesive protein (MAP)-based adhesives. Cha et al. produced rfp-1 MAP (AKP-SYPPTYK) using recombinant protein expression techniques. This MAP was used to form hydrogels by coordinating crosslinking with Fe^{3+} ions or covalent crosslinking with $NaIO_4$ [187]. Lu et al. designed a hybrid molecular adhesive by combining CsgA protein, an amyloid-based adhesive protein from *E. coli*, with mussel adhesive proteins [188].

In addition to mussel protein-based adhesives, research has also explored polysaccharide-based adhesion technologies. Deng et al. proposed a wet tissue adhesive based on the coacervates formed from low-molecular-weight methacrylated chitosan (CSMA) and HA (Figure 10). The fabricated coacervates can be applied to wet tissue surfaces and photo-crosslinked to form in situ double-network hydrogels, which enhances cohesion and ensures long-lasting adhesion. After immersion in PBS for 24 h, the hydrogel burst pressure increased to approximately 623 mmHg due to dynamic bond reorganization and low swelling ability [146].

Despite these advances in coacervation-based research, a complete tissue adhesive that is non-toxic, biocompatible, biodegradable, user-friendly, scalable, and capable of mimicking natural adhesion has yet to be developed. Future research needs to investigate how the chemical structure of the polymer backbone, hydrophilicity/hydrophobicity, combinations of amino acids, and charged side groups influence the adhesive strength. In addition, the application of tissue adhesives should expand beyond basic tissue bonding to include drug delivery, tissue grafting, wound healing, and tissue regeneration, providing new opportunities for biomedical advancement.

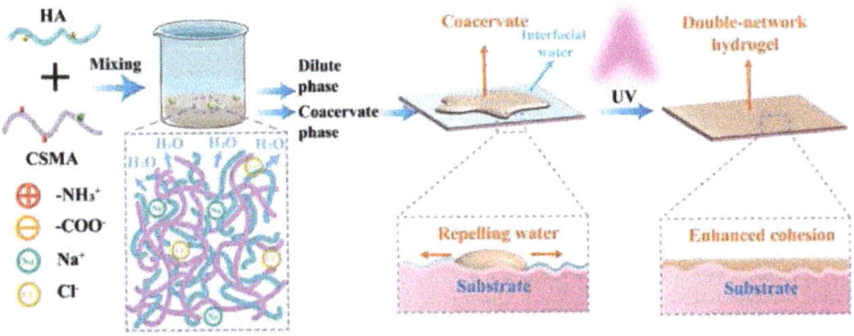

Figure 10. Schematic illustration of coacervation for adhesive technology: HA and low-molecular-weight methacrylated chitosan-formed coacervation. The coacervate was coated on the interfacial water of the substrate and treated by UV to form double-network hydrogel. Double-network hydrogel enhanced the cohesion of the substrate compared to a blank substrate. (Reused from ref. [146]. Copyright © 2025 American Chemical Society).

6.5. Cellular Mimicking

Coacervation occurs when oppositely charged polymers interact in water, resulting in LLPS and the formation of highly concentrated liquid droplets [189]. Although this phase separation is similar to the behavior of oil and water, the droplets represent a distinct aqueous phase within the water, often referred to as a "second liquid" [190]. In the past, Oparin observed these properties of coacervates and proposed that such structures may have functioned as protocells in the prebiotic era, prior to the emergence of life [191]. However, this hypothesis was largely dismissed at the time because coacervates lack membranes, whereas real cells are clearly defined by their membrane structures. Recently, this idea has regained significant attention due to the discovery of membraneless organelles such as the nucleoli [192], P granules, and stress granules [193–195]. These findings have revitalized research in the field, supporting the design of synthetic cells and artificial organelles and extending the applications of coacervation to areas such as origin-of-life studies, neurodegenerative diseases, cellular compartmentalization, and protein-based drug delivery [194,196].

For example, Jia et al. generated polyester coacervates based on prebiotically accessible α-hydroxy acids. The formation of these coacervates demonstrated their ability to partition proteins and RNA in a manner consistent with origin-of-life conditions [197]. Longo et al. further demonstrated that coacervation can facilitate the conversion of abiotic ornithine residues into arginine, thereby enabling the synthesis of proteins capable of binding dsDNA [198].

In another study, Donau et al. proposed a coacervate-based model to understand the behavior of membraneless organelles. In their study, they prepared and used active coacervates containing RNA that were driven by ATP-dependent reactions. The results showed that these coacervates exhibit dynamic properties, including emergence, decay, building block exchange, and the concentration of functional RNA in its active folded state. Thus, the study presented these fuel-regulated coacervate droplets driven by a reaction cycle as a useful model for investigating the mechanisms of compositional control and functionality in membraneless organelles [83].

Recent studies have explored not only membraneless protocell models but also strategies to control the membranization and demembranization of protocells, as well as approaches to understanding tissue-level multicompartment systems and biochemical reaction networks through prototissue models. Zhou and colleagues proposed a strategy to regulate the membrane formation (membranization) and membrane removal (demembranization) of coacervate protocells, demonstrating the potential to enhance their functionality and dynamic reconfiguration capabilities [150] (Figure 11). Membraneless coacervates were transformed into membranized structures through coating with terpolymer-based nanoparticles, and the membrane was subsequently removed using the anionic polysaccharide carboxymethyl dextran. This process allowed for structural reconfiguration and controlled permeability toward biomolecules, ultimately leading to the development of synthetic protein-containing protocells with hierarchical and asymmetric membrane architectures.

Meanwhile, Hu and colleagues presented a binary droplet-based protocell network formed from coacervates and aqueous two-phase systems (ATPSs) to mimic prototissues. This network facilitated the spatial self-sorting of biomacromolecules, thereby enhancing biological reactions and material extraction processes. Furthermore, the study highlighted the importance of the dynamic nature of coacervation in organizing intracellular environments and regulating biochemical processes [151].

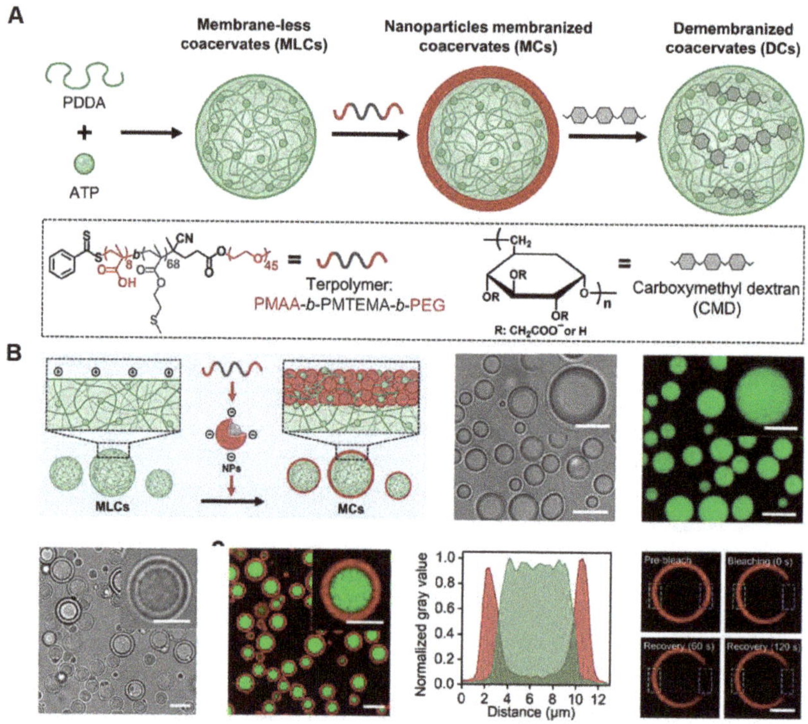

Figure 11. (**A**) Controlled demembranization of membranized coacervate droplets. The membranization of coacervates is achieved by introducing a terpolymer (PMAA-*b*-PMTEMA-*b*-PEG) into MLCs composed of PDDA and ATP. Following this, the demembranization of MCs is triggered by the addition of CMD, which causes the dissociation of the outer membrane and results in DCs integrated with CMD. (**B**) Preparation and characterization of (NR-loaded) NP membranized coacervates. The schematic illustration shows the formation of MCs through the addition of (NR-loaded) NPs. In the presence and absence of NR, the terpolymer initially self-assembles into anionic (NR-loaded) NPs, followed by the membranization on the surface of the cationic MLCs, accompanied by a small partial dissociation of the coacervate phase and the redistribution of the coacervate components into the NP membrane. Bright field and fluorescence field CLSM images of MLCs and MCs, respectively. The scale bar represents 10 μm, and the scale bar in the inserted image represents 5 μm. (Adjusted from ref. [150]. Copyright © 2025 American Chemical Society).

7. Fundamental Challenges of Coacervation

Coacervate systems encounter notable stability issues in biological environments. The main limitation arises from their sensitivity to variations in ionic strength, pH, or temperature [107,199]. This characteristic, while allowing for controlled release properties, creates significant obstacles for systemic delivery routes where the coacervate must traverse various biological environments with different ionic compositions and pH levels. To overcome the inherent limitations while preserving the beneficial properties that make coacervates attractive for drug delivery applications, covalent crosslinking strategies, click chemistry, and multivalent hydrogen bonding have been demonstrated [199–201]. The effectiveness of these crosslinking strategies has been validated through dynamic light scattering studies, which demonstrated significant improvements in stability against both salt and pH changes [199]. Post-coacervation crosslinking uses thiolene click chemistry, which enables the formation of covalent crosslinks in aqueous media without exposure to heat or organic solvents [200]. Visual confirmation through optical microscopy has demonstrated that, while both the crosslinked and uncrosslinked coacervates maintain droplet structure at a pH of 6.5, only the crosslinked coacervates preserve their droplet morphology at a pH of 2.0. The enhanced stability achieved through this crosslinking approach extends beyond pH resistance to include improved performance in various biological conditions such as the stomach or inflammatory sites. A dopamine-containing multi-hydrogen bonded peptide complex coacervate exhibiting stability in a pH range of 1–11 and various salt concentration ranges has been developed [201]. These peptide coacervates demonstrated high drug encapsulation efficiency and trypsin-induced release characteristics, showing great potential for oral drug delivery applications.

Recent research has shown that coacervates can function as transient protective structures in vivo or serve as smart platforms for stimulus responsive drug release [202,203]. For example, Zhao et al. developed an oral drug delivery strategy using nanoparticle-assembled bioadhesive coacervates. In a mouse model, the orally administered drug-loaded coacervates significantly alleviated the symptoms of inflammatory bowel disease, restored gut microbiota diversity, reduced systemic drug exposure, and improved therapeutic efficacy in acute colitis models [204]. This system can be administered orally, providing a more patient-friendly alternative to traditional drug delivery methods such as enemas or injections. Additionally, by releasing the drug directly into the gastrointestinal tract in a controlled manner, it minimizes systemic exposure to the drug, thereby reducing the risk of the adverse effects associated with the long-term use of corticosteroids and other drugs that may cause serious side effects, while maintaining therapeutic efficacy. In other studies, Wang et al. and Park et al. utilized coacervation-based hydrogels for wound healing applications and demonstrated enhanced tissue regeneration in mouse models, with wound healing and infection control following treatment. Specifically, the self-healing ability of the hydrogel after being damaged means that it can maintain its functionality over time, reducing the need for frequent dressing changes [163,205]. This can lead to improved patient comfort and reduced healthcare costs associated with the wound management. As a result, coacervation has gained increasing attention in various in vivo applications. The favorable results suggest strong potential for clinical use.

8. Future Perspective for Coacervation

Despite encouraging in vitro results, the clinical translation of coacervate-based delivery systems continues to face significant challenges. In particular, their structural instability under physiological conditions remains a major limitation for systemic administration routes such as intravenous or oral delivery [201]. In vivo, coacervates are exposed to various destabilizing factors, including proteolytic enzymes, immune responses, fluctuations in ionic strength, and interactions with the extracellular matrix, all of which can induce phase separation breakdown or premature disassembly. To address these challenges, systematic in vivo studies are essential to evaluate the long-term stability, functional retention, target specificity, and immunogenicity of coacervation [18].

One promising direction is the development of stimuli-responsive coacervates that can remain stable during circulation but disassemble selectively at the target site [206]. For instance, complex coacervate core micelles incorporating reversible crosslinking have shown enhanced control over phase stability and release behavior. Although polyethylenimine is frequently used to generate small, stable coacervates, concerns remain regarding its cytotoxicity and limited transfection efficiency in vivo [203]. In response, low-molecular-weight polyethylenimine grafted with biocompatible polymers such as chitosan has demonstrated improved buffering capacity and cellular uptake with reduced cytotoxic effects, making it a promising alternative [202].

Another critical consideration for clinical trials is the scalability and reproducibility of manufacturing processes. Recent advances in microfluidic-based fabrication, particularly for lipid nanoparticle production in mRNA vaccine platforms, offer a blueprint for adapting similar technologies to coacervate systems. These platforms enable precise control over particle size, composition, and encapsulation efficiency, while maintaining the batch-to-batch consistency required for regulatory approval [9,152].

In addition to engineering advances, gaining mechanistic insight into how coacervates behave in complex biological environments is crucial. Techniques such as confocal laser scanning microscopy, dynamic light scattering, isothermal titration calorimetry, and cryo-electron microscopy are increasingly employed to study phase behavior, structural integrity, and biomolecular interactions under physiologically relevant conditions [207–209].

9. Conclusions

This review highlights biomolecule-based coacervation as a flexible and dynamic platform that offers a wide range of functions in the biomedical and biotechnological fields, including drug encapsulation, self-assembly, and the development of functional materials. Biomolecule-based coacervates provide a favorable environment for handling sensitive biological molecules such as proteins, RNA, and enzymes, owing to their inherent biocompatibility, tunable composition, and mild assembly conditions. Recent studies have demonstrated the applicability of these systems in diverse areas such as the controlled release of therapeutics, gene editing platforms (e.g., Cas9, siRNA, and mRNA), wound healing scaffolds, and protocell-based synthetic biological architectures. However, several challenges remain before these coacervate systems can be fully translated into clinical or industrial settings. Major hurdles include mechanical instability under physiological conditions, insufficient droplet uniformity and reproducibility, and a limited understanding of long-term biodegradability and biocompatibility. The lack of a well-established regulatory framework for therapeutic applications is also a significant barrier. To overcome these challenges, integration with advanced technologies such as microfluidics, AI-assisted material design, and 3D bioprinting is required to enable precise spatial control, large-scale production, and clinical translation. In parallel, convergence with systems biology, omics-based analysis, and high-throughput screening platforms will be essential to elucidate the mechanisms of interaction between the coacervates and cells, and to develop application-specific design strategies. With continued technological refinement and interdisciplinary collaboration, biomolecule-based coacervation is expected to play a central role in a variety of fields, including precision medicine, synthetic cell engineering, next-generation therapeutics, functional food and vaccine platforms, and the construction of spatially controlled biological environments.

Author Contributions: Conceptualization, D.H.K., M.-R.K. and S.P.P.; investigation, D.H.K., M.-R.K., D.Y.C. and S.P.P.; resources, D.H.K., D.Y.C. and M.-R.K.; writing—original draft preparation, D.H.K.; writing—review and editing, D.H.K. and M.-R.K.; visualization, D.H.K.; supervision, S.P.P.; project administration, S.P.P.; and funding acquisition, S.P.P. All authors have read and agreed to the published version of the manuscript.

Funding: This research was funded by National Research Foundation of Korea (NRF) grants, funded by the Korean government (MSIT) (grant numbers NRF-RS-2021-NR060107 and NRF-RS-2021-NR059450). This work was also supported by the National Research Foundation of Korea (NRF), funded by the Korean Ministry of Education (NRF-RS-2021-NR065961).

Institutional Review Board Statement: Not applicable.

Informed Consent Statement: Not applicable.

Data Availability Statement: Dataset available on request from the authors.

Conflicts of Interest: The authors declare no conflicts of interest.

Abbreviations

The following abbreviations are used in this manuscript:

ApoEV	Apoptotic extracellular vesicle
ATP	Adenosine 5′-triphosphate
BMP-2	Bone morphogenetic protein 2
BMSC	Bone marrow mesenchymal stem cell
CLSM	Confocal laser scanning microscope
CMC	Carboxymethyl cellulose
CMCS	Carboxymethyl chitosan
CMD	Carboxymethyl dextran
CS	Chitosan
DC	Demembranized coacervate
ECM	Extracellular matrix
ELP	Elastin-like polypeptide
FG	Fish gelatin
GA	Arabic gum, gum arabic
GAG	Glycosaminoglycan
HA	Hyaluronic acid
HMWC	High-molecular-weight chitosan
HMWG	High-molecular-weight cationized gelatin
HRV	Rhinovirus
IGF-1	Insulin-like growth factor-1
LLPS	Liquid–liquid phase separation
LMWC	Low-molecular-weight chitosan
LMWG	Low-molecular-weight cationized gelatin
MAP	Mussel adhesive protein
MC	Nanoparticle membranized coacervate
MLC	Membraneless coacervate
OVA	Ovalbumin
PDDA	Poly(diallyldimethylammonium chloride)
PLL	Poly-L-lysine
PP	Pea protein
PPV	Procine parvovirus
QHECE	Hydroxyethylcellulose ethoxylate
RNP	Ribonucleoprotein

ROS	Reactive oxygen species
SA	Sodium alginate
SPI	Soy protein isolate
SPP	Soluble pea protein
SSPS	Soluble soybean polysaccharide
WPI	Whey protein isolate

References

1. Zhou, L.; Shi, H.; Li, Z.; He, C. Recent Advances in Complex Coacervation Design from Macromolecular Assemblies and Emerging Applications. *Macromol. Rapid Commun.* **2020**, *41*, 2000149. [CrossRef] [PubMed]
2. Kruyt, H.; Bungenberg de Jong, H. *Chemistry–Coacervation (Partial Miscibility in Colloid Systems)*; Koninklijke Nederlandse Akademie van Wetenschappen: Amsterdam, The Netherlands, 1929; pp. 849–856.
3. Nakashima, K.K.; Vibhute, M.A.; Spruijt, E. Biomolecular Chemistry in Liquid Phase Separated Compartments. *Front. Mol. Biosci.* **2019**, *6*, 21. [CrossRef]
4. Abdelhamid, M.A.A.; Khalifa, H.O.; Ki, M.-R.; Pack, S.P. Nanoengineered Silica-Based Biomaterials for Regenerative Medicine. *Int. J. Mol. Sci.* **2024**, *25*, 6125. [CrossRef] [PubMed]
5. Kim, K.H.; Ki, M.-R.; Nguyen, T.K.M.; Min, K.H.; Pack, S.P. In vivo self-assembly of an intact functional cage protein: Intracellular generation of chimeric ferritins without disassembly-involved damage. *J. Ind. Eng. Chem.* **2025**, *141*, 67–71. [CrossRef]
6. Abdelhamid, M.A.A.; Son, R.G.; Ki, M.-R.; Pack, S.P. Biosilica-coated carbonic anhydrase displayed on Escherichia coli: A novel design approach for efficient and stable biocatalyst for CO_2 sequestration. *Int. J. Biol. Macromol.* **2024**, *277*, 134058. [CrossRef] [PubMed]
7. Kayitmazer, A.B.; Seeman, D.; Minsky, B.B.; Dubin, P.L.; Xu, Y. Protein–polyelectrolyte interactions. *Soft Matter* **2013**, *9*, 2553–2583. [CrossRef]
8. Banani, S.F.; Lee, H.O.; Hyman, A.A.; Rosen, M.K. Biomolecular condensates: Organizers of cellular biochemistry. *Nat. Rev. Mol. Cell Biol.* **2017**, *18*, 285–298. [CrossRef]
9. Yewdall, N.A.; André, A.A.M.; Lu, T.; Spruijt, E. Coacervates as models of membraneless organelles. *Curr. Opin. Colloid Interface Sci.* **2021**, *52*, 101416. [CrossRef]
10. Alberti, S.; Gladfelter, A.; Mittag, T. Considerations and Challenges in Studying Liquid-Liquid Phase Separation and Biomolecular Condensates. *Cell* **2019**, *176*, 419–434. [CrossRef] [PubMed]
11. Croguennec, T.; Tavares, G.M.; Bouhallab, S. Heteroprotein complex coacervation: A generic process. *Adv. Colloid Interface Sci.* **2017**, *239*, 115–126. [CrossRef] [PubMed]
12. Devi, N.; Sarmah, M.; Khatun, B.; Maji, T.K. Encapsulation of active ingredients in polysaccharide-protein complex coacervates. *Adv. Colloid Interface Sci.* **2017**, *239*, 136–145. [CrossRef] [PubMed]
13. Moschakis, T.; Biliaderis, C.G. Biopolymer-based coacervates: Structures, functionality and applications in food products. *Curr. Opin. Colloid Interface Sci.* **2017**, *28*, 96–109. [CrossRef]
14. Shin, Y.; Brangwynne, C.P. Liquid phase condensation in cell physiology and disease. *Science* **2017**, *357*, eaaf4382. [CrossRef] [PubMed]
15. Kwon, I.; Kato, M.; Xiang, S.; Wu, L.; Theodoropoulos, P.; Mirzaei, H.; Han, T.; Xie, S.; Corden, J.L.; McKnight, S.L. Phosphorylation-regulated binding of RNA polymerase II to fibrous polymers of low-complexity domains. *Cell* **2013**, *155*, 1049–1060. [CrossRef]
16. Lin, Y.; Protter, D.S.; Rosen, M.K.; Parker, R. Formation and maturation of phase-separated liquid droplets by RNA-binding proteins. *Mol. Cell* **2015**, *60*, 208–219. [CrossRef]
17. Molliex, A.; Temirov, J.; Lee, J.; Coughlin, M.; Kanagaraj, A.P.; Kim, H.J.; Mittag, T.; Taylor, J.P. Phase separation by low complexity domains promotes stress granule assembly and drives pathological fibrillization. *Cell* **2015**, *163*, 123–133. [CrossRef]
18. Ban, E.; Kim, A. Coacervates: Recent developments as nanostructure delivery platforms for therapeutic biomolecules. *Int. J. Pharm.* **2022**, *624*, 122058. [CrossRef]
19. Blocher, W.C.; Perry, S.L. Complex coacervate-based materials for biomedicine. *Wiley Interdiscip. Rev. Nanomed. Nanobiotechnology* **2017**, *9*, e1442. [CrossRef]
20. Timilsena, Y.P.; Akanbi, T.O.; Khalid, N.; Adhikari, B.; Barrow, C.J. Complex coacervation: Principles, mechanisms and applications in microencapsulation. *Int. J. Biol. Macromol.* **2019**, *121*, 1276–1286. [CrossRef] [PubMed]
21. Overbeek, J.T.; Voorn, M.J. Phase separation in polyelectrolyte solutions. Theory of complex coacervation. *J. Cell. Comp. Physiol.* **1957**, *49*, 7–26. [CrossRef]
22. Gooch, J.W. Flory-Huggins Theory. In *Encyclopedic Dictionary of Polymers*; Springer Science & Business Media: New York, NY, USA, 2011; p. 315.
23. Flory, P.J. *Principles of Polymer Chemistry*; Cornell University Press: Ithaca, NY, USA, 1953.

24. Black, J.R. Debye-Hückel Equation. In *Encyclopedia of Geochemistry*; Encyclopedia of Earth Sciences Series; Springer: Cham, Switzerland, 2016; pp. 1–3.
25. Michaeli, I.; Overbeek, J.T.G.; Voorn, M. Phase separation of polyelectrolyte solutions. *J. Polym. Sci.* **1957**, *23*, 443–450. [CrossRef]
26. Veis, A. A review of the early development of the thermodynamics of the complex coacervation phase separation. *Adv. Colloid Interface Sci.* **2011**, *167*, 2–11. [CrossRef]
27. Veis, A.; Aranyi, C. Phase separation in polyelectrolyte systems. I. Complex coacervates of gelatin. *J. Phys. Chem.* **1960**, *64*, 1203–1210. [CrossRef]
28. Tainaka, K.I. Effect of counterions on complex coacervation. *Biopolym. Orig. Res. Biomol.* **1980**, *19*, 1289–1298. [CrossRef]
29. Sing, C.E. Development of the modern theory of polymeric complex coacervation. *Adv. Colloid Interface Sci.* **2017**, *239*, 2–16. [CrossRef]
30. Astoricchio, E.; Alfano, C.; Rajendran, L.; Temussi, P.A.; Pastore, A. The Wide World of Coacervates: From the Sea to Neurodegeneration. *Trends Biochem. Sci.* **2020**, *45*, 706–717. [CrossRef]
31. Furlani, F.; Donati, I.; Marsich, E.; Sacco, P. Characterization of chitosan/hyaluronan complex coacervates assembled by varying polymers weight ratio and chitosan physical-chemical composition. *Colloids Interfaces* **2020**, *4*, 12. [CrossRef]
32. Furlani, F.; Parisse, P.; Sacco, P. On the formation and stability of chitosan/hyaluronan-based complex coacervates. *Molecules* **2020**, *25*, 1071. [CrossRef]
33. Kim, S.; Huang, J.; Lee, Y.; Dutta, S.; Yoo, H.Y.; Jung, Y.M.; Jho, Y.; Zeng, H.; Hwang, D.S. Complexation and coacervation of like-charged polyelectrolytes inspired by mussels. *Proc. Natl. Acad. Sci. USA* **2016**, *113*, E847–E853. [CrossRef]
34. Najafi, S.; Lin, Y.; Longhini, A.P.; Zhang, X.; Delaney, K.T.; Kosik, K.S.; Fredrickson, G.H.; Shea, J.-E.; Han, S. Liquid–liquid phase separation of Tau by self and complex coacervation. *Protein Sci.* **2021**, *30*, 1393–1407. [CrossRef] [PubMed]
35. Ki, M.-R.; Nguyen, T.K.M.; Park, T.-I.; Park, H.-M.; Pack, S.P. Biomimetic Silica Particles with Self-Loading BMP-2 Knuckle Epitope Peptide and Its Delivery for Bone Regeneration. *Pharmaceutics* **2023**, *15*, 1061. [CrossRef] [PubMed]
36. Cai, H.; Gabryelczyk, B.; Manimekalai, M.S.S.; Grüber, G.; Salentinig, S.; Miserez, A. Self-coacervation of modular squid beak proteins—A comparative study. *Soft Matter* **2017**, *13*, 7740–7752. [CrossRef] [PubMed]
37. Jing, H.; Du, X.; Mo, L.; Wang, H. Self-coacervation of carboxymethyl chitosan as a pH-responsive encapsulation and delivery strategy. *Int. J. Biol. Macromol.* **2021**, *192*, 1169–1177. [CrossRef]
38. Kaushik, P.; Rawat, K.; Bohidar, H.B. Heat-induced coacervation of elastin and its possible thermoreversibility. *Colloid Polym. Sci.* **2019**, *297*, 947–956. [CrossRef]
39. Gulão, E.d.S.; de Souza, C.J.F.; Andrade, C.T.; Garcia-Rojas, E.E. Complex coacervates obtained from peptide leucine and gum arabic: Formation and characterization. *Food Chem.* **2016**, *194*, 680–686. [CrossRef]
40. Black, K.A.; Priftis, D.; Perry, S.L.; Yip, J.; Byun, W.Y.; Tirrell, M. Protein Encapsulation via Polypeptide Complex Coacervation. *ACS Macro Lett.* **2014**, *3*, 1088–1091. [CrossRef]
41. Sing, C.E.; Perry, S.L. Recent progress in the science of complex coacervation. *Soft Matter* **2020**, *16*, 2885–2914. [CrossRef]
42. Xu, Z.; Wang, W.; Cao, Y.; Xue, B. Liquid-liquid phase separation: Fundamental physical principles, biological implications, and applications in supramolecular materials engineering. *Supramol. Mater.* **2023**, *2*, 100049. [CrossRef]
43. Yan, Z.; Liu, J.; Li, C.; Ren, J.; Wang, Z.; Zhang, R.; Liu, X. Heteroprotein complex coacervation of ovalbumin and lysozyme: Phase behavior, microstructure and processing properties. *Food Hydrocoll.* **2023**, *144*, 109013. [CrossRef]
44. Li, L.; Srivastava, S.; Andreev, M.; Marciel, A.B.; de Pablo, J.J.; Tirrell, M.V. Phase Behavior and Salt Partitioning in Polyelectrolyte Complex Coacervates. *Macromolecules* **2018**, *51*, 2988–2995. [CrossRef]
45. Kuroyanagi, S.; Shimada, N.; Fujii, S.; Furuta, T.; Harada, A.; Sakurai, K.; Maruyama, A. Highly Ordered Polypeptide with UCST Phase Separation Behavior. *J. Am. Chem. Soc.* **2019**, *141*, 1261–1268. [CrossRef] [PubMed]
46. Zhao, M.; Xia, X.; Mao, J.; Wang, C.; Dawadi, M.B.; Modarelli, D.A.; Zacharia, N.S. Composition and property tunable ternary coacervate: Branched polyethylenimine and a binary mixture of a strong and weak polyelectrolyte. *Mol. Syst. Des. Eng.* **2019**, *4*, 110–121. [CrossRef]
47. Perry, S.L.; Leon, L.; Hoffmann, K.Q.; Kade, M.J.; Priftis, D.; Black, K.A.; Wong, D.; Klein, R.A.; Pierce, C.F., 3rd; Margossian, K.O.; et al. Chirality-selected phase behaviour in ionic polypeptide complexes. *Nat. Commun.* **2015**, *6*, 6052. [CrossRef]
48. Pramanik, U.; Das, A.; Brown, E.; Struckman, H.L.; Wang, H.; Stealey, S.; Sprunger, M.L.; Wasim, A.; Fascetti, J.; Mondal, J.; et al. Histidine-rich enantiomeric peptide coacervates enhance antigen sequestration and presentation to T cells. *Chem. Sci.* **2025**, *16*, 7523–7536. [CrossRef]
49. Vieregg, J.R.; Lueckheide, M.; Marciel, A.B.; Leon, L.; Bologna, A.J.; Rivera, J.R.; Tirrell, M.V. Oligonucleotide-Peptide Complexes: Phase Control by Hybridization. *J. Am. Chem. Soc.* **2018**, *140*, 1632–1638. [CrossRef]
50. Nichols, M.K.; Kumar, R.K.; Bassindale, P.G.; Tian, L.; Barnes, A.C.; Drinkwater, B.W.; Patil, A.J.; Mann, S. Fabrication of Micropatterned Dipeptide Hydrogels by Acoustic Trapping of Stimulus-Responsive Coacervate Droplets. *Small* **2018**, *14*, 1800739. [CrossRef]

51. Huang, G.-Q.; Han, X.-N.; Xiao, J.-X. Glutaraldehyde-crosslinked O-carboxymethyl chitosan–gum Arabic coacervates: Characteristics versus complexation acidity. *J. Dispers. Sci. Technol.* **2017**, *38*, 1607–1612. [CrossRef]
52. Kayitmazer, A.B.; Koksal, A.F.; Kilic Iyilik, E. Complex coacervation of hyaluronic acid and chitosan: Effects of pH, ionic strength, charge density, chain length and the charge ratio. *Soft Matter* **2015**, *11*, 8605–8612. [CrossRef] [PubMed]
53. Doshi, N.; Venema, P.; van der Linden, E.; de Vries, R. The generality of pH-induced liquid-liquid phase separation in plant proteins extends to commercial legume flours. *Food Hydrocoll.* **2025**, *162*, 110927. [CrossRef]
54. Jamshidian, H.; Rafe, A. Complex coacervate of wheat germ protein/high methoxy pectin in encapsulation of d-limonene. *Chem. Biol. Technol. Agric.* **2024**, *11*, 60. [CrossRef]
55. Varanko, A.K.; Su, J.C.; Chilkoti, A. Elastin-like polypeptides for biomedical applications. *Annu. Rev. Biomed. Eng.* **2020**, *22*, 343–369. [CrossRef]
56. Poudyal, R.R.; Pir Cakmak, F.; Keating, C.D.; Bevilacqua, P.C. Physical Principles and Extant Biology Reveal Roles for RNA-Containing Membraneless Compartments in Origins of Life Chemistry. *Biochemistry* **2018**, *57*, 2509–2519. [CrossRef]
57. Aumiller, W.M., Jr.; Keating, C.D. Experimental models for dynamic compartmentalization of biomolecules in liquid organelles: Reversible formation and partitioning in aqueous biphasic systems. *Adv. Colloid Interface Sci.* **2017**, *239*, 75–87. [CrossRef]
58. Adhikari, S.; Prabhu, V.M.; Muthukumar, M. Lower critical solution temperature behavior in polyelectrolyte complex coacervates. *Macromolecules* **2019**, *52*, 6998–7004. [CrossRef]
59. Priftis, D.; Laugel, N.; Tirrell, M. Thermodynamic Characterization of Polypeptide Complex Coacervation. *Langmuir* **2012**, *28*, 15947–15957. [CrossRef]
60. Fu, X.; Xing, C.; Sun, J. Tunable LCST/UCST-Type Polypeptoids and Their Structure–Property Relationship. *Biomacromolecules* **2020**, *21*, 4980–4988. [CrossRef]
61. Nie, J.; Zhang, X.; Hu, Z.; Wang, W.; Schroer, M.A.; Ren, J.; Svergun, D.; Chen, A.; Yang, P.; Zeng, A.-P. A globular protein exhibits rare phase behavior and forms chemically regulated orthogonal condensates in cells. *Nat. Commun.* **2025**, *16*, 2449. [CrossRef]
62. Kaushik, P.; Rawat, K.; Aswal, V.K.; Kohlbrecher, J.; Bohidar, H.B. Mixing ratio dependent complex coacervation versus bicontinuous gelation of pectin with in situ formed zein nanoparticles. *Soft Matter* **2018**, *14*, 6463–6475. [CrossRef] [PubMed]
63. Zhang, J.; Du, H.; Ma, N.; Zhong, L.; Ma, G.; Pei, F.; Chen, H.; Hu, Q. Effect of ionic strength and mixing ratio on complex coacervation of soy protein isolate/Flammulina velutipes polysaccharide. *Food Sci. Hum. Wellness* **2023**, *12*, 183–191. [CrossRef]
64. Wei, Y.; Huang, Y.; Wen, C.; Wei, K.; Peng, L.; Wei, X. Theabrownin/whey protein isolate complex coacervate strengthens C_2C_{12} cell proliferation via modulation of energy metabolism and mitochondrial apoptosis. *Int. J. Biol. Macromol.* **2024**, *283*, 137686. [CrossRef]
65. Cui, Y.; Chen, K.; Chen, K.; Li, Y.; Jiang, L. The complex coacervation of gum Arabic and krill protein isolate and their application for Antarctic krill oil encapsulation. *Carbohydr. Polym.* **2025**, *348*, 122831. [CrossRef]
66. Li, L.; Lai, B.; Yan, J.-N.; Yambazi, M.H.; Wang, C.; Wu, H.-T. Characterization of complex coacervation between chia seed gum and whey protein isolate: Effect of pH, protein/polysaccharide mass ratio and ionic strength. *Food Hydrocoll.* **2024**, *148*, 109445. [CrossRef]
67. Mirlohi, K.; Blocher McTigue, W.C. Coacervation for biomedical applications: Innovations involving nucleic acids. *Soft Matter* **2025**, *21*, 8–26. [CrossRef] [PubMed]
68. Sabadini, J.B.; Oliveira, C.L.P.; Loh, W. Assessing the Structure and Equilibrium Conditions of Complex Coacervate Core Micelles by Varying Their Shell Composition and Medium Ionic Strength. *Langmuir* **2024**, *40*, 2015–2027. [CrossRef] [PubMed]
69. Soussi Hachfi, R.; Hamon, P.; Rousseau, F.; Famelart, M.-H.; Bouhallab, S. Ionic Strength Dependence of the Complex Coacervation between Lactoferrin and β-Lactoglobulin. *Foods* **2023**, *12*, 1040. [CrossRef]
70. Xiao, K.; Yang, Y.; Xu, X.; Szymanowski, J.E.S.; Zhou, Y.; Sigmon, G.E.; Burns, P.C.; Liu, T. Coacervate Formation in Dilute Aqueous Solutions of Inorganic Molecular Clusters with Simple Divalent Countercations. *Inorg. Chem.* **2024**, *63*, 15331–15339. [CrossRef]
71. Holkar, A.; Gao, S.; Villaseñor, K.; Lake, M.; Srivastava, S. Quantitative turbidimetric characterization of stabilized complex coacervate dispersions. *Soft Matter* **2024**, *20*, 5060–5070. [CrossRef]
72. Sathyavageeswaran, A.; Bonesso Sabadini, J.; Perry, S.L. Self-Assembling Polypeptides in Complex Coacervation. *Acc. Chem. Res.* **2024**, *57*, 386–398. [CrossRef] [PubMed]
73. Onuchic, P.L.; Milin, A.N.; Alshareedah, I.; Deniz, A.A.; Banerjee, P.R. Divalent cations can control a switch-like behavior in heterotypic and homotypic RNA coacervates. *Sci. Rep.* **2019**, *9*, 12161. [CrossRef] [PubMed]
74. Warnakulasuriya, S.N.; Nickerson, M.T. Review on plant protein–polysaccharide complex coacervation, and the functionality and applicability of formed complexes. *J. Sci. Food Agric.* **2018**, *98*, 5559–5571. [CrossRef]
75. Boeynaems, S.; Holehouse, A.S.; Weinhardt, V.; Kovacs, D.; Van Lindt, J.; Larabell, C.; Van Den Bosch, L.; Das, R.; Tompa, P.S.; Pappu, R.V.; et al. Spontaneous driving forces give rise to protein– RNA condensates with coexisting phases and complex material properties. *Proc. Natl. Acad. Sci. USA* **2019**, *116*, 7889–7898. [CrossRef]

76. Rousi, Z.; Malhiac, C.; Fatouros, D.G.; Paraskevopoulou, A. Complex coacervates formation between gelatin and gum Arabic with different arabinogalactan protein fraction content and their characterization. *Food Hydrocoll.* **2019**, *96*, 577–588. [CrossRef]
77. Simon, J.R.; Eghtesadi, S.A.; Dzuricky, M.; You, L.; Chilkoti, A. Engineered ribonucleoprotein granules inhibit translation in protocells. *Mol. Cell* **2019**, *75*, 66–75.e65. [CrossRef]
78. Gracia, P.; Polanco, D.; Tarancón-Díez, J.; Serra, I.; Bracci, M.; Oroz, J.; Laurents, D.V.; García, I.; Cremades, N. Molecular mechanism for the synchronized electrostatic coacervation and co-aggregation of alpha-synuclein and tau. *Nat. Commun.* **2022**, *13*, 4586. [CrossRef]
79. Allahyartorkaman, M.; Chan, T.-H.; Chen, E.H.L.; Ng, S.-T.; Chen, Y.-A.; Wen, J.-K.; Ho, M.-R.; Yen, H.-Y.; Kuan, Y.-S.; Kuo, M.-H.; et al. Phosphorylation-Induced Self-Coacervation versus RNA-Assisted Complex Coacervation of Tau Proteins. *J. Am. Chem. Soc.* **2025**, *147*, 10172–10187. [CrossRef]
80. Liu, J.; Chai, J.; Zhang, T.; Yuan, Y.; Saini, R.K.; Xu, M.; Li, S.; Shang, X. Phase behavior, thermodynamic and rheological properties of ovalbumin/dextran sulfate: Effect of biopolymer ratio and salt concentration. *Food Hydrocoll.* **2021**, *118*, 106777. [CrossRef]
81. Archut, A.; Klost, M.; Drusch, S.; Kastner, H. Complex coacervation of pea protein and pectin: Contribution of different protein fractions to turbidity. *Food Hydrocoll.* **2023**, *134*, 108032. [CrossRef]
82. Park, K.S.; Park, T.-I.; Lee, J.E.; Hwang, S.-Y.; Choi, A.; Pack, S.P. Aptamers and Nanobodies as New Bioprobes for SARS-CoV-2 Diagnostic and Therapeutic System Applications. *Biosensors* **2024**, *14*, 146. [CrossRef]
83. Donau, C.; Späth, F.; Sosson, M.; Kriebisch, B.A.K.; Schnitter, F.; Tena-Solsona, M.; Kang, H.-S.; Salibi, E.; Sattler, M.; Mutschler, H.; et al. Active coacervate droplets as a model for membraneless organelles and protocells. *Nat. Commun.* **2020**, *11*, 5167. [CrossRef] [PubMed]
84. Douliez, J.P.; Martin, N.; Gaillard, C.; Beneyton, T.; Baret, J.C.; Mann, S.; Beven, L. Catanionic coacervate droplets as a surfactant-based membrane-free protocell model. *Angew. Chem. Int. Ed.* **2017**, *56*, 13689–13693. [CrossRef] [PubMed]
85. Samanta, A.; Sabatino, V.; Ward, T.R.; Walther, A. Functional and morphological adaptation in DNA protocells via signal processing prompted by artificial metalloenzymes. *Nat. Nanotechnol.* **2020**, *15*, 914–921. [CrossRef]
86. Liu, W.; Lupfer, C.; Samanta, A.; Sarkar, A.; Walther, A. Switchable hydrophobic pockets in DNA protocells enhance chemical conversion. *J. Am. Chem. Soc.* **2023**, *145*, 7090–7094. [CrossRef] [PubMed]
87. Morita, M.; Sakamoto, T.; Nomura, S.M.; Murata, S.; Yanagisawa, M.; Takinoue, M. Liquid DNA Coacervates form Porous Capsular Hydrogels via Viscoelastic Phase Separation on Microdroplet Interface. *Adv. Mater. Interfaces* **2024**, *11*, 2300898. [CrossRef]
88. van Haren, M.H.I.; Helmers, N.S.; Verploegen, L.; Beckers, V.A.C.; Spruijt, E. Shape transformations in peptide–DNA coacervates driven by enzyme-catalyzed deacetylation. *Soft Matter* **2024**, *20*, 9493–9502. [CrossRef]
89. Liu, Q.; Wan, K.; Shang, Y.; Wang, Z.G.; Zhang, Y.; Dai, L.; Wang, C.; Wang, H.; Shi, X.; Liu, D.; et al. Cofactor-free oxidase-mimetic nanomaterials from self-assembled histidine-rich peptides. *Nat. Mater.* **2021**, *20*, 395–402. [CrossRef]
90. Ma, L.; Fang, X.; Wang, C. Peptide-based coacervates in therapeutic applications. *Front. Bioeng. Biotechnol.* **2023**, *10*, 1100365. [CrossRef]
91. Sheehan, F.; Sementa, D.; Jain, A.; Kumar, M.; Tayarani-Najjaran, M.; Kroiss, D.; Ulijn, R.V. Peptide-Based Supramolecular Systems Chemistry. *Chem. Rev.* **2021**, *121*, 13869–13914. [CrossRef]
92. Zheng, Y.; Yu, L.; Zou, Y.; Yang, Y.; Wang, C. Steric Dependence of Chirality Effect in Surface-Mediated Peptide Assemblies Identified with Scanning Tunneling Microscopy. *Nano Lett.* **2019**, *19*, 5403–5409. [CrossRef]
93. Zhu, J.; Xia, K.; Yu, W.; Wang, Y.; Hua, J.; Liu, B.; Gong, Z.; Wang, J.; Xu, A.; You, Z.; et al. Sustained release of GDF5 from a designed coacervate attenuates disc degeneration in a rat model. *Acta Biomater.* **2019**, *86*, 300–311. [CrossRef]
94. Sun, Y.; Lau, S.Y.; Lim, Z.W.; Chang, S.C.; Ghadessy, F.; Partridge, A.; Miserez, A. Phase-separating peptides for direct cytosolic delivery and redox-activated release of macromolecular therapeutics. *Nat. Chem.* **2022**, *14*, 274–283. [CrossRef] [PubMed]
95. Min, K.H.; Shin, J.W.; Ki, M.-R.; Kim, S.H.; Kim, K.H.; Pack, S.P. Bio-inspired formation of silica particles using the silica-forming peptides found by silica-binding motif sequence, RRSSGGRR. *Process Biochem.* **2021**, *111*, 262–269. [CrossRef]
96. Fisher, R.S.; Cheng, Y.; Goessling, L.; Obermeyer, A.C. Formation and Gelation of Elastin-like Polypeptide Complex Coacervates. *bioRxiv* **2025**, bioRxiv:2025.2001.2015.633264. [CrossRef]
97. Joshi, P.U.; Decker, C.; Zeng, X.; Sathyavageeswaran, A.; Perry, S.L.; Heldt, C.L. Design Rules for the Sequestration of Viruses into Polypeptide Complex Coacervates. *Biomacromolecules* **2024**, *25*, 741–753. [CrossRef] [PubMed]
98. Xu, N.; Feng, Y.; Wan, H.; Li, Z.; Sun, K.; Ye, S. Construction and characterization of probiotic intestinal-targeted delivery system based on complex coacervation and double-emulsion structure. *Food Hydrocoll.* **2025**, *160*, 110814. [CrossRef]
99. Yang, Z.; Guo, J.; Song, K.; Wan, Z.; Guo, J.; Yang, X. Plant protein-based complex coacervation via protein deamidation. *Food Hydrocoll.* **2025**, *159*, 110660. [CrossRef]
100. Aktaş, H.; Custodio-Mendoza, J.; Szpicer, A.; Pokorski, P.; Samborska, K.; Kurek, M.A. Polysaccharide-potato protein coacervates for enhanced anthocyanin bioavailability and stability. *Int. J. Biol. Macromol.* **2024**, *282*, 136829. [CrossRef]

101. Li, G.-Y.; Chen, Q.-H.; Su, C.-R.; Wang, H.; He, S.; Liu, J.; Nag, A.; Yuan, Y. Soy protein-polysaccharide complex coacervate under physical treatment: Effects of pH, ionic strength and polysaccharide type. *Innov. Food Sci. Emerg. Technol.* **2021**, *68*, 102612. [CrossRef]
102. Yuan, Y.; Kong, Z.-Y.; Sun, Y.-E.; Zeng, Q.-Z.; Yang, X.-Q. Complex coacervation of soy protein with chitosan: Constructing antioxidant microcapsule for algal oil delivery. *LWT* **2017**, *75*, 171–179. [CrossRef]
103. Wang, Y.; Liu, J.; Ding, Y.; Zheng, X.; Jiang, Y.; Tang, K. Fabrication and characterization of fish gelatin/soluble soybean polysaccharide edible blend films through complex coacervation. *Food Hydrocoll.* **2024**, *155*, 110226. [CrossRef]
104. de Kruif, C.G.; Weinbreck, F.; de Vries, R. Complex coacervation of proteins and anionic polysaccharides. *Curr. Opin. Colloid Interface Sci.* **2004**, *9*, 340–349. [CrossRef]
105. Zheng, J.; Van der Meeren, P.; Sun, W. New insights into protein–polysaccharide complex coacervation: Dynamics, molecular parameters, and applications. *Aggregate* **2024**, *5*, e449. [CrossRef]
106. Fraccia, T.P.; Martin, N. Non-enzymatic oligonucleotide ligation in coacervate protocells sustains compartment-content coupling. *Nat. Commun.* **2023**, *14*, 2606. [CrossRef]
107. Johnson, N.R.; Wang, Y. Coacervate delivery systems for proteins and small molecule drugs. *Expert Opin. Drug Deliv.* **2014**, *11*, 1829–1832. [CrossRef]
108. Roy, P.S. Complex Coacervate-Based Materials for Biomedicine: Recent Advancements and Future Prospects. *Ind. Eng. Chem. Res.* **2024**, *63*, 5414–5487. [CrossRef]
109. Pathak, J.; Priyadarshini, E.; Rawat, K.; Bohidar, H.B. Complex coacervation in charge complementary biopolymers: Electrostatic versus surface patch binding. *Adv. Colloid Interface Sci.* **2017**, *250*, 40–53. [CrossRef]
110. Nezamdoost-Sani, N.; Amiri, S.; Mousavi Khaneghah, A. The application of the coacervation technique for microencapsulation bioactive ingredients: A critical review. *J. Agric. Food Res.* **2024**, *18*, 101431. [CrossRef]
111. Han, Y.; Kim, D.H.; Pack, S.P. Marine-Derived Bioactive Ingredients in Functional Foods for Aging: Nutritional and Therapeutic Perspectives. *Mar. Drugs* **2024**, *22*, 496. [CrossRef]
112. Ki, M.-R.; Youn, S.; Kim, D.H.; Pack, S.P. Natural Compounds for Preventing Age-Related Diseases and Cancers. *Int. J. Mol. Sci.* **2024**, *25*, 7530. [CrossRef] [PubMed]
113. Muhoza, B.; Yuyang, H.; Uriho, A.; Harindintwali, J.D.; Liu, Q.; Li, Y. Spray-and freeze-drying of microcapsules prepared by complex coacervation method: A review. *Food Hydrocoll.* **2023**, *140*, 108650. [CrossRef]
114. Kwant, A.N.; Es Sayed, J.S.; Kamperman, M.; Burgess, J.K.; Slebos, D.-J.; Pouwels, S.D. Sticky Science: Using Complex Coacervate Adhesives for Biomedical Applications. *Adv. Healthc. Mater.* **2025**, *14*, 2402340. [CrossRef] [PubMed]
115. Stewart, R.J.; Wang, C.S.; Shao, H. Complex coacervates as a foundation for synthetic underwater adhesives. *Adv. Colloid Interface Sci.* **2011**, *167*, 85–93. [CrossRef]
116. Cook, A.B.; Novosedlik, S.; van Hest, J.C.M. Complex Coacervate Materials as Artificial Cells. *Acc. Mater. Res.* **2023**, *4*, 287–298. [CrossRef]
117. Chen, Q.; Zhao, C.; Ma, X.; Yan, W.; Wang, F. Preparation and characterization of walnut oil microcapsules by complex coacervation with sodium alginate and chitosan. *LWT* **2025**, *222*, 117630. [CrossRef]
118. Barajas-Álvarez, P.; Haro-González, J.N.; González-Ávila, M.; Espinosa-Andrews, H. Gum Arabic/Chitosan Coacervates for Encapsulation and Protection of Lacticaseibacillus rhamnosus in Storage and Gastrointestinal Environments. *Probiotics Antimicrob. Proteins* **2024**, *16*, 2073–2084. [CrossRef]
119. Hernández-Nava, R.; López-Malo, A.; Palou, E.; Ramírez-Corona, N.; Jiménez-Munguía, M.T. Encapsulation of oregano essential oil (*Origanum vulgare*) by complex coacervation between gelatin and chia mucilage and its properties after spray drying. *Food Hydrocoll.* **2020**, *109*, 106077. [CrossRef]
120. Chen, G.; Dong, S.; Chen, Y.; Gao, Y.; Zhang, Z.; Li, S.; Chen, Y. Complex coacervation of zein-chitosan via atmospheric cold plasma treatment: Improvement of encapsulation efficiency and dispersion stability. *Food Hydrocoll.* **2020**, *107*, 105943. [CrossRef]
121. Xue, F.; Zhao, X.; Li, C.; Adhikari, B. Modification of plum seed protein isolate via enzymatic hydrolysis, polyphenol conjugation and polysaccharide complexation to enhance emulsification and encapsulation of essential oils. *Int. J. Biol. Macromol.* **2025**, *306*, 141812. [CrossRef]
122. Zhang, J.; Wang, Z.; Wu, X.; Piao, S.; Zhang, Q.; Zhou, D. Covalent modulation of zein surface potential by gallic acid to enhance the formation of electrostatic-driven ternary antioxidant complex coacervates with chitosan. *Food Chem.* **2025**, *475*, 143233. [CrossRef] [PubMed]
123. Li, H.; Zong, Y.; Chen, W.; Zhao, Y.; Geng, J.; He, Z.; Du, R. Microencapsulation of Deer Oil in Soy Protein Isolate–Chitosan Complex Coacervate—Preparation, Characterization, and Simulated Digestion. *Foods* **2025**, *14*, 181. [CrossRef] [PubMed]
124. Zheng, J.; Gao, Q.; Ge, G.; Sun, W.; Van der Meeren, P.; Zhao, M. Encapsulation behavior of curcumin in heteroprotein complex coacervates and precipitates fabricated from β-conglycinin and lysozyme. *Food Hydrocoll.* **2022**, *133*, 107964. [CrossRef]

125. Wang, K.-L.; Yu, B.-K.; Zhao, H.-F.; Liu, Y.-X.; Wu, C.-Y.; Zhang, Y.-H.; Mu, Z.-S. Preparation and characterization of microcapsules for tuna oil by maillard reaction products of whey protein isolate and Arabic gum via complex coacervation. *Food Chem.* **2025**, *475*, 143269. [CrossRef]
126. Huang, J.; Liu, D.; Wang, Q.; Xu, M.; An, S.; Chu, L.; Yan, T. Preparation and characterization of resveratrol-loaded microcapsules with whey protein and flaxseed gum by membrane emulsification and complex coacervation methods. *Int. J. Biol. Macromol.* **2025**, *306*, 141783. [CrossRef]
127. Wen, P.; Huang, H.; Zhang, R.; Zheng, H.; Liang, T.; Zhuang, C.; Wu, Q.; Wang, J.; Liu, F.; Zhang, K.; et al. Coacervate vesicles assembled by liquid–liquid phase separation improve delivery of biopharmaceuticals. *Nat. Chem.* **2025**, *17*, 279–288. [CrossRef]
128. Delaporte, A.; Paraskevopoulou, A.; Grisel, M.; Gore, E. Animal-free coacervates: The combination of fungal chitosan-gum Arabic for the encapsulation of lipophilic compounds. *Int. J. Biol. Macromol.* **2025**, *299*, 140003. [CrossRef] [PubMed]
129. Ban, E.; Park, M.; Kim, Y.; Park, J.; Kim, A. Enhanced delivery of anti-inflammatory miRNA-497 to dermal fibroblasts using cationized gelatin-sodium alginate coacervates. *J. Drug Deliv. Sci. Technol.* **2024**, *97*, 105767. [CrossRef]
130. Yim, W.; Jin, Z.; Chang, Y.-C.; Brambila, C.; Creyer, M.N.; Ling, C.; He, T.; Li, Y.; Retout, M.; Penny, W.F.; et al. Polyphenol-stabilized coacervates for enzyme-triggered drug delivery. *Nat. Commun.* **2024**, *15*, 7295. [CrossRef]
131. Chenglong, W.; Shuhan, X.; Jiayi, Y.; Wencai, G.; Guoxiong, X.; Hongjing, D. Dextran-based coacervate nanodroplets as potential gene carriers for efficient cancer therapy. *Carbohydr. Polym.* **2020**, *231*, 115687. [CrossRef]
132. Li, Y.; Zhang, Y.; Dai, W.; Zhang, Q. Enhanced oral absorption and anti-inflammatory activity of ellagic acid via a novel type of case in nanosheets constructed by simple coacervation. *Int. J. Pharm.* **2021**, *594*, 120131. [CrossRef] [PubMed]
133. Sun, Y.; Xu, X.; Chen, L.; Chew, W.L.; Ping, Y.; Miserez, A. Redox-Responsive Phase-Separating Peptide as a Universal Delivery Vehicle for CRISPR/Cas9 Genome Editing Machinery. *ACS Nano* **2023**, *17*, 16597–16606. [CrossRef] [PubMed]
134. Wang, Z.; Zhang, X.; Han, M.; Jiao, X.; Zhou, J.; Wang, X.; Su, R.; Wang, Y.; Qi, W. An ultra pH-responsive peptide nanocarrier for cancer gene therapy. *J. Mater. Chem. B* **2023**, *11*, 8974–8984. [CrossRef]
135. Zhang, J.; Hong, L.; Han, H.; Zhang, Y.; Zhu, W.; Xu, Y.; Xu, Z.; Yang, Y.; Qian, X. Highly Stable, Excellent Foliar Adhesion and Anti-Photodegradation Nucleic Acid-Peptide Coacervates for Broad Agrochemicals Delivery. *Small* **2025**, *21*, 2500044. [CrossRef] [PubMed]
136. Jo, H.; Gajendiran, M.; Kim, K. Development of Polymer Coacersome Structure with Enhanced Colloidal Stability for Therapeutic Protein Delivery. *Macromol. Biosci.* **2019**, *19*, e1900207. [CrossRef]
137. Gharanjig, H.; Najaf Zadeh, H.; Stevens, C.; Abhayawardhana, P.; Huber, T.; Nazmi, A.R. Development of Extrudable Hydrogels Based on Carboxymethyl Cellulose–Gelatin Complex Coacervates. *Gels* **2025**, *11*, 51. [CrossRef]
138. Wang, Z.; Yu, X.; Song, L.; Jiao, J.; Prakash, S.; Dong, X. Encapsulation of β-carotene in gelatin-gum Arabic-sodium carboxymethyl-cellulose complex coacervates: Enhancing surimi gel properties and exploring 3D printing potential. *Int. J. Biol. Macromol.* **2024**, *278*, 134129. [CrossRef]
139. Karabıyık Acar, Ö.; Bedir, S.; Kayitmazer, A.B.; Kose, G.T. Chondro-inductive hyaluronic acid/chitosan coacervate-based scaffolds for cartilage tissue engineering. *Int. J. Biol. Macromol.* **2021**, *188*, 300–312. [CrossRef]
140. George Joy, J.; Son, H.K.; Lee, S.-J.; Kim, J.-C. Glucose-responsive soluble microneedle prepared with "Gelatin—Hydroxyethylcellulose ethoxylate complex coacervate" for the transdermal drug delivery. *J. Ind. Eng. Chem.* **2024**, *134*, 432–447. [CrossRef]
141. Wu, M.; Deng, D.; Peng, D.; Tan, C.; Lv, J.; Zhang, W.; Liu, Y.; Tian, H.; Zhao, Y. Development of low-molecular-weight polysaccharide-based wound dressings for full-thickness cutaneous wound healing via coacervate formation. *Carbohydr. Polym.* **2025**, *348*, 122851. [CrossRef]
142. Xie, J.; Bi, J.; Liu, X.; Blecker, C.; Jacquet, N.; Lyu, J. Development of soy protein isolate-chelator soluble pectin composite gels as extrusion-based 3D food printing inks: Effects of mingling strategy. *Food Hydrocoll.* **2025**, *162*, 110904. [CrossRef]
143. Jiang, L.; Dong, J.; Jiang, M.; Tan, W.; Zeng, Y.; Liu, X.; Wang, P.; Jiang, H.; Zhou, J.; Liu, X.; et al. 3D-printed multifunctional bilayer scaffold with sustained release of apoptotic extracellular vesicles and antibacterial coacervates for enhanced wound healing. *Biomaterials* **2025**, *318*, 123196. [CrossRef]
144. Cho, H.; Kim, J.; Kim, S.; Jung, Y.C.; Wang, Y.; Kang, B.-J.; Kim, K. Dual delivery of stem cells and insulin-like growth factor-1 in coacervate-embedded composite hydrogels for enhanced cartilage regeneration in osteochondral defects. *J. Control. Release* **2020**, *327*, 284–295. [CrossRef] [PubMed]
145. Hou, J.; Liu, Y.; Ma, Y.; Zhang, H.; Xia, N.; Li, H.; Wang, Z.; Rayan, A.M.; Ghamry, M.; Mohamed, T.A. High internal phase Pickering emulsions stabilized by egg yolk-carboxymethyl cellulose as an age-friendly dysphagia food: Tracking the dynamic transition from co-solubility to coacervates. *Carbohydr. Polym.* **2024**, *342*, 122430. [CrossRef] [PubMed]
146. Deng, D.; Peng, D.; Lv, J.; Zhang, W.; Tian, H.; Wang, T.; Wu, M.; Zhao, Y. Double-Network Hydrogel Based on Methacrylated Chitosan/Hyaluronic Acid Coacervate for Enhanced Wet-Tissue Adhesion. *Biomacromolecules* **2025**, *26*, 2317–2330. [CrossRef]
147. Galland, P.; Iqbal, M.H.; Favier, D.; Legros, M.; Schaaf, P.; Boulmedais, F.; Vahdati, M. Tuning the underwater adhesiveness of antibacterial polysaccharides complex coacervates. *J. Colloid Interface Sci.* **2024**, *661*, 196–206. [CrossRef]

148. Ji, F.; Li, Y.; Zhao, H.; Wang, X.; Li, W. Solvent-Exchange Triggered Solidification of Peptide/POM Coacervates for Enhancing the On-Site Underwater Adhesion. *Molecules* **2024**, *29*, 681. [CrossRef]
149. Yun, J.; Woo, H.T.; Lee, S.; Cha, H.J. Visible light-induced simultaneous bioactive amorphous calcium phosphate mineralization and in situ crosslinking of coacervate-based injectable underwater adhesive hydrogels for enhanced bone regeneration. *Biomaterials* **2025**, *315*, 122948. [CrossRef]
150. Zhou, Y.; Maitz, M.F.; Zhang, K.; Voit, B.; Appelhans, D. Dynamic and Diverse Coacervate Architectures by Controlled Demembranization. *J. Am. Chem. Soc.* **2025**, *147*, 12239–12250. [CrossRef]
151. Hu, J.; Li, J.; Liu, J.; Huang, Y.; Zhu, M.; Chen, C.; Ji, W.; Huang, X. Alternating Binary Droplets-Based Protocell Networks Driven by Heterogeneous Liquid–Liquid Phase Separation. *Angew. Chem. Int. Ed.* **2025**, *64*, e202422175. [CrossRef] [PubMed]
152. Deng, N.-N.; Huck, W.T.S. Microfluidic Formation of Monodisperse Coacervate Organelles in Liposomes. *Angew. Chem.* **2017**, *129*, 9868–9872. [CrossRef]
153. Williams, D.S.; Koga, S.; Hak, C.R.C.; Majrekar, A.; Patil, A.J.; Perriman, A.W.; Mann, S. Polymer/nucleotide droplets as bio-inspired functional micro-compartments. *Soft Matter* **2012**, *8*, 6004–6014. [CrossRef]
154. Drobot, B.; Iglesias-Artola, J.M.; Le Vay, K.; Mayr, V.; Kar, M.; Kreysing, M.; Mutschler, H.; Tang, T.Y.D. Compartmentalised RNA catalysis in membrane-free coacervate protocells. *Nat. Commun.* **2018**, *9*, 3643. [CrossRef]
155. Zhao, M.; Zacharia, N.S. Protein encapsulation via polyelectrolyte complex coacervation: Protection against protein denaturation. *J. Chem. Phys.* **2018**, *149*, 163326. [CrossRef] [PubMed]
156. Koga, S.; Williams, D.S.; Perriman, A.W.; Mann, S. Peptide–nucleotide microdroplets as a step towards a membrane-free protocell model. *Nat. Chem.* **2011**, *3*, 720–724. [CrossRef] [PubMed]
157. Spruijt, E.; Sprakel, J.; Cohen Stuart, M.A.; van der Gucht, J. Interfacial tension between a complex coacervate phase and its coexisting aqueous phase. *Soft Matter* **2010**, *6*, 172–178. [CrossRef]
158. Altenburg, W.J.; Yewdall, N.A.; Vervoort, D.F.M.; van Stevendaal, M.H.M.E.; Mason, A.F.; van Hest, J.C.M. Programmed spatial organization of biomacromolecules into discrete, coacervate-based protocells. *Nat. Commun.* **2020**, *11*, 6282. [CrossRef]
159. Pippa, N.; Karayianni, M.; Pispas, S.; Demetzos, C. Complexation of cationic-neutral block polyelectrolyte with insulin and in vitro release studies. *Int. J. Pharm.* **2015**, *491*, 136–143. [CrossRef]
160. Schuster, B.S.; Reed, E.H.; Parthasarathy, R.; Jahnke, C.N.; Caldwell, R.M.; Bermudez, J.G.; Ramage, H.; Good, M.C.; Hammer, D.A. Controllable protein phase separation and modular recruitment to form responsive membraneless organelles. *Nat. Commun.* **2018**, *9*, 2985. [CrossRef]
161. Yewdall, N.A.; Buddingh, B.C.; Altenburg, W.J.; Timmermans, S.B.P.E.; Vervoort, D.F.M.; Abdelmohsen, L.K.E.A.; Mason, A.F.; van Hest, J.C.M. Physicochemical Characterization of Polymer-Stabilized Coacervate Protocells. *ChemBioChem* **2019**, *20*, 2643–2652. [CrossRef] [PubMed]
162. Margossian, K.O.; Brown, M.U.; Emrick, T.; Muthukumar, M. Coacervation in polyzwitterion-polyelectrolyte systems and their potential applications for gastrointestinal drug delivery platforms. *Nat. Commun.* **2022**, *13*, 2250. [CrossRef] [PubMed]
163. Park, U.; Lee, M.S.; Jeon, J.; Lee, S.; Hwang, M.P.; Wang, Y.; Yang, H.S.; Kim, K. Coacervate-mediated exogenous growth factor delivery for scarless skin regeneration. *Acta Biomater.* **2019**, *90*, 179–191. [CrossRef] [PubMed]
164. Kishimura, A.; Koide, A.; Osada, K.; Yamasaki, Y.; Kataoka, K. Encapsulation of Myoglobin in PEGylated Polyion Complex Vesicles Made from a Pair of Oppositely Charged Block Ionomers: A Physiologically Available Oxygen Carrier. *Angew. Chem. Int. Ed.* **2007**, *46*, 6085–6088. [CrossRef]
165. Kim, Y.H.; Lee, K.; Li, S. Nucleic Acids Based Polyelectrolyte Complexes: Their Complexation Mechanism, Morphology, and Stability. *Chem. Mater.* **2021**, *33*, 7923–7943. [CrossRef]
166. Zhao, M.; Eghtesadi, S.A.; Dawadi, M.B.; Wang, C.; Huang, S.; Seymore, A.E.; Vogt, B.D.; Modarelli, D.A.; Liu, T.; Zacharia, N.S. Partitioning of Small Molecules in Hydrogen-Bonding Complex Coacervates of Poly(acrylic acid) and Poly(ethylene glycol) or Pluronic Block Copolymer. *Macromolecules* **2017**, *50*, 3818–3830. [CrossRef]
167. Zhao, M.; Zacharia, N.S. Sequestration of Methylene Blue into Polyelectrolyte Complex Coacervates. *Macromol. Rapid Commun.* **2016**, *37*, 1249–1255. [CrossRef]
168. Obermeyer, A.C.; Mills, C.E.; Dong, X.-H.; Flores, R.J.; Olsen, B.D. Complex coacervation of supercharged proteins with polyelectrolytes. *Soft Matter* **2016**, *12*, 3570–3581. [CrossRef]
169. Sarmento, B.; Ribeiro, A.; Veiga, F.; Sampaio, P.; Neufeld, R.; Ferreira, D. Alginate/Chitosan Nanoparticles are Effective for Oral Insulin Delivery. *Pharm. Res.* **2007**, *24*, 2198–2206. [CrossRef]
170. Sakiyama-Elbert, S.E.; Hubbell, J.A. Controlled release of nerve growth factor from a heparin-containing fibrin-based cell ingrowth matrix. *J. Control. Release* **2000**, *69*, 149–158. [CrossRef]
171. Malmo, J.; Sørgård, H.; Vårum, K.M.; Strand, S.P. siRNA delivery with chitosan nanoparticles: Molecular properties favoring efficient gene silencing. *J. Control. Release* **2012**, *158*, 261–268. [CrossRef] [PubMed]
172. Strand, S.P.; Lelu, S.; Reitan, N.K.; de Lange Davies, C.; Artursson, P.; Vårum, K.M. Molecular design of chitosan gene delivery systems with an optimized balance between polyplex stability and polyplex unpacking. *Biomaterials* **2010**, *31*, 975–987. [CrossRef]

173. Li, M.-F.; Chen, L.; Xu, M.-Z.; Zhang, J.-L.; Wang, Q.; Zeng, Q.-Z.; Wei, X.-C.; Yuan, Y. The formation of zein-chitosan complex coacervated particles: Relationship to encapsulation and controlled release properties. *Int. J. Biol. Macromol.* **2018**, *116*, 1232–1239. [CrossRef]
174. Lee, A.L.Z.; Voo, Z.X.; Chin, W.; Ono, R.J.; Yang, C.; Gao, S.; Hedrick, J.L.; Yang, Y.Y. Injectable Coacervate Hydrogel for Delivery of Anticancer Drug-Loaded Nanoparticles in vivo. *ACS Appl. Mater. Interfaces* **2018**, *10*, 13274–13282. [CrossRef]
175. Huei, G.O.S.; Muniyandy, S.; Sathasivam, T.; Veeramachineni, A.K.; Janarthanan, P. Iron cross-linked carboxymethyl cellulose–gelatin complex coacervate beads for sustained drug delivery. *Chem. Pap.* **2016**, *70*, 243–252. [CrossRef]
176. Min, K.H.; Kim, D.H.; Kim, K.H.; Seo, J.H.; Pack, S.P. Biomimetic Scaffolds of Calcium-Based Materials for Bone Regeneration. *Biomimetics* **2024**, *9*, 511. [CrossRef] [PubMed]
177. Kim, D.H.; Min, K.H.; Pack, S.P. Efficient Bioactive Surface Coatings with Calcium Minerals: Step-Wise Biomimetic Transformation of Vaterite to Carbonated Apatite. *Biomimetics* **2024**, *9*, 402. [CrossRef] [PubMed]
178. Raj, S.; Kumar Sharma, P.; Malviya, R. Pharmaceutical and tissue engineering applications of polyelectrolyte complexes. *Curr. Smart Mater.* **2018**, *3*, 21–31. [CrossRef]
179. Wei, W.; Ma, Y.; Yao, X.; Zhou, W.; Wang, X.; Li, C.; Lin, J.; He, Q.; Leptihn, S.; Ouyang, H. Advanced hydrogels for the repair of cartilage defects and regeneration. *Bioact. Mater.* **2021**, *6*, 998–1011. [CrossRef]
180. Waite, J.H.; Andersen, N.H.; Jewhurst, S.; Sun, C. Mussel Adhesion: Finding the Tricks Worth Mimicking. *J. Adhes.* **2005**, *81*, 297–317. [CrossRef]
181. Hwang, D.S.; Zeng, H.; Lu, Q.; Israelachvili, J.; Waite, J.H. Adhesion mechanism in a DOPA-deficient foot protein from green mussels. *Soft Matter* **2012**, *8*, 5640–5648. [CrossRef] [PubMed]
182. Wei, W.; Tan, Y.; Martinez Rodriguez, N.R.; Yu, J.; Israelachvili, J.N.; Waite, J.H. A mussel-derived one component adhesive coacervate. *Acta Biomater.* **2014**, *10*, 1663–1670. [CrossRef] [PubMed]
183. Dompé, M.; Vahdati, M.; van Ligten, F.; Cedano-Serrano, F.J.; Hourdet, D.; Creton, C.; Zanetti, M.; Bracco, P.; van der Gucht, J.; Kodger, T.; et al. Enhancement of the Adhesive Properties by Optimizing the Water Content in PNIPAM-Functionalized Complex Coacervates. *ACS Appl. Polym. Mater.* **2020**, *2*, 1722–1730. [CrossRef]
184. Winslow, B.D.; Shao, H.; Stewart, R.J.; Tresco, P.A. Biocompatibility of adhesive complex coacervates modeled after the sandcastle glue of Phragmatopoma californica for craniofacial reconstruction. *Biomaterials* **2010**, *31*, 9373–9381. [CrossRef] [PubMed]
185. Yang, B.; Jin, S.; Park, Y.; Jung, Y.M.; Cha, H.J. Coacervation of Interfacial Adhesive Proteins for Initial Mussel Adhesion to a Wet Surface. *Small* **2018**, *14*, e1803377. [CrossRef]
186. Lim, S.; Choi, Y.S.; Kang, D.G.; Song, Y.H.; Cha, H.J. The adhesive properties of coacervated recombinant hybrid mussel adhesive proteins. *Biomaterials* **2010**, *31*, 3715–3722. [CrossRef]
187. Kim, B.J.; Oh, D.X.; Kim, S.; Seo, J.H.; Hwang, D.S.; Masic, A.; Han, D.K.; Cha, H.J. Mussel-Mimetic Protein-Based Adhesive Hydrogel. *Biomacromolecules* **2014**, *15*, 1579–1585. [CrossRef]
188. Zhong, C.; Gurry, T.; Cheng, A.A.; Downey, J.; Deng, Z.; Stultz, C.M.; Lu, T.K. Strong underwater adhesives made by self-assembling multi-protein nanofibres. *Nat. Nanotechnol.* **2014**, *9*, 858–866. [CrossRef]
189. Biesheuvel, P.M.; Cohen Stuart, M.A. Electrostatic Free Energy of Weakly Charged Macromolecules in Solution and Intermacromolecular Complexes Consisting of Oppositely Charged Polymers. *Langmuir* **2004**, *20*, 2785–2791. [CrossRef]
190. Hyman, A.A.; Weber, C.A.; Jülicher, F. Liquid-liquid phase separation in biology. *Annu. Rev. Cell Dev. Biol.* **2014**, *30*, 39–58. [CrossRef]
191. Oparin, A.I. *The Origin of Life*; Dover Pubns: Garden City, NY, USA, 1953.
192. Spector, D.L. Nuclear domains. *J. Cell Sci.* **2001**, *114*, 2891–2893. [CrossRef] [PubMed]
193. Nott, T.J.; Petsalaki, E.; Farber, P.; Jervis, D.; Fussner, E.; Plochowietz, A.; Craggs, T.D.; Bazett-Jones, D.P.; Pawson, T.; Forman-Kay, J.D.; et al. Phase Transition of a Disordered Nuage Protein Generates Environmentally Responsive Membraneless Organelles. *Mol. Cell* **2015**, *57*, 936–947. [CrossRef] [PubMed]
194. Berry, J.; Weber, S.C.; Vaidya, N.; Haataja, M.; Brangwynne, C.P. RNA transcription modulates phase transition-driven nuclear body assembly. *Proc. Natl. Acad. Sci. USA* **2015**, *112*, E5237–E5245. [CrossRef]
195. Hubstenberger, A.; Noble, S.L.; Cameron, C.; Evans, T.C. Translation repressors, an RNA helicase, and developmental cues control RNP phase transitions during early development. *Dev. Cell* **2013**, *27*, 161–173. [CrossRef]
196. Yewdall, N.A.; Mason, A.F.; van Hest, J.C.M. The hallmarks of living systems: Towards creating artificial cells. *Interface Focus* **2018**, *8*, 20180023. [CrossRef]
197. Jia, T.Z.; Chandru, K.; Hongo, Y.; Afrin, R.; Usui, T.; Myojo, K.; Cleaves, H.J. Membraneless polyester microdroplets as primordial compartments at the origins of life. *Proc. Natl. Acad. Sci. USA* **2019**, *116*, 15830–15835. [CrossRef]
198. Longo, L.M.; Despotović, D.; Weil-Ktorza, O.; Walker, M.J.; Jabłońska, J.; Fridmann-Sirkis, Y.; Varani, G.; Metanis, N.; Tawfik, D.S. Primordial emergence of a nucleic acid-binding protein via phase separation and statistical ornithine-to-arginine conversion. *Proc. Natl. Acad. Sci. USA* **2020**, *117*, 15731–15739. [CrossRef] [PubMed]

199. Kembaren, R.; Kleijn, J.M.; Borst, J.W.; Kamperman, M.; Hofman, A.H. Enhanced stability of complex coacervate core micelles following different core-crosslinking strategies. *Soft Matter* **2022**, *18*, 3052–3062. [CrossRef]
200. Zhao, M.; Cho, S.-H.; Wu, X.; Mao, J.; Vogt, B.D.; Zacharia, N.S. Covalently crosslinked coacervates: Immobilization and stabilization of proteins with enhanced enzymatic activity. *Soft Matter* **2024**, *20*, 7623–7633. [CrossRef]
201. Chen, S.; Zou, G.; Guo, Q.; Qian, X.; Li, H.; Gao, H.; Yu, J. Extreme pH Tolerance in Peptide Coacervates Mediated by Multivalent Hydrogen Bonds for Enzyme-Triggered Oral Drug Delivery. *J. Am. Chem. Soc.* **2025**, *147*, 9704–9715. [CrossRef]
202. Li, C.-L.; Qin, F.; Li, R.-r.; Zhuan, J.; Zhu, H.-Y.; Wang, Y.; Wang, K. Preparation and In Vivo Expression of CS-PEI/pCGRP Complex for Promoting Fracture Healing. *Int. J. Polym. Sci.* **2019**, *2019*, 9432194. [CrossRef]
203. Santos, J.L.; Ren, Y.; Vandermark, J.; Archang, M.M.; Williford, J.-M.; Liu, H.-W.; Lee, J.; Wang, T.-H.; Mao, H.-Q. Continuous Production of Discrete Plasmid DNA-Polycation Nanoparticles Using Flash Nanocomplexation. *Small* **2016**, *12*, 6214–6222. [CrossRef]
204. Zhao, P.; Xia, X.; Xu, X.; Leung, K.K.C.; Rai, A.; Deng, Y.; Yang, B.; Lai, H.; Peng, X.; Shi, P.; et al. Nanoparticle-assembled bioadhesive coacervate coating with prolonged gastrointestinal retention for inflammatory bowel disease therapy. *Nat. Commun.* **2021**, *12*, 7162. [CrossRef] [PubMed]
205. Wang, L.; Duan, L.; Liu, G.; Sun, J.; Shahbazi, M.-A.; Kundu, S.C.; Reis, R.L.; Xiao, B.; Yang, X. Bioinspired Polyacrylic Acid-Based Dressing: Wet Adhesive, Self-Healing, and Multi-Biofunctional Coacervate Hydrogel Accelerates Wound Healing. *Adv. Sci.* **2023**, *10*, 2207352. [CrossRef]
206. Buscail, L.; Bournet, B.; Vernejoul, F.; Cambois, G.; Lulka, H.; Hanoun, N.; Dufresne, M.; Meulle, A.; Vignolle-Vidoni, A.; Ligat, L.; et al. First-in-man phase 1 clinical trial of gene therapy for advanced pancreatic cancer: Safety, biodistribution, and preliminary clinical findings. *Mol. Ther.* **2015**, *23*, 779–789. [CrossRef]
207. Marvin, L.; Paiva, W.; Gill, N.; Morales, M.A.; Halpern, J.M.; Vesenka, J.; Balog, E.R.M. Flow imaging microscopy as a novel tool for high-throughput evaluation of elastin-like polymer coacervates. *PLoS ONE* **2019**, *14*, e0216406. [CrossRef]
208. Pei, Y.; Zheng, Y.; Li, Z.; Liu, J.; Zheng, X.; Tang, K.; Kaplan, D.L. Ethanol-induced coacervation in aqueous gelatin solution for constructing nanospheres and networks: Morphology, dynamics and thermal sensitivity. *J. Colloid Interface Sci.* **2021**, *582*, 610–618. [CrossRef] [PubMed]
209. Peng, Q.; Wang, T.; Yang, D.; Peng, X.; Zhang, H.; Zeng, H. Recent advances in coacervation and underlying noncovalent molecular interaction mechanisms. *Prog. Polym. Sci.* **2024**, *153*, 101827. [CrossRef]

biomolecules

MDPI

Review

Current Modalities in Soft-Tissue Reconstruction and Vascularized Adipose Engineering

Jessica C. El-Mallah [1,*], Connie Wen [2], Olivia Waldron [1], Neekita R. Jikaria [1], Mohammad Hossein Asgardoon [1], Kevin Schlidt [1], Dana Goldenberg [1], Summer Horchler [1], Mary E. Landmesser [1], Ji Ho Park [1], Urara Hasegawa [3], Yong Wang [2,4] and Dino J. Ravnic [1,4,*]

1 Department of Surgery, Penn State Milton S. Hershey Medical Center, 500 University Drive, Hershey, PA 17033, USA; owaldron@pennstatehealth.psu.edu (O.W.); njikaria@pennstatehealth.psu.edu (N.R.J.); masgardoon@pennstatehealth.psu.edu (M.H.A.); kschlidt@lifebridgehealth.org (K.S.); dana.goldenberg96@gmail.com (D.G.); summer.horchler@gmail.com (S.H.); mlandmesser@pennstatehealth.psu.edu (M.E.L.); jpark9@pennstatehealth.psu.edu (J.H.P.)

2 Department of Biomedical Engineering, The Pennsylvania State University, 201 Old Main, University Park, PA 16802, USA; cqw5733@psu.edu (C.W.); yxw30@psu.edu (Y.W.)

3 Department of Materials Science and Engineering, The Pennsylvania State University, 201 Old Main, University Park, PA 16802, USA; uph5002@psu.edu

4 Huck Institutes of the Life Sciences, The Pennsylvania State University, 201 Old Main, University Park, PA 16802, USA

* Correspondence: jcollins10@pennstatehealth.psu.edu (J.C.E.-M.); dravnic@pennstatehealth.psu.edu (D.J.R.); Tel.: +001-814-441-5632 (J.C.E.-M.); +001-717-531-1019 (D.J.R.); Fax: +001-717-531-4339 (J.C.E.-M. & D.J.R.)

Abstract: Soft-tissue loss resulting from trauma or oncologic resection is a significant problem worldwide. Surgical reconstruction using adipose tissue has long been the gold-standard solution. However, these surgeries are often highly morbid, not always feasible in patients with insufficient adipose, and can have unpredictable results. Engineered soft-tissue replacements present a promising alternative. Many cell types, such as adipose-derived stem cells, have been recognized as a viable starting platform upon which new avenues in tissue engineering can be built. Additionally, efforts to develop scaffolds that can mimic the native extracellular matrix have been made with varying success. However, the suboptimal vascularization of engineered replacements is still a major limiting factor for achieving clinical translation. The current research explores the integration of all these techniques, including the use of growth factors, bioactive molecules, and advanced microsurgical techniques to enhance the vascularization process. This translational review covers the clinically standard methods of soft-tissue reconstruction and dives into emerging engineering techniques to develop vascularized adipose alternatives.

Keywords: soft-tissue reconstruction; adipose tissue engineering; biocompatible scaffolds; vascularized tissue engineering

1. Introduction

Fat, or adipose, is a form of loose connective tissue derived from the mesoderm. It is composed mainly of adipocytes but houses a variety of cell types, including preadipocytes, stem cells, endothelial cells (ECs), pericytes, fibroblasts, macrophages, and immune cells [1,2]. Adipose is abundant and largely dispensable. Due to its ubiquitous nature, it has been used extensively for soft-tissue reconstruction throughout the body [3,4]. Unfortunately, not all individuals possess enough adipose for reconstructive applications, and engineering platforms have been developed to mitigate these insufficiencies.

1.1. Adipose Development

Adipose development, or adipogenesis, is the process by which mesenchymal stem cells (MSCs) differentiate into mature adipocytes. This process is regulated by a complex transcriptional cascade. While over two dozen relevant transcription factors have been noted, PPAR-γ is the master regulator, as no other pro-adipogenic factors can function in its absence [5–13]. Induced by C/EBP-β and δ proteins, PPAR-γ works with C/EBP-α to establish adipocyte maturity (Figure 1) [14–19]. Understanding adipogenesis at the molecular level is central to adipose tissue engineering, as manipulating these pathways enables the directed differentiation of stem cells into functional adipocytes. This knowledge facilitates the development of biomimetic scaffolds and culture conditions that recapitulate native adipose tissue architecture and function, advancing strategies for soft-tissue reconstruction and regenerative therapies.

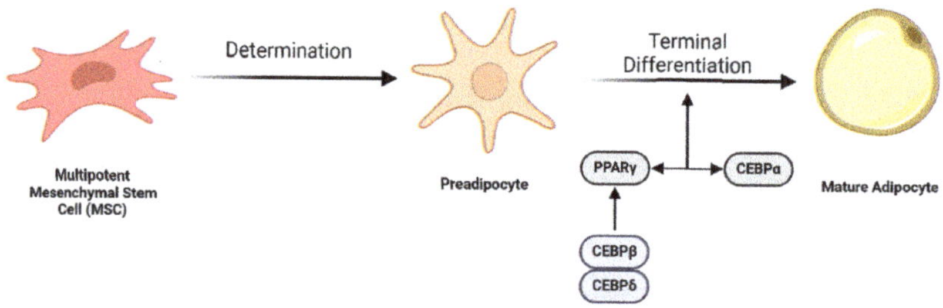

Figure 1. Overview of adipogenesis. MSCs serve as adipocyte precursors. The initial determination phase involves the conversion of an MSC to a preadipocyte. PPAR-γ and C/EBPα activate one another through a variety of signaling molecules and thus promote the terminal differentiation phase. The preadipocyte then undergoes terminal differentiation to a mature adipocyte. (Image created with Biorender.com).

1.2. Adipose Angiogenesis

Adipocyte angiogenesis, the formation of new blood vessels within adipose tissue in response to hypoxia and growth demands, is essential for supporting adipose expansion and function. This process is regulated by adipocyte-derived adipokines such as vascular endothelial growth factor (VEGF), which promotes neovascularization, and platelet-derived growth factor (PDGF), which contributes to vessel maturation and adipocyte development [20,21]. During adipose hyperplasia, microvascular proliferation occurs at the leading edge of the fat pat, where preadipocytes residing within the mural-cell compartment are found clustered along the expanding vasculature. This vascularization process differs from the process that takes place when adipose hypertrophy occurs (Figure 2) [22]. This spatial and functional coupling of angiogenesis and adipogenesis demonstrates the importance of vascularization in adipose tissue formation. In adipose tissue engineering, this relationship, which is leveraged by incorporating angiogenic cues and which supports vascular networks within scaffolds, is critical for sustaining cell viability, promoting integration with the host tissue, and enhancing the regenerative potential of engineered constructs.

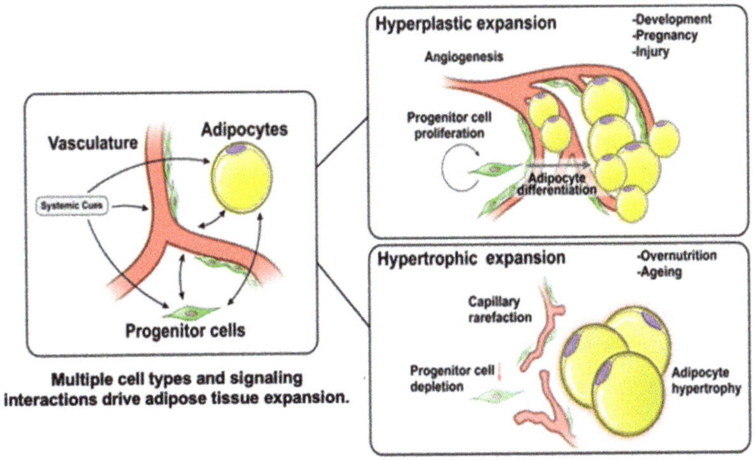

Figure 2. Microvascular growth in adipocyte hyperplasia vs. hypertrophy. The diagram illustrates adipose expansion with hyperplasia or hypertrophy. Adapted from Corvera et al. under the terms of the Creative Commons Attribution (CC BY) license (CC BY 4.0 Deed | Attribution 4.0 International | Creative Commons) [23].

2. Soft-Tissue Reconstruction

Soft-tissue loss is common with aging, traumatic injury, and oncologic resection. Over the past hundred years, surgeons have used autologous adipose grafts and flaps to correct these deficiencies. The ease of adipose harvest and its omnipresence has resulted in the vast majority of plastic surgeons utilizing these methods to treat virtually any anatomic site [24].

2.1. Fat-Graft Principles

Grafts lack an intrinsic vascular network, and embedded cells are reliant on diffusion from the recipient wound bed until neovascularization occurs. Because of this, only grafted adipose within a 250 μm diameter reliably survives [25]. Therefore, angiogenesis into the graft is critical and initially involves capillary inosculation from the recipient, which takes three to seven days [26]. This delay causes core necrosis, as oxygen cannot diffuse into the center of a thick graft. Consequently, this portion is dependent on intrinsic progenitor cells to induce EC transformation and capillary development [27]. Unfortunately, when the recipient site has dysfunctional microcirculation, such as in poorly controlled diabetes or irradiation, this angiogenic process is significantly impaired.

2.2. Fat Grafts in Clinical Care

Autologous fat grafting has become an increasingly versatile tool, offering both volumetric enhancement and regenerative potential (Figure 3A). Clinically, fat grafts are commonly employed for soft-tissue augmentation in aesthetic procedures as well as in reconstructive settings such as in postmastectomy breast reconstruction, the correction of contour deformities, and the treatment of radiation-induced fibrosis [28]. The minimally invasive nature of fat grafting combined with the use of autologous tissue reduces the risk of immunogenic reactions and foreign-body responses while offering natural-appearing and durable results. Initially, there were concerns that the intrinsic adipocyte-derived stem cells (ASCs) within grafts would increase the risk of cancer propagation. However, while the relationship between fat grafting and cancer is complex, studies have demonstrated this technique's safety and efficacy [29]. A major issue with fat grafting in breast reconstruction is that previous irradiation often requires repeated sessions of grafting. This is secondary to the detrimental effects of radiation on the microvasculature and oxygen diffusion in the recipient [30]. This highlights the importance of optimizing the recipient site prior to grafting.

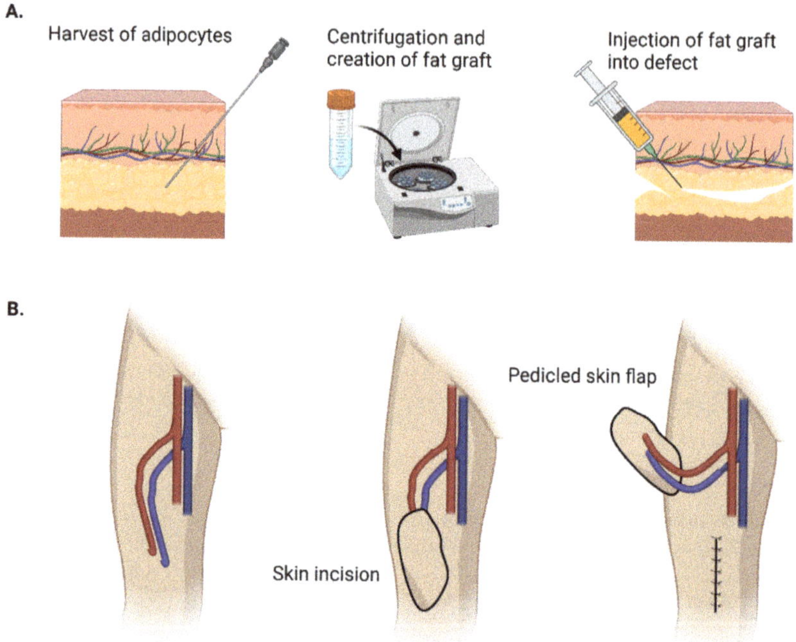

Figure 3. Fat grafting vs. adipose flap for soft-tissue reconstruction. (**A**) Fat-grafting technique where fat is harvested with a cannula, centrifuged, and then injected into a defect. (**B**) A pedicled flap that can be rotated into a soft-tissue defect with its accompanying vascular pedicle. (Image created with Biorender.com).

2.3. Graft and Recipient-Site Preparation

Two fundamental factors related to adipose retention are the graft–recipient interface and interstitial-fluid pressure limitations. Due to oxygen diffusion limitations, the "viable zone" of a graft extends 200 μm from the periphery of grafted tissue, as per the cell survival theory [31]. For this reason, thin surgical cannulas have been developed to deliver fat droplets in micro-ribbon form to increase the graft–microvasculature contact area [31]. Where adequate dispersion of the microdroplets is not achieved, grafts will be prone to hypoxia and the formation of necrotic cysts. Moreover, over-grafting profoundly increases the interstitial osmotic pressure, impairing normal capillary fluid dynamics, blood flow, and oxygen delivery.

Adipocytes are also susceptible to mechanical forces. Different harvesting techniques place varying degrees of mechanical stress on fat grafts, affecting their viability and function [32]. Common harvesting modalities include hand-held syringe aspiration, suction-assisted lipectomy, and ultrasound-assisted lipectomy [33]. Several studies have investigated the effect of cannula diameter on graft survival. Unfortunately, while our understanding has been expanded, no universal approach to graft preparation and delivery has been accepted. Hence, a variety of experimental techniques have been explored [34].

Preconditioning the recipient site through external volume expansion, for instance, uses vacuum-assisted devices to enhance vascularity and the graft capacity. Khouri et al. were able to demonstrate improved graft survival rates with lower rates of appreciable necrosis in patients pre-treated with the Brava vacuum-based expander device [35,36]. However, widespread clinical adoption of this technique has been limited by the cumbersome volume expansion process, significant complication profile, and marginal long-term viable graft retention improvement [35,36].

Other innovations, including the use of alloplastic materials to stimulate local inflammation and angiogenesis, angiogenic growth factor delivery (such as VEGF or the stromal vascular fraction (SVF)), ischemic preconditioning, and microneedling have shown promise in preclinical and early clinical studies [37–39]. These techniques aim to optimize the recipient bed through neovascular enhancement and the reduction of fat absorption. More studies are needed to test their long-term clinical efficacy and practicality in an operative setting.

2.4. Adipose-Flap Principles

While fat grafting is routinely performed, surgeons and scientists continue to seek complementary and alternative options to mitigate the unpredictable results [40,41]. Specifically, due to inadequate vascularization, only around 50% of the grafted volume is maintained long-term [42,43]. This leads to patients undergoing multiple surgeries. Furthermore, fat grafting is unsuitable for voluminous defects [44,45]. For these reasons, adipose-flap surgery (Figure 3B) has become commonplace.

Flap surgery is defined as the transfer of vascularized tissue together with its feeding artery and draining vein (vascular pedicle). Since flaps carry their own blood supply, they can be of any thickness and are suitable for wound reconstruction of any depth. Autologous flaps are broadly defined as pedicled or free depending on whether their vascular pedicle remains intact or needs to be divided and reconnected to the main vascular system, respectively. Pedicled flaps, therefore, are only suitable for wounds that exist in proximity to their donor site. Because the vascularized adipose is being moved a short distance without any disturbance in blood perfusion, the approach is technically easy. Free flaps allow for the movement of tissue from further distances; however, this approach is technically challenging, requiring specialized expertise and equipment that not all centers can provide. Furthermore, complications with microsurgical reconstruction are common and include the devastating loss of a flap secondary to thrombosis (up to 10%) as well as donor-site injuries such as scarring, wound dehiscence, seromas, hernias, muscle weakness, and paresthesia [46–51]. These inherent problems lead to patient frustration, re-operation, increased morbidity, and significantly higher costs [52,53]. This has led to the emergence of engineered alternatives.

3. Adipose Engineering

The goal of tissue engineering is to assemble functional constructs that restore, maintain, or improve damaged tissues or whole organs. Fung introduced the term in 1985, and the seminal paper was published in 1993 by Langer and Vacanti [54]. Evolved from the field of biomaterials, it refers to the practice of combining cells, scaffolds (artificial ECMs), and biologically active molecules into functional tissues. The methodologies and combinations available have grown exponentially over the past three decades. However, translatable scale-up has been largely prevented by the issue of vascularization. Developing vascularized adipose tissue (VAT) is especially complex because it requires cell sources that support both adipogenesis and angiogenesis [55]. Vascularized adipose engineering would be a welcome addition to the surgeon's armamentarium for soft-tissue reconstruction and be of significant benefit to patients. Various cell types can be combined with scaffolds and growth factors for vascularized adipose engineering (Table 1).

Table 1. Cell types utilized in adipose tissue engineering.

Cell Type	Source	Differentiation Potential	Qualities Relevant to Adipose Engineering
Adipose-derived stem cells (ASCs)	Adipose tissue, stromal vascular fraction (SVF)	Differentiate into adipocytes, endothelial cells, pericyte-like cells	- easy to harvest - low immunogenicity - upregulate vascularization
	Adult somatic cells (e.g., skin fibroblasts)	Potential to differentiate into various cell lines, including endothelial cells	- low immunogenicity - risk of tumorigenicity - high cost - time-consuming to generate and differentiate cells
Human umbilical vein endothelial cells (HUVECs)	Umbilical cord blood vessels	Endothelial cells	- ease of isolation - abundant cell harvest
Endothelial progenitor cells (EPCs)	Circulating blood, bone marrow, umbilical cord blood, adipose tissue	Endothelial cells	- accessible source - low immunogenicity - minimal ethical concerns - limited expansion capacity
Human adipose-derived microvascular endothelial cells (hAMECs)	Adipose tissue, SVF	Endothelial cells	- low cell abundance - mimic the native endothelial cells in adipose tissue - limited expansion capacity

3.1. Stem Cell Applications in Adipose Engineering

Stem cells have gained significant traction in regenerative medicine for their ability to self-renew and to differentiate into multiple cell lineages. Because mature adipocytes are terminally differentiated and mechanically fragile, they are a poor option for VAT bioengineering. MSCs—adult stem cells isolated from various types of tissue, including skeletal muscle, peripheral blood, dermis, synovial membrane, and adipose—are a better source [56].

Among the most widely utilized stem cell types are adipose-derived stem cells (ASCs), which are easily harvested in large quantities and which display robust adipogenic and angiogenic potential [57,58]. ASCs have been incorporated into various biomaterials, including hydrogels, electrospun scaffolds, and decellularized matrices, to support tissue regeneration [59–63]. They secrete pro-angiogenic factors as well as extracellular vesicles to further promote vascular ingrowth [64]. The ASC donor site influences the cellular yield, with subcutaneous depots such as the thigh providing higher ASC counts and superior adipogenic potential than the abdomen, waist, or inner knee, for example [65].

Induced pluripotent stem cells (iPSCs) are stem cells obtained from somatic cells through the ectopic expression of pluripotency transcription factors that have characteristics similar to those of embryonic stem cells (ESCs) [66]. They offer a promising avenue for regenerative medicine and disease modeling as they bypass the ethical concerns associated with human ESCs. iPSCs can be produced in large numbers and directed to differentiate into vascular lineages, providing a scalable source for engineering perfusable tissue constructs [67]. While its autologous use remains limited by cost and logistical hurdles, the development of HLA-matched iPSC lines holds promise for off-the-shelf applications. These cells hold the potential to address challenges associated with immunogenic rejection and the substantial quantity of autologous ECs needed to populate a pre-vascularized scaffold for clinical use [68].

To vascularize engineered adipose tissue, ECs are frequently co-cultured with ASCs or iPSCs. Human umbilical vein endothelial cells (HUVECs) are a commonly utilized cell type in VAT engineering. These cells can be readily obtained from umbilical cords, making them a convenient and abundant cell source for in vitro studies, and they can form functional vascular networks to support angiogenesis and tissue perfusion in engineered adipose constructs [69–71]. When seeded alongside ASCs within decellularized scaffolds, HUVECs support vessel-like structure formation and have demonstrated effective integration following implantation in small-animal models [72]. While HUVECs are often used for in vitro studies, their clinical translation is limited due to their potential for immunogenicity [73]. Additionally, maintaining the HUVEC phenotype and functionality over extended culture periods is challenging given the high rate of apoptosis following multiple cell-culture passages [74].

Possibly more clinically translatable EC sources include endothelial progenitor cells (EPCs) and human adipose microvascular endothelial cells (hAMECs). EPCs, isolated from peripheral blood, have been co-cultured with ASCs to form microvascular networks within collagen and dermal scaffolds [75]. hAMECs, although only a small fraction of the SVF, can be enriched and used to create complex vascular networks when paired with ASCs. These cells have demonstrated synergistic effects, including more mature vessel formation and improved scaffold integration in vivo. Scaffold-free models using hAMECs and stem cells have also shown promising outcomes, highlighting a path toward fully human, immunologically compatible adipose tissue constructs suitable for reconstructive applications [76].

3.2. *Scaffolds*

The ECM is a major component of native tissues. It provides cells with mechanical and structural support through networks of collagen, reticular and elastin fibers, and glycosaminoglycans (GAGs) [77]. Cells attach to the ECM by interacting with receptors such as integrin receptors [78]. Moreover, the ECM serves as a reservoir for sequestering and releasing growth factors and signaling molecules that affect cell proliferation, differentiation, and other cellular activities through signal transduction. To recapitulate these natural ECM functions, extensive studies have been conducted to engineer scaffolds from different materials such as biopolymers, synthetic polymers, and decellularized ECM (dECM).

3.2.1. Biopolymer-Based Scaffolds

Biopolymers are proteins or polysaccharides that are derived from animals or plants. These materials generally degrade enzymatically or hydrolytically and have low toxicity in the human body. In addition, many biopolymers, especially those produced in mammals, contain bioactive motifs that can interact with cells and growth factors to enhance cellular attachment and proliferation. Thus far, different biopolymers have been used for adipose tissue engineering, including collagen, gelatin, fibrin gel, and alginate.

Collagen is the most widely used biopolymer for engineering cell scaffolds. Generally, collagen scaffolds are prepared by lyophilizing acidic collagen solutions. These scaffolds have sponge-like structures with interconnected pores that are suited for cell penetration and oxygen delivery. One of the challenges in using collagen sponges is their poor mechanical strength compared with natural ECM and their uncontrolled degradation rate. One way to address these limitations is to crosslink collagen fibers chemically via crosslinking agents such as glutaraldehyde, genipin, and hexamethylene diisocyanate [79]. Kimura and colleagues reported that glutaraldehyde-crosslinked collagen sponges incorporated with preadipocytes and FGF could support fat tissue formation [80,81]. Despite the promise of this approach, crosslinking agents are generally too short to bridge collagen fibers, resulting in low crosslinking efficiency and, therefore, insufficient mechanical properties. In addition, the potential toxicity of residual crosslinking agents poses a concern [82]. To address these issues, biopolymer-based crosslinkers have been used. Davidenko et al. crosslinked collagen sponges with hyaluronic acid (HA) via the carbodiimide/*N*-hydroxysuccinimide coupling reaction, which significantly improved the dissolution resistance of the collagen sponge [83]. Zhu et al. used the same chemistry to develop porous collagen–chitosan scaffolds [84]. The combination of collagen and chitosan improved the mechanical properties of the scaffold. It was also demonstrated that the porous collagen–chitosan scaffold promoted the adhesion and proliferation of ASCs and maintained cell pluripotency [84].

Gelatin, obtained by the partial hydrolysis of collagen, is another important biopolymer in adipose tissue engineering. It maintains many of the biological functions of collagen as it contains the Arg-Gly-Asp (RGD) cell-adhesion peptide motif as well as matrix metallopeptidase (MMP)-sensitive sequences. Unlike collagen, it is soluble in water at temperatures above 30–35 °C, allowing for the facile modification of gelatin with different functional groups. Among its derivatives, gelatin methacrylamide (GelMA) has frequently been used to prepare hydrogels. GelMA can be crosslinked in the presence of a photoinitiator upon UV light irradiation. Due to the relatively mild reaction conditions, the photopolymerization of GelMA can be performed in the presence of cells, enabling their encapsulation within the hydrogel network. It has been reported that both ASCs and mature adipocytes can be encapsulated in GelMA hydrogels without affecting cell viability [85]. In addition, vascularized adipose tissue-like constructs can be generated by co-culturing ASCs and HUVECs within a GelMA hydrogel [86]. While GelMA is a promising material, phototoxicity induced by UV exposure can be a potential issue for practical applications. For this reason, other crosslinking chemistries have also been exploited to engineer gelatin-based hydrogels. For example, maleimide-functionalized gelatin (GelMAL) has been crosslinked with a dithiol crosslinker via a Michael-type addition reaction, which does not require photoinitiation [87]. It was shown that hematopoietic stem cells encapsulated in a GelMAL hydrogel generated a much lower level of reactive oxygen species (ROS) compared with those in a GelMA hydrogel, indicating that the Michael-type addition is better suited for encapsulation of cells in hydrogels because of the reduction in cell damage by ROS compared with UV-light-initiated photopolymerization. Furthermore, like collagen, gelatin contains amino and carboxyl groups that can be used for crosslinking reactions with other biopolymers. For example, gelatin was reacted with HA via the carbodiimide coupling reaction followed by a cryogelation process to fabricate a porous scaffold [88]. This scaffold exhibited mechanical properties similar to those of adipose tissue and stimulated the adipogenesis of ASCs seeded in the scaffold.

Fibrin gel is a mesh-like fibrous protein network found in blood clots. This material can be prepared by mixing fibrinogen and thrombin. Fibrin gel can bind to different growth factors as well as fibronectin, and it is degraded enzymatically by the action of plasmin [89]. These biological functions make fibrin gel an attractive biomaterial for tissue engineering. Wittman et al. demonstrated that a fibrin gel containing cells from the SVF formed vascularized adipose tissue in vivo [90]. Fibrin gels were also used to co-culture ASCs and ECs derived from peripheral blood, which led to vessel-like structure formation within the hydrogel [91]. While fibrin gel holds promise, it shows relatively fast degradation (generally, within a few days in the body), hampering its long-term applications [92]. This issue can be addressed by combining fibrin gel with other biopolymers. For example, a composite of fibrin gel and collagen microfibers was used to generate a pre-vascularized adipose tissue construct from ASCs and HUVECs. This tissue construct maintained its volume with a high cell-survival rate over three months after subcutaneous implantation [93].

Alginate is an anionic polysaccharide obtained from brown seaweed. Due to its mild gelation condition, which only requires the addition of divalent cations such as Ca^{2+}, this natural polymer has been used in many biomedical applications, including wound dressing and as a cell carrier [94]. Unlike protein-based biopolymers, alginate does not contain any cell-adhesion motifs. Therefore, conjugation of functional groups such as RGD peptides is often required to support cell attachment and growth [95]. Yoo et al. homogenously mixed adipose tissues with ionically crosslinked alginate gels to generate an alginate–fat scaffold. The adipose tissue within the alginate–fat scaffold remained viable and secreted adipokines in vitro. More importantly, the alginate–fat scaffold preserved the volume of adipose tissue in vivo [96]. Since alginate is practically non-degradable in the body, alginate hydrogels are quite stable. While these gels dissociate gradually by releasing Ca^{2+} ions, dissociation rates are generally slow, which prevents cell migration and vascularization. To make alginate gels susceptible to hydrolysis, partially oxidized alginate has been developed. Kim et al. reported that in vivo injection of oxidized alginate hydrogel containing pre-differentiated human ASCs resulted in the formation of adipose tissue within ten weeks [97].

3.2.2. Synthetic Scaffolds

While biopolymers have been used extensively in tissue engineering due to their bioactivity, biocompatibility, and degradability, some of their drawbacks include difficulties in fine-tuning material properties such as mechanical strength, viscoelasticity, biodegradability; high costs; batch-to-batch variability; and immunogenicity. In addition, animal-derived materials have the potential risk of transmitting infectious diseases. In contrast, synthetic polymers can be manufactured reproducibly and tailored for specific applications to fulfill the required material properties. In general, synthetic materials are bio-inert, and introducing bioactive motifs is often required to ensure sufficient cell ingrowth. Thus far, various synthetic polymers have been used as scaffold materials in adipose tissue engineering, including polyethylene glycol (PEG) and its derivatives, as well as biodegradable plastics such as polyglycolic acid (PGA) and poly(lactic-co-glycolic) acid (PLGA).

PEG is a highly water-soluble polymer commonly used for engineering drug–polymer conjugates, surface coating biomedical devices, and in scaffolds for tissue engineering. Generally, PEG polymers are chemically crosslinked to generate a hydrated 3D-network structure (hydrogel). Brandl et al. reported that enzymatically degradable PEG hydrogels can promote adipogenesis in 3T3-L1 preadipocytes [98]. To confer biodegradability and cell-adhesion capability, a collagenase-sensitive peptide sequence, as well as an integrin-binding motif, was incorporated into the PEG hydrogel network structure. It was found that the PEG hydrogels containing these peptide sequences enhanced lipid synthesis from differentiating adipocytes. One of the drawbacks of chemically crosslinked PEG hydrogels is the potential toxicity of the residual reactive functional groups within a PEG hydrogel network reacting with surrounding tissues. In addition, these systems generally require complicated administration procedures such as light irradiation and the mixing of two or more components [99]. To circumvent these issues, researchers have explored the use of thermally induced gelling systems (thermogels) based on PEG-based amphiphilic block copolymers, which are liquid at room temperature but which transform into hydrogels at body temperature [100]. Because of the unique gelling mechanism using body heat to induce a sol–gel transition without the need for additional toxic chemicals, thermogels have great potential in tissue engineering applications. Vashi et al. demonstrated that PEG–polypropylene oxide–PEG amphiphilic triblock copolymers (Pluronic F127) mixed with type I collagen could serve as an injectable scaffold for supporting adipogenic differentiation of bone marrow-derived MSCs [101].

Aliphatic polyesters, such as PGA and PLGA, are semi-crystalline/glassy polymers that degrade upon hydrolysis of their ester linkages. Due to their biocompatibility, these polymers are a popular material choice in tissue engineering. In addition, these polymers have excellent processability, allowing the fabrication of different sizes and shapes of scaffolds using common manufacturing techniques such as electrospinning and 3D printing. Weiser et al. used PGA fiber meshes to culture 3T3-L1 adipocytes under adipogenic conditions [102]. Subcutaneous implantation of these cell–PGA mesh constructs led to the formation of vascularized mature adipose tissues in vivo. Xu et al. reported that the implantation of a porous PLGA scaffold seeded with ASCs in a laminectomy defect resulted in the restoration of epidural fat without scar tissue formation [103]. Additionally, Patrick et al. conducted in vivo studies using preadipocyte-seeded PLGA scaffolds in rats with successful adipose tissue development [104]. PLGA scaffolds have also demonstrated successful fat regeneration in a rabbit model. However, these are often prone to a foreign-body response, with complications such as fibrous encapsulation and inflammatory reactions. Aliphatic polycarbonates are another class of material that is of interest to biomedical engineers as these materials are degradable and resorbable [105]. Poly(trimethylene carbonate) (PTMC) is one such material that is being explored [106]. PTMC is a flexible, non-toxic scaffold that does not form acidic degradation products. Jain et al. used a 3D-printing technique to fabricate a scaffold made of poly(L-lactide-co-trimethylene carbonate) (PLATMC), which was further coated with polydopamine (PDA) for increased hydrophilicity. This scaffold augmented ASC proliferation and differentiation (Figure 4) [107].

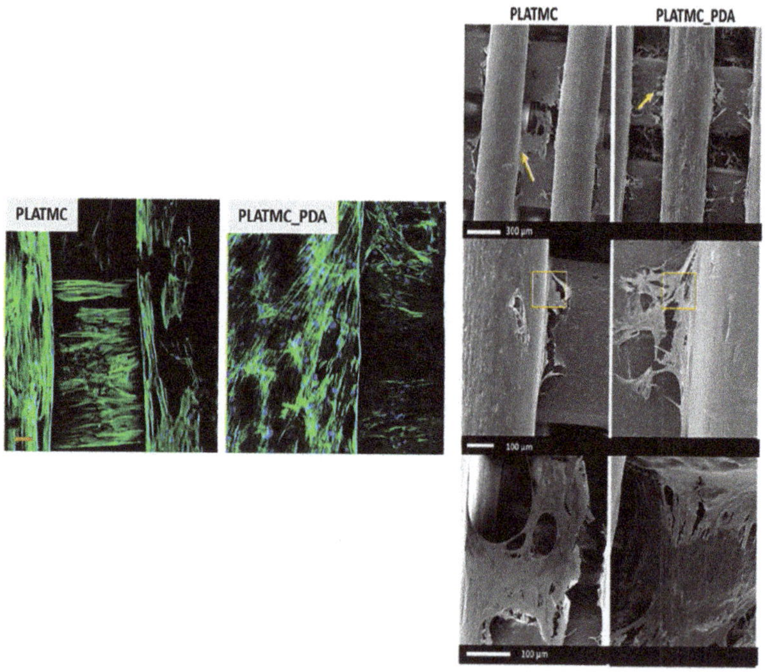

(A) (B)

Figure 4. ASC response to PLATMC 3D-printed scaffolds with or without PDA coating. (**A**) Confocal images demonstrating augmented ASC distribution with the actin scaffold depicted in green and nuclei staining in blue, scale bar 50 μm. (**B**) Scanning electron microscopy images demonstrating the cell protrusion and distribution of ASCs along the PLATMC scaffolds [107]. Adapted from Jain et al. under the terms of the Creative Commons Attribution (CC BY) license (CC BY 4.0 Deed | Attribution 4.0 International | Creative Commons).

For the successful formation of tissue-like constructs, an appropriate scaffold design is critical. Porosity is an important factor for allowing efficient cell ingrowth, sufficient nutrient and oxygen supply, and waste elimination. Pore size and interconnectivity have a significant influence on angiogenesis. It has been reported that large pores of 50–150 μm permitted mature vascularized tissue formation throughout the scaffold [108]. In addition, mechanical compatibility, degradability, and biological functionalities of scaffolds can affect adipose tissue formation.

3.2.3. Decellularized ECM

Decellularized ECM (dECM) has been used in tissue engineering. The goal of decellularization is to remove all immunogenic components, such as nucleic acids, while retaining biologically active components of the ECM to provide a microenvironment for stem cell growth and differentiation after transplantation. The process of decellularization involves treating tissues with high concentrations of salts, enzymes such as trypsin, and non-ionic detergents like Triton X-100.

Adipose-tissue-derived decellularized extracellular matrix (DAM) in combination with a scaffold can be used to induce the development of adipose tissue and capillary formation [109]. DAM can be extracted from wasted adipose tissue and is composed of ECM components such as collagen, laminin, fibronectin, elastin, GAGs, and other biologically active macromolecules [110]. The fibrillar collagen and glycoproteins within the DAM provide structural stretch resistance and resilience, allowing for the dynamic remodeling of stem cells. It also contains growth factors such as VEGF, bFGF, and TGF-B, which play an important role in soft-tissue regeneration [63]. Stem cells can be seeded on the DAM and injected or transplanted into subcutaneous tissue to promote adipogenesis and angiogenesis [111]. Following co-culture of a DAM with ASCs, the DAM was demonstrated to express the adipogenic markers PPAR-γ and C/eBP-α [112]. Cell-tracking techniques have verified that this ASC/DAM combination promotes adipogenesis originating from the host [113]. These results have also been confirmed in vitro, with increased regeneration of adipocytes within DAM constructs, and further studies have confirmed the biocompatibility of the DAM with surrounding tissues [114–116]. Wang et al. used decellularized human adipose tissues and processed them into an injectable hydrogel for seeding with human ASCs [111]. The viability and proliferation of the ASCs were confirmed in vitro. The in vivo results showed that the dECM stimulated host-cell infiltration and neovascularization, accompanied by the formation of new adipose tissue, demonstrating the feasibility of applying this system to adipose tissue engineering. Notably, the ASC-seeded dECM did not elicit an immunogenic response [111].

The availability of adipose tissue can limit decellularization. Thus, decellularization of other tissues (i.e., placental tissue) has also been investigated for adipose tissue engineering. Flynn et al. perfused the placenta with different formulations of detergent solutions and treated it with enzymatic digestion [117]. The decellularized placenta preserved the original architecture and vascular network, and histological and immunohistochemical analyses demonstrated the successful removal of immunogenic cellular components. The ASCs attached to the decellularized placental ECM, suggesting that other types of tissues can be decellularized for adipose tissue engineering. However, dECMs have limitations related to their mechanical properties, degradation kinetics, and suboptimal cellular environments and the time-consuming nature of constructing these matrices.

3.2.4. Adipose Collagen Fragments

Although acellular dermal matrices provide a framework, the decellularization process eliminates adipokines. To combat this, Xu et al. utilized adipose collagen fragments (ACFs) to capture adipokines and to functionalize these molecules to an acellular adipose matrix. Through this model, they identified the differentiation abilities of adipokines on human ASCs by evaluating the structure of neo-adipocytes and neo-adipose tissue. The adipose collagen fragments contained a diverse set of adipokines and were rich in angiogenic proteins that were able to create mature, functional, and highly vascularized adipose tissue when released in the presence of acellular adipose or dermal matrices [118].

3.3. *Growth Factors/Biologics*

Upon transplantation, the graft experiences a hostile hypoxic environment that induces growth factor and cytokine secretions that influence the newly grafted preadipocytes, adipocytes, and ASCs, as well as surrounding native adipocytes, to engage [119]. Methods have been devised to enhance and optimize this natural process by selecting specific growth factors to introduce into cell cultures, scaffolds, or grafts to promote viability. Recent advancements have utilized placental membranes to extract growth factors to create conditioned cell culture media. These membranes have abundant growth factors, including PDGF, FGF, epidermal growth factor (EGF), keratinocyte growth factor (KGF), PIGF, interleukin-4 (IL-4), transforming growth factor (TGF-β), VEGF, and tissue inhibitor metalloproteinases (TIMPs) [120]. Magana et al. found that in the presence of such factors, preadipocytes had higher cell viability in hypoxic environments when compared with normal conditions, indicating a synergistic effect between the two. Further analysis identified higher expression of VEGF-A after seven days in the hypoxic environment, suggesting these growth factors trigger angiogenesis under hypoxic conditions [121].

Furthermore, growth factors that have been paired with biodegradable scaffolds and strategies to augment their slow and controlled release from scaffolds have been extensively studied [122]. Some techniques include the use of heparin and fibronectin-binding domains to augment scaffold degradation kinetics, scaffold layering, covalent linking, and encapsulation [123]. Song et al. used decellularized adipose tissue crosslinked with heparin to encapsulate VEGF for controlled release. They found improved tissue vascularization with the benefit of a biocompatible and stable scaffold in vitro [124]. Other approaches utilize the layer-by-layer technique, alternating scaffold polymers with VEGF to allow for sequential delivery. Khanna et al. developed a polycaprolactone (PCL) scaffold with alternating layers of heparin, VEGF, and MMP-2s. The early release of VEGF followed by ECM degradation by the MMPs and heparin release improved long-term graft integration by reducing thrombogenesis [123]. Similarly, researchers used acellular adipose matrices functionalized with specific adipose-derived growth factors, including VEGF, HGF, and stromal cell-derived factor-1 (SDF-1), to induce angiogenic potential [125].

3.3.1. Extracellular Vesicles

EVs are lipid-bound vesicles naturally secreted by cells that contain proteins, lipids, and nucleic acids for intercellular communication. Within VAT engineering, EVs have gained popularity for their likely role in angiogenesis [126,127]. Additionally, these molecules pose low risk for immune rejection. Studies have demonstrated that ASC-derived EVs can promote fat-graft survival through the enhancement of angiogenesis in addition to increasing graft volume retention [128–130]. They are also useful in the repair and regeneration of tissues but are quickly degraded. Consequently, researchers have paired them with hydrogel scaffolds for targeted delivery [131].

3.3.2. Platelet-Rich Plasma

Other cell types can also augment scaffolds and serve similar functions to growth factors. A systematic review by Vyas et al. recognized platelet-rich plasma (PRP) and ASCs to be the most efficacious in promoting graft survival in vivo [132]. Li et al. demonstrated that the combination of ASCs and PRP not only promoted graft survival but maintained tissue volume in mice [133]. Sasaki et al. demonstrated a statistically significant difference between mean graft retention with PRP supplementation in fat grafting for anterior mid-face grafts compared with fat alone [134]. Furthermore, Gentile et al. demonstrated maintenance of tissue volume and shape in breast reconstruction with the addition of PRP to autologous fat transfers [135].

3.4. *Approaches to Engineering Vascularized Adipose Tissue*

Two principal approaches in VAT engineering have emerged: top–down and bottom–up [136]. The top–down approach involves seeding cells onto porous scaffolds, stimulating cell proliferation with growth factors, and cultivating the construct in a supportive environment [137,138]. The bottom–up approach (modular) utilizes individual cells or cell agglomerates, such as spheroids, organoids, and cell sheets, which are then assembled into a more complex structure to mimic native tissue [139]. This approach offers greater control over the tissue architecture.

3.4.1. Top–Down Approach

In the top–down approach, 3D pre-shaped constructs can be seeded with ASCs and ECs to form mature VAT [140]. Additionally, scaffolds can deliver complementary growth factors and biologics that support VAT. Zhang et al. demonstrated that when scaffolds were integrated with human ASCs as well as microspheres that release VEGF, the constructs showed neovascularization and persistent adipose tissue and ECM formation in rats [141]. Additionally, there has been growing interest in the use of nanotechnology in vascularized adipose engineering. Nanotechnology refers to the use of nano-sized particles (drugs, proteins, etc.) that can be placed within or around scaffolds or other implantable materials to deliver molecules. These particles increase the surface area, allowing for a more widespread therapeutic effect, and can be targeted to specific tissues. The top–down approach poses many limitations, including slow vascularization, diffusion limitations, low cell densities, and a non-uniform cell distribution. For this reason, many researchers have started focusing on a bottom–up approach instead.

3.4.2. Bottom–Up Approach

Modular engineering often utilizes cell clusters such as spheroids or organoids. Spheroids are a 3D cell cluster formed by exposing cells to a non-adherent environment. Spheroids exhibit close cell compaction, creating oxygen and nutrient gradients similar to that in natural tissues, thus providing the ability to mimic in vivo conditions [142]. ASCs have been cultured in 3D spheroids with success [143]. Similarly to fat grafts, spheroids can only be grown to up to 400 μm in diameter, as an increased size results in limited diffusion, leading to a necrotic core. However, recent research has shown them to have some pro-angiogenic qualities, and they are also amenable to be used as a bioink, allowing for controlled placement to maximize oxygen diffusion and vascularization [144]. Challenges persist, including achieving the ideal spheroid size and compactness, the capacity for fusion, and the high cell density required for mimicking native tissues.

Organoids, or "mini-organs", are self-organizing in vitro cell cultures that differentiate into functional cell types with the ability to grow in a 3D environment. Generated using ESCs, iPSCs, or adult stem cells, organoids can mimic any tissue [145]. There have been several studies using ASCs to create organoids through adipogenesis and vascularization [146]. Strobel et al. aimed to incorporate vascular structures into adipose organoids by differentiating human MSCs into preadipocytes and mixing them with microvessels [147]. However, like spheroids, they suffer from oxygen diffusion limitations and slow vascular inosculation to the recipient.

3.5. Initiating Perfusion

Despite some success in VAT engineering, the same problem that plagues autologous fat grafts persists—the inability to provide prompt oxygen delivery upon implantation. The microcirculation within VAT needs to integrate with that of the recipient as quickly as possible to prevent necrosis. Essentially, the recipient macrovasculature needs to establish continuity with the embedded adipose microvasculature. There have been some promising advances in microsurgery to improve perfusion.

3.5.1. Arteriovenous Loops

One technique described extensively in the literature to promote recipient-site angiogenesis is the use of arteriovenous loops (AVLs) (Figure 5A). AVLs are deliberate arterio-venous fistulas that are created with a grafted venous interposition. AVLs can stimulate capillary formation into the surrounding matrix. It has been demonstrated that AVLs induce angiogenesis by modifying blood flow dynamics; increased flow in the AVL causes higher wall shear stress leading to the formation of new vessels [148,149]. Further analysis demonstrated that elevated shear stress leads to angiogenesis by inhibiting the klf2 gene, which is responsible for the endothelial to mesenchymal transition, and increasing the expression of pro-inflammatory, pro-angiogenic macrophages and connexin 43 (a gap-junction protein), which is typically negligible in veins [150,151]. Despite their technical complexity, AVLs have been safely utilized in plastic surgery for vascular reconstruction and subsequent flap transfer with good clinical outcomes [152]. By combining an AVL with a fat graft, the fat receives a pedicled blood supply, and angiogenesis into the graft is stimulated. The vascularized tissue can then be either left in situ or transplanted to a distal site for soft-tissue reconstruction. This model thereby mimics an autologous flap. Debels et al. combined fat grafts with an AVL in an isolation chamber and demonstrated vascularized adipogenesis [153]. The adipocytes that remained within the isolation chamber appeared to be the product of adipogenesis rather than adipocyte survival, suggesting that the success of fat grafts may be based on new adipocyte development. Similarly, Henn et al. combined an injectable nanofiber hydrogel with an AVL to engineer a soft-tissue flap, demonstrating adipose vascularization within the isolation chamber [154]. This hydrogel combination could offer an off-the-shelf injectable scaffold that produces a flap with biomechanical properties similar to that of human fat. However, AVLs are technically cumbersome to perform, and simpler microsurgical approaches have been trialed as well.

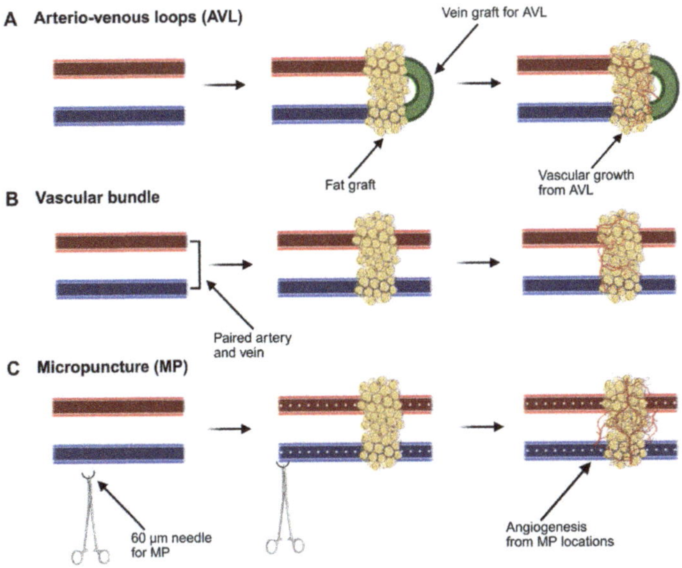

Figure 5. Schematic of microsurgical techniques to improve vascularization. (**A**) An AVL with a fat graft demonstrating the vascularization from a venous graft. (**B**) Fat-graft deposition over a vascular bundle showing microvascular formation. (**C**) Transmural MPs into a vascular bundle followed by placement of a fat graft demonstrating augmented vascularization. Image created with biorender.com.

3.5.2. Vascular Bundles

Another avenue for engineering VAT is with its own vascular pedicle. Here, fat grafts are placed in direct continuity with an underlying arterial and venous macrovasculature (Figure 5B). Vascular bundles are simple and effective techniques for tissue or construct vascularization. Previous studies have demonstrated the pro-angiogenic effects of vascular bundles with and without anastomoses in silk-scaffold vascularization [155]. Furthermore, Tanaka et al. demonstrated adipose tissue growth utilizing this technique in vivo [156,157]. In this study, the groins of rabbits were implanted with a tissue-growth chamber containing a vascular pedicle bundle with a collagen sponge, PRP, and bFGF. At the 12-week timepoint, adequate vascular tissue had developed sufficiently to transfer this tissue as an adipose flap with the vascular pedicle outside of the chamber [157]. A technique like this would offer the ability to spontaneously generate an autologous adipose flap for transfer without the donor-site morbidity that is currently encountered with flap tissue. Lu et al. further expanded on this model by utilizing an adipose tissue extract in combination with the chamber model to further promote tissue growth and vascularization with growth factors [158]. Tissue engineering chambers not only play a promising role in soft-tissue reconstruction but also serve as a mechanistic model for understanding tissue growth. While promising, both AVL and vascular bundles still suffer from a lack of rapidity.

3.5.3. Micropuncture

Our group has been using an experimental microsurgery technique termed micropuncture (MP) to stimulate cell extravasation and rapid microvascular formation out of a macrovascular bundle (Figure 5C). Sprouting angiogenesis is a complex and sequential process that requires disruption of the basement membrane. Normally, this is the rate-limiting step in angiogenesis, as there needs to be a substantial buildup of inflammatory cells, MMPs, and cytokines. In microsurgery, we routinely use needles that have diameters in the capillary range, so we sought to explore whether a 60 μm needle could be used to purposely disrupt the vessel basement membrane and rapidly stimulate microvascular outgrowth. To date, we have demonstrated that MPs can expedite adjacent hydrogel scaffold vascularization, with the induced microvasculature demonstrating sustainability for up to one month [159]. Angiogenesis in the MP cohort is induced by increased infiltration of ECs and macrophages and increased expression of VEGF-receptor 2 and Tie-2, which are involved in vascular remodeling [159,160]. We have performed preliminary studies incorporating MPs into the femoral vascular bundle with autologous adipose tissue to determine whether this can improve fat-graft vascularization (Figure 6A,B). In gross analysis, samples undergoing MP demonstrate increased microvasculature formation when compared with the non-MP control (Figure 6C–F). In the next phase, we aim to study the rapid vascularization of an engineered adipose replacement graft with MP.

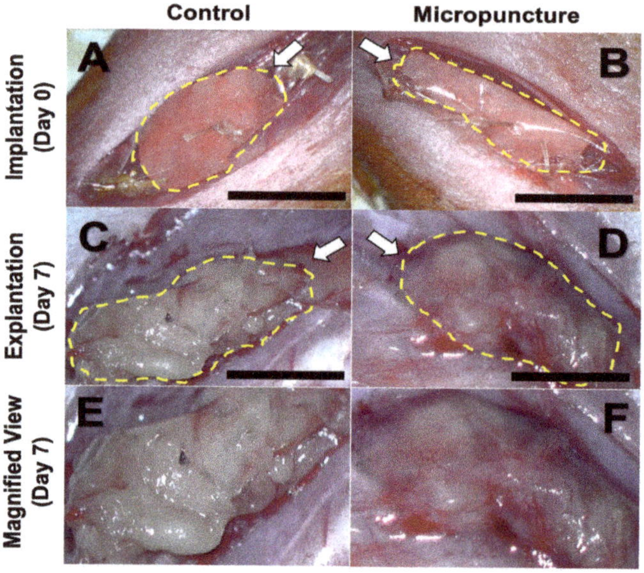

Figure 6. Fat graft vascularization after 7 days. Equal amounts of autologous adipose (yellow outline) was loaded onto a silicone sheet that circumferentially wrapped the femoral vessels (**A**,**B**). After 7 days, FGs were analyzed in situ, with control non-MP fat grafts remaining avascular and undergoing liquefaction necrosis (**C**,**E**) while MP grafts have evidence of robust vessel ingrowth (**D**,**F**). Arrows show the direction of the underlying femoral vessels. Scale bar = 10 mm.

4. Conclusions

Adipose is an abundantly available tissue in the human body with a wide variety of clinical and engineering applications. Soft-tissue reconstructive efforts to date have primarily focused on fat grafting and adipose flaps. Grafts are limited by oxygen diffusion, and are thus only suitable for the reconstruction of small defects, while flaps carry substantial donor-site morbidity. Recently, there have been a multitude of tissue engineering efforts to develop vascularized adipose tissue. Once successful, our landscape of soft-tissue reconstruction will profoundly change for the benefit of patient care.

Author Contributions: J.C.E.-M. and D.J.R. conceived the idea for this review and determined the content to be included. O.W., C.W., N.R.J., M.H.A., M.E.L., D.G., S.H., J.H.P. and K.S. all contributed to the writing and editing of this manuscript. U.H., Y.W. and D.J.R. revised the manuscript. All authors have read and agreed to the published version of the manuscript.

Funding: We acknowledge the funding provided through NIH-R01EB035568 and the Dorothy Foehr Huck and J. Lloyd Huck Endowed Chair (D.J.R. and Y.W.).

Institutional Review Board Statement: Not applicable.

Informed Consent Statement: Not applicable.

Data Availability Statement: All content of this article is available and has been shared.

Acknowledgments: There were no contributors to this article who did not meet the criteria for authorship.

Conflicts of Interest: The authors declare no conflict of interest.

References

1. Berry, D.C.; Stenesen, D.; Zeve, D.; Graff, J.M. The developmental origins of adipose tissue. *Development* **2013**, *140*, 3939–3949. [CrossRef] [PubMed]
2. Lenz, M.; Arts, I.C.W.; Peeters, R.L.M.; de Kok, T.M.; Ertaylan, G. Adipose tissue in health and disease through the lens of its building blocks. *Sci. Rep.* **2020**, *10*, 10433. [CrossRef] [PubMed]
3. Teresa Minjung, O.; Chan, K.; Brennan, T.; Roden, D.; Shamouelian, D.; Chung, H.Y.; Waner, M. Autologous Fat Grafting Restores Soft-tissue Contour Deformities after Vascular Anomaly Surgery. *Plast. Reconstr. Surg. Glob. Open* **2019**, *7*, e2196. [CrossRef]
4. Abu-Ghname, A.; Perdanasari, A.T.; Reece, E.M. Principles and Applications of Fat Grafting in Plastic Surgery. *Semin. Plast. Surg.* **2019**, *33*, 147–154. [CrossRef]
5. Mota de Sa, P.; Richard, A.J.; Hang, H.; Stephens, J.M. Transcriptional Regulation of Adipogenesis. *Compr. Physiol.* **2017**, *7*, 635–674. [CrossRef]
6. Farmer, S.R. Transcriptional control of adipocyte formation. *Cell Metab.* **2006**, *4*, 263–273. [CrossRef]
7. Siersbaek, R.; Mandrup, S. Transcriptional networks controlling adipocyte differentiation. *Cold Spring Harb. Symp. Quant. Biol.* **2011**, *76*, 247–255. [CrossRef]
8. Rosen, E.D.; Walkey, C.J.; Puigserver, P.; Spiegelman, B.M. Transcriptional regulation of adipogenesis. *Genes. Dev.* **2000**, *14*, 1293–1307. [CrossRef]
9. Tontonoz, P.; Hu, E.; Spiegelman, B.M. Stimulation of adipogenesis in fibroblasts by PPAR gamma 2, a lipid-activated transcription factor. *Cell* **1994**, *79*, 1147–1156. [CrossRef]
10. Barak, Y.; Nelson, M.C.; Ong, E.S.; Jones, Y.Z.; Ruiz-Lozano, P.; Chien, K.R.; Koder, A.; Evans, R.M. PPAR gamma is required for placental, cardiac, and adipose tissue development. *Mol. Cell* **1999**, *4*, 585–595. [CrossRef]
11. Lefterova, M.I.; Haakonsson, A.K.; Lazar, M.A.; Mandrup, S. PPARgamma and the global map of adipogenesis and beyond. *Trends Endocrinol. Metab.* **2014**, *25*, 293–302. [CrossRef] [PubMed]
12. Lee, J.E.; Ge, K. Transcriptional and epigenetic regulation of PPARgamma expression during adipogenesis. *Cell Biosci.* **2014**, *4*, 29. [CrossRef]
13. Imai, T.; Takakuwa, R.; Marchand, S.; Dentz, E.; Bornert, J.M.; Messaddeq, N.; Wendling, O.; Mark, M.; Desvergne, B.; Wahli, W.; et al. Peroxisome proliferator-activated receptor gamma is required in mature white and brown adipocytes for their survival in the mouse. *Proc. Natl. Acad. Sci. USA* **2004**, *101*, 4543–4547. [CrossRef]
14. Linhart, H.G.; Ishimura-Oka, K.; DeMayo, F.; Kibe, T.; Repka, D.; Poindexter, B.; Bick, R.J.; Darlington, G.J. C/EBPalpha is required for differentiation of white, but not brown, adipose tissue. *Proc. Natl. Acad. Sci. USA* **2001**, *98*, 12532–12537. [CrossRef]
15. Wu, Z.; Rosen, E.D.; Brun, R.; Hauser, S.; Adelmant, G.; Troy, A.E.; McKeon, C.; Darlington, G.J.; Spiegelman, B.M. Cross-regulation of C/EBP alpha and PPAR gamma controls the transcriptional pathway of adipogenesis and insulin sensitivity. *Mol. Cell* **1999**, *3*, 151–158. [CrossRef]
16. Tamori, Y.; Masugi, J.; Nishino, N.; Kasuga, M. Role of peroxisome proliferator-activated receptor-gamma in maintenance of the characteristics of mature 3T3-L1 adipocytes. *Diabetes* **2002**, *51*, 2045–2055. [CrossRef]
17. Wu, Z.; Bucher, N.L.; Farmer, S.R. Induction of peroxisome proliferator-activated receptor gamma during the conversion of 3T3 fibroblasts into adipocytes is mediated by C/EBPbeta, C/EBPdelta, and glucocorticoids. *Mol. Cell. Biol.* **1996**, *16*, 4128–4136. [CrossRef]
18. Cao, Z.; Umek, R.M.; McKnight, S.L. Regulated expression of three C/EBP isoforms during adipose conversion of 3T3-L1 cells. *Genes. Dev.* **1991**, *5*, 1538–1552. [CrossRef]
19. Lefterova, M.I.; Zhang, Y.; Steger, D.J.; Schupp, M.; Schug, J.; Cristancho, A.; Feng, D.; Zhuo, D.; Stoeckert, C.J., Jr.; Liu, X.S.; et al. PPARgamma and C/EBP factors orchestrate adipocyte biology via adjacent binding on a genome-wide scale. *Genes. Dev.* **2008**, *22*, 2941–2952. [CrossRef]

20. Kawai, T.; Autieri, M.V.; Scalia, R. Adipose tissue inflammation and metabolic dysfunction in obesity. *Am. J. Physiol. Cell Physiol.* **2021**, *320*, C375–C391. [CrossRef]
21. Fantuzzi, G. Adipose tissue, adipokines, and inflammation. *J. Allergy Clin. Immunol.* **2005**, *115*, 911–919; quiz 920. [CrossRef] [PubMed]
22. Han, J.; Lee, J.E.; Jin, J.; Lim, J.S.; Oh, N.; Kim, K.; Chang, S.I.; Shibuya, M.; Kim, H.; Koh, G.Y. The spatiotemporal development of adipose tissue. *Development* **2011**, *138*, 5027–5037. [CrossRef] [PubMed]
23. Corvera, S.; Solivan-Rivera, J.; Yang Loureiro, Z. Angiogenesis in adipose tissue and obesity. *Angiogenesis* **2022**, *25*, 439–453. [CrossRef]
24. Strong, A.L.; Adidharma, W.; Brown, O.H.; Cederna, P.S. Fat Grafting Subjectively Improves Facial Skin Elasticity and Hand Function of Scleroderma Patients. *Plast. Reconstr. Surg. Glob. Open* **2021**, *9*, e3373. [CrossRef]
25. Pu, L.L. Mechanisms of Fat Graft Survival. *Ann. Plast. Surg.* **2016**, *77* (Suppl. S1), S84–S86. [CrossRef]
26. Evans, B.G.A.; Gronet, E.M.; Saint-Cyr, M.H. How Fat Grafting Works. *Plast. Reconstr. Surg. Glob. Open* **2020**, *8*, e2705. [CrossRef]
27. Gan, F.; Liu, L.; Zhou, Q.; Huang, W.; Huang, X.; Zhao, X. Effects of adipose-derived stromal cells and endothelial progenitor cells on adipose transplant survival and angiogenesis. *PLoS ONE* **2022**, *17*, e0261498. [CrossRef]
28. Shauly, O.; Gould, D.J.; Ghavami, A. Fat Grafting: Basic Science, Techniques, and Patient Management. *Plast. Reconstr. Surg. Glob. Open* **2022**, *10*, e3987. [CrossRef]
29. Valente, D.S.; Ely, P.B.; Kieling, L.; Konzen, A.T.; Steffen, L.P.; Lazzaretti, G.S.; Zanella, R.K. Breast fat grafting and cancer: A systematic review of the science behind enhancements and concerns. *Transl. Breast Cancer Res.* **2024**, *5*, 14. [CrossRef]
30. Delaporte, T.; Delay, E.; Toussoun, G.; Delbaere, M.; Sinna, R. Breast volume reconstruction by lipomodeling technique: About 15 consecutive cases. *Ann. Chir. Plast. Esthet.* **2009**, *54*, 303–316. [CrossRef]
31. Khouri, R.K., Jr.; Khouri, R.E.; Lujan-Hernandez, J.R.; Khouri, K.R.; Lancerotto, L.; Orgill, D.P. Diffusion and perfusion: The keys to fat grafting. *Plast. Reconstr. Surg. Glob. Open* **2014**, *2*, e220. [CrossRef] [PubMed]
32. Shih, L.; Davis, M.J.; Winocour, S.J. The Science of Fat Grafting. *Semin. Plast. Surg.* **2020**, *34*, 5–10. [CrossRef]
33. Strong, A.L.; Cederna, P.S.; Rubin, J.P.; Coleman, S.R.; Levi, B. The Current State of Fat Grafting: A Review of Harvesting, Processing, and Injection Techniques. *Plast. Reconstr. Surg.* **2015**, *136*, 897–912. [CrossRef]
34. Bagheri, S.C.; Bohluli, B.; Consky, E.K. Current Techniques in Fat Grafting. *Atlas Oral Maxillofac. Surg. Clin. N. Am.* **2018**, *26*, 7–13. [CrossRef]
35. Fuentes-Felix, C. BRAVA: Results did not meet expectations. *Aesthet. Surg. J.* **2003**, *23*, 42. [CrossRef]
36. Oranges, C.M.; Striebel, J.; Tremp, M.; Madduri, S.; Kalbermatten, D.F.; Schaefer, D.J. The Impact of Recipient Site External Expansion in Fat Grafting Surgical Outcomes. *Plast. Reconstr. Surg. Glob. Open* **2018**, *6*, e1649. [CrossRef]
37. Topcu, A.; Aydin, O.E.; Unlu, M.; Barutcu, A.; Atabey, A. Increasing the viability of fat grafts by vascular endothelial growth factor. *Arch. Facial Plast. Surg.* **2012**, *14*, 270–276. [CrossRef]
38. Gassman, A.A.; Lewis, M.S.; Bradley, J.P.; Lee, J.C. Remote Ischemic Preconditioning Improves the Viability of Donor Lipoaspirate during Murine Fat Transfer. *Plast. Reconstr. Surg.* **2015**, *136*, 495–502. [CrossRef]
39. Samdal, F.; Skolleborg, K.C.; Berthelsen, B. The effect of preoperative needle abrasion of the recipient site on survival of autologous free fat grafts in rats. *Scand. J. Plast. Reconstr. Surg. Hand Surg.* **1992**, *26*, 33–36. [CrossRef]
40. Leong, D.T.; Hutmacher, D.W.; Chew, F.T.; Lim, T.C. Viability and adipogenic potential of human adipose tissue processed cell population obtained from pump-assisted and syringe-assisted liposuction. *J. Dermatol. Sci.* **2005**, *37*, 169–176. [CrossRef]
41. Simonacci, F.; Bertozzi, N.; Grieco, M.P.; Grignaffini, E.; Raposio, E. Procedure, applications, and outcomes of autologous fat grafting. *Ann. Med. Surg.* **2017**, *20*, 49–60. [CrossRef] [PubMed]
42. Bauer, C.A.; Valentino, J.; Hoffman, H.T. Long-term result of vocal cord augmentation with autogenous fat. *Ann. Otol. Rhinol. Laryngol.* **1995**, *104*, 871–874. [CrossRef]
43. Nguyen, A.; Pasyk, K.A.; Bouvier, T.N.; Hassett, C.A.; Argenta, L.C. Comparative study of survival of autologous adipose tissue taken and transplanted by different techniques. *Plast. Reconstr. Surg.* **1990**, *85*, 378–386; discussion 387–379. [CrossRef]
44. Delay, E.; Garson, S.; Tousson, G.; Sinna, R. Fat injection to the breast: Technique, results, and indications based on 880 procedures over 10 years. *Aesthet. Surg. J.* **2009**, *29*, 360–376. [CrossRef]
45. Schipper, J.A.M.; Tuin, A.J.; Loonen, T.G.J.; Dijkstra, P.U.; Spijkervet, F.K.L.; Schepers, R.H.; Jansma, J. Volume and patient satisfaction, 5 years of follow up after facial fat grafting. *J. Plast. Reconstr. Aesthet. Surg.* **2025**, *102*, 231–237. [CrossRef]
46. Mehrara, B.J.; Santoro, T.D.; Arcilla, E.; Watson, J.P.; Shaw, W.W.; Da Lio, A.L. Complications after microvascular breast reconstruction: Experience with 1195 flaps. *Plast. Reconstr. Surg.* **2006**, *118*, 1100–1109. [CrossRef]
47. Chiu, Y.H.; Chang, D.H.; Perng, C.K. Vascular Complications and Free Flap Salvage in Head and Neck Reconstructive Surgery: Analysis of 150 Cases of Reexploration. *Ann. Plast. Surg.* **2017**, *78*, S83–S88. [CrossRef]
48. Pan, X.L.; Chen, G.X.; Shao, H.W.; Han, C.M.; Zhang, L.P.; Zhi, L.Z. Effect of heparin on prevention of flap loss in microsurgical free flap transfer: A meta-analysis. *PLoS ONE* **2014**, *9*, e95111. [CrossRef]

49. Senchenkov, A.; Lemaine, V.; Tran, N.V. Management of perioperative microvascular thrombotic complications—The use of multiagent anticoagulation algorithm in 395 consecutive free flaps. *J. Plast. Reconstr. Aesthet. Surg.* **2015**, *68*, 1293–1303. [CrossRef]
50. Chang, E.I.; Chang, E.I.; Soto-Miranda, M.A.; Zhang, H.; Nosrati, N.; Robb, G.L.; Chang, D.W. Comprehensive analysis of donor-site morbidity in abdominally based free flap breast reconstruction. *Plast. Reconstr. Surg.* **2013**, *132*, 1383–1391. [CrossRef]
51. Atisha, D.M.; Tessiatore, K.M.; Rushing, C.N.; Dayicioglu, D.; Pusic, A.; Hwang, S. A National Snapshot of Patient-Reported Outcomes Comparing Types of Abdominal Flaps for Breast Reconstruction. *Plast. Reconstr. Surg.* **2019**, *143*, 667–677. [CrossRef] [PubMed]
52. Las, D.E.; de Jong, T.; Zuidam, J.M.; Verweij, N.M.; Hovius, S.E.; Mureau, M.A. Identification of independent risk factors for flap failure: A retrospective analysis of 1530 free flaps for breast, head and neck and extremity reconstruction. *J. Plast. Reconstr. Aesthet. Surg.* **2016**, *69*, 894–906. [CrossRef]
53. Largo, R.D.; Selber, J.C.; Garvey, P.B.; Chang, E.I.; Hanasono, M.M.; Yu, P.; Butler, C.E.; Baumann, D.P. Outcome Analysis of Free Flap Salvage in Outpatients Presenting with Microvascular Compromise. *Plast. Reconstr. Surg.* **2018**, *141*, 20e–27e. [CrossRef]
54. Langer, R.; Vacanti, J.P. Tissue engineering. *Science* **1993**, *260*, 920–926. [CrossRef]
55. Peirsman, A.; Nguyen, H.T.; Van Waeyenberge, M.; Ceballos, C.; Bolivar, J.; Kawakita, S.; Vanlauwe, F.; Tirpakova, Z.; Van Dorpe, S.; Van Damme, L.; et al. Vascularized adipose tissue engineering: Moving towards soft tissue reconstruction. *Biofabrication* **2023**, *15*, 32003. [CrossRef]
56. Bianco, P. "Mesenchymal" stem cells. *Annu. Rev. Cell Dev. Biol.* **2014**, *30*, 677–704. [CrossRef]
57. Si, Z.; Wang, X.; Sun, C.; Kang, Y.; Xu, J.; Wang, X.; Hui, Y. Adipose-derived stem cells: Sources, potency, and implications for regenerative therapies. *Biomed. Pharmacother.* **2019**, *114*, 108765. [CrossRef]
58. Bacakova, L.; Zarubova, J.; Travnickova, M.; Musilkova, J.; Pajorova, J.; Slepicka, P.; Kasalkova, N.S.; Svorcik, V.; Kolska, Z.; Motarjemi, H.; et al. Stem cells: Their source, potency and use in regenerative therapies with focus on adipose-derived stem cells—A review. *Biotechnol. Adv.* **2018**, *36*, 1111–1126. [CrossRef]
59. Huang, Q.; Zou, Y.; Arno, M.C.; Chen, S.; Wang, T.; Gao, J.; Dove, A.P.; Du, J. Hydrogel scaffolds for differentiation of adipose-derived stem cells. *Chem. Soc. Rev.* **2017**, *46*, 6255–6275. [CrossRef]
60. Liu, Y.; Liu, Y.; Wu, M.; Zou, R.; Mao, S.; Cong, P.; Hou, M.; Jin, H.; Zhao, Y.; Bao, Y. Adipose-derived mesenchymal stem cell-loaded beta-chitin nanofiber hydrogel promote wound healing in rats. *J. Mater. Sci. Mater. Med.* **2022**, *33*, 12. [CrossRef]
61. Pacelli, S.; Chakravarti, A.R.; Modaresi, S.; Subham, S.; Burkey, K.; Kurlbaum, C.; Fang, M.; Neal, C.A.; Mellott, A.J.; Chakraborty, A.; et al. Investigation of human adipose-derived stem-cell behavior using a cell-instructive polydopamine-coated gelatin-alginate hydrogel. *J. Biomed. Mater. Res. A* **2021**, *109*, 2597–2610. [CrossRef] [PubMed]
62. Razavi, S.; Karbasi, S.; Morshed, M.; Zarkesh Esfahani, H.; Golozar, M.; Vaezifar, S. Cell Attachment and Proliferation of Human Adipose-Derived Stem Cells on PLGA/Chitosan Electrospun Nano-Biocomposite. *Cell J.* **2015**, *17*, 429–437. [CrossRef]
63. Yang, J.Z.; Qiu, L.H.; Xiong, S.H.; Dang, J.L.; Rong, X.K.; Hou, M.M.; Wang, K.; Yu, Z.; Yi, C.G. Decellularized adipose matrix provides an inductive microenvironment for stem cells in tissue regeneration. *World J. Stem Cells* **2020**, *12*, 585–603. [CrossRef]
64. Hong, K.Y.; Bae, H.; Park, I.; Park, D.Y.; Kim, K.H.; Kubota, Y.; Cho, E.S.; Kim, H.; Adams, R.H.; Yoo, O.J.; et al. Perilipin+ embryonic preadipocytes actively proliferate along growing vasculatures for adipose expansion. *Development* **2015**, *142*, 2623–2632. [CrossRef]
65. Tsekouras, A.; Mantas, D.; Tsilimigras, D.I.; Moris, D.; Kontos, M.; Zografos, G.C. Comparison of the Viability and Yield of Adipose-Derived Stem Cells (ASCs) from Different Donor Areas. *In Vivo* **2017**, *31*, 1229–1234. [CrossRef]
66. Dani, V.; Yao, X.; Dani, C. Transplantation of fat tissues and iPSC-derived energy expenditure adipocytes to counteract obesity-driven metabolic disorders: Current strategies and future perspectives. *Rev. Endocr. Metab. Disord.* **2022**, *23*, 103–110. [CrossRef]
67. Patsch, C.; Challet-Meylan, L.; Thoma, E.C.; Urich, E.; Heckel, T.; O'Sullivan, J.F.; Grainger, S.J.; Kapp, F.G.; Sun, L.; Christensen, K.; et al. Generation of vascular endothelial and smooth muscle cells from human pluripotent stem cells. *Nat. Cell Biol.* **2015**, *17*, 994–1003. [CrossRef]
68. Kong, A.M.; Yap, K.K.; Lim, S.Y.; Marre, D.; Pebay, A.; Gerrand, Y.W.; Lees, J.G.; Palmer, J.A.; Morrison, W.A.; Mitchell, G.M. Bio-engineering a tissue flap utilizing a porous scaffold incorporating a human induced pluripotent stem cell-derived endothelial cell capillary network connected to a vascular pedicle. *Acta Biomater.* **2019**, *94*, 281–294. [CrossRef]
69. Koduru, S.V.; Leberfinger, A.N.; Pasic, D.; Forghani, A.; Lince, S.; Hayes, D.J.; Ozbolat, I.T.; Ravnic, D.J. Cellular Based Strategies for Microvascular Engineering. *Stem Cell Rev. Rep.* **2019**, *15*, 218–240. [CrossRef]
70. Koivunotko, E.; Snirvi, J.; Merivaara, A.; Harjumaki, R.; Rautiainen, S.; Kelloniemi, M.; Kuismanen, K.; Miettinen, S.; Yliperttula, M.; Koivuniemi, R. Angiogenic Potential of Human Adipose-Derived Mesenchymal Stromal Cells in Nanofibrillated Cellulose Hydrogel. *Biomedicines* **2022**, *10*, 2584. [CrossRef]
71. Sarkanen, J.R.; Vuorenpaa, H.; Huttala, O.; Mannerstrom, B.; Kuokkanen, H.; Miettinen, S.; Heinonen, T.; Ylikomi, T. Adipose stromal cell tubule network model provides a versatile tool for vascular research and tissue engineering. *Cells Tissues Organs* **2012**, *196*, 385–397. [CrossRef] [PubMed]

72. Zhang, Q.; Johnson, J.A.; Dunne, L.W.; Chen, Y.; Iyyanki, T.; Wu, Y.; Chang, E.I.; Branch-Brooks, C.D.; Robb, G.L.; Butler, C.E. Decellularized skin/adipose tissue flap matrix for engineering vascularized composite soft tissue flaps. *Acta Biomater.* **2016**, *35*, 166–184. [CrossRef] [PubMed]
73. Kocherova, I.; Bryja, A.; Mozdziak, P.; Angelova Volponi, A.; Dyszkiewicz-Konwinska, M.; Piotrowska-Kempisty, H.; Antosik, P.; Bukowska, D.; Bruska, M.; Izycki, D.; et al. Human Umbilical Vein Endothelial Cells (HUVECs) Co-Culture with Osteogenic Cells: From Molecular Communication to Engineering Prevascularised Bone Grafts. *J. Clin. Med.* **2019**, *8*, 1602. [CrossRef] [PubMed]
74. Prasad Chennazhy, K.; Krishnan, L.K. Effect of passage number and matrix characteristics on differentiation of endothelial cells cultured for tissue engineering. *Biomaterials* **2005**, *26*, 5658–5667. [CrossRef]
75. Yoder, M.C. Human endothelial progenitor cells. *Cold Spring Harb. Perspect. Med.* **2012**, *2*, a006692. [CrossRef]
76. Freiman, A.; Shandalov, Y.; Rozenfeld, D.; Shor, E.; Segal, S.; Ben-David, D.; Meretzki, S.; Egozi, D.; Levenberg, S. Adipose-derived endothelial and mesenchymal stem cells enhance vascular network formation on three-dimensional constructs in vitro. *Stem Cell Res. Ther.* **2016**, *7*, 5. [CrossRef]
77. Yue, B. Biology of the extracellular matrix: An overview. *J. Glaucoma* **2014**, *23*, S20–S23. [CrossRef]
78. Frantz, C.; Stewart, K.M.; Weaver, V.M. The extracellular matrix at a glance. *J. Cell Sci.* **2010**, *123*, 4195–4200. [CrossRef]
79. Fan, L.; Ren, Y.; Emmert, S.; Vuckovic, I.; Stojanovic, S.; Najman, S.; Schnettler, R.; Barbeck, M.; Schenke-Layland, K.; Xiong, X. The Use of Collagen-Based Materials in Bone Tissue Engineering. *Int. J. Mol. Sci.* **2023**, *24*, 3744. [CrossRef]
80. Kimura, Y.; Inamoto, T.; Tabata, Y. Adipose tissue formation in collagen scaffolds with different biodegradabilities. *J. Biomater. Sci. Polym. Ed.* **2010**, *21*, 463–476. [CrossRef]
81. Kimura, Y.; Ozeki, M.; Inamoto, T.; Tabata, Y. Adipose tissue engineering based on human preadipocytes combined with gelatin microspheres containing basic fibroblast growth factor. *Biomaterials* **2003**, *24*, 2513–2521. [CrossRef] [PubMed]
82. Yang, G.; Xiao, Z.; Long, H.; Ma, K.; Zhang, J.; Ren, X.; Zhang, J. Assessment of the characteristics and biocompatibility of gelatin sponge scaffolds prepared by various crosslinking methods. *Sci. Rep.* **2018**, *8*, 1616. [CrossRef]
83. Davidenko, N.; Campbell, J.J.; Thian, E.S.; Watson, C.J.; Cameron, R.E. Collagen-hyaluronic acid scaffolds for adipose tissue engineering. *Acta Biomater.* **2010**, *6*, 3957–3968. [CrossRef]
84. Zhu, Y.; Liu, T.; Song, K.; Jiang, B.; Ma, X.; Cui, Z. Collagen-chitosan polymer as a scaffold for the proliferation of human adipose tissue-derived stem cells. *J. Mater. Sci. Mater. Med.* **2009**, *20*, 799–808. [CrossRef]
85. Albrecht, F.B.; Schmidt, F.F.; Volz, A.C.; Kluger, P.J. Bioprinting of 3D Adipose Tissue Models Using a GelMA-Bioink with Human Mature Adipocytes or Human Adipose-Derived Stem Cells. *Gels* **2022**, *8*, 611. [CrossRef]
86. Cheng, M.H.; Chang, C.W.; Wang, J.; Bupphathong, S.; Huang, W.; Lin, C.H. 3D-Bioprinted GelMA Scaffold with ASCs and HUVECs for Engineering Vascularized Adipose Tissue. *ACS Appl. Bio Mater.* **2024**, *7*, 406–415. [CrossRef]
87. Gilchrist, A.E.; Serrano, J.F.; Ngo, M.T.; Hrnjak, Z.; Kim, S.; Harley, B.A.C. Encapsulation of murine hematopoietic stem and progenitor cells in a thiol-crosslinked maleimide-functionalized gelatin hydrogel. *Acta Biomater.* **2021**, *131*, 138–148. [CrossRef]
88. Chang, K.H.; Liao, H.T.; Chen, J.P. Preparation and characterization of gelatin/hyaluronic acid cryogels for adipose tissue engineering: In vitro and in vivo studies. *Acta Biomater.* **2013**, *9*, 9012–9026. [CrossRef]
89. Janmey, P.A.; Winer, J.P.; Weisel, J.W. Fibrin gels and their clinical and bioengineering applications. *J. R. Soc. Interface* **2009**, *6*, 1–10. [CrossRef]
90. Wittmann, K.; Dietl, S.; Ludwig, N.; Berberich, O.; Hoefner, C.; Storck, K.; Blunk, T.; Bauer-Kreisel, P. Engineering vascularized adipose tissue using the stromal-vascular fraction and fibrin hydrogels. *Tissue Eng. Part. A* **2015**, *21*, 1343–1353. [CrossRef]
91. Holnthoner, W.; Hohenegger, K.; Husa, A.M.; Muehleder, S.; Meinl, A.; Peterbauer-Scherb, A.; Redl, H. Adipose-derived stem cells induce vascular tube formation of outgrowth endothelial cells in a fibrin matrix. *J. Tissue Eng. Regen. Med.* **2015**, *9*, 127–136. [CrossRef] [PubMed]
92. Ahmed, T.A.; Griffith, M.; Hincke, M. Characterization and inhibition of fibrin hydrogel-degrading enzymes during development of tissue engineering scaffolds. *Tissue Eng.* **2007**, *13*, 1469–1477. [CrossRef]
93. Louis, F.; Sowa, Y.; Irie, S.; Higuchi, Y.; Kitano, S.; Mazda, O.; Matsusaki, M. Injectable Prevascularized Mature Adipose Tissues (iPAT) to Achieve Long-Term Survival in Soft Tissue Regeneration. *Adv. Healthc. Mater.* **2022**, *11*, e2201440. [CrossRef]
94. Lee, K.Y.; Mooney, D.J. Alginate: Properties and biomedical applications. *Prog. Polym. Sci.* **2012**, *37*, 106–126. [CrossRef]
95. Kang, S.W.; Cha, B.H.; Park, H.; Park, K.S.; Lee, K.Y.; Lee, S.H. The effect of conjugating RGD into 3D alginate hydrogels on adipogenic differentiation of human adipose-derived stromal cells. *Macromol. Biosci.* **2011**, *11*, 673–679. [CrossRef]
96. Yoo, B.; Kim, S.; Shin, B.H.; Lee, M.H.; Choy, Y.B.; Lee, K.; Heo, C.Y.; Koh, W.G. Preparation of alginate hydrogel with human-derived adipose tissue to improve fat graft survival and adipogenesis. *J. Ind. Eng. Chem.* **2021**, *95*, 148–155. [CrossRef]
97. Kim, W.S.; Mooney, D.J.; Arany, P.R.; Lee, K.; Huebsch, N.; Kim, J. Adipose tissue engineering using injectable, oxidized alginate hydrogels. *Tissue Eng. Part. A* **2012**, *18*, 737–743. [CrossRef]
98. Brandl, F.P.; Seitz, A.K.; Tessmar, J.K.; Blunk, T.; Gopferich, A.M. Enzymatically degradable poly(ethylene glycol) based hydrogels for adipose tissue engineering. *Biomaterials* **2010**, *31*, 3957–3966. [CrossRef]

99. Salvatore, L.; Natali, M.L.; Brunetti, C.; Sannino, A.; Gallo, N. An Update on the Clinical Efficacy and Safety of Collagen Injectables for Aesthetic and Regenerative Medicine Applications. *Polymers* **2023**, *15*, 1020. [CrossRef]
100. Van Nieuwenhove, I.; Tytgat, L.; Ryx, M.; Blondeel, P.; Stillaert, F.; Thienpont, H.; Ottevaere, H.; Dubruel, P.; Van Vlierberghe, S. Soft tissue fillers for adipose tissue regeneration: From hydrogel development toward clinical applications. *Acta Biomater.* **2017**, *63*, 37–49. [CrossRef]
101. Vashi, A.V.; Keramidaris, E.; Abberton, K.M.; Morrison, W.A.; Wilson, J.L.; O'Connor, A.J.; Cooper-White, J.J.; Thompson, E.W. Adipose differentiation of bone marrow-derived mesenchymal stem cells using Pluronic F-127 hydrogel in vitro. *Biomaterials* **2008**, *29*, 573–579. [CrossRef] [PubMed]
102. Weiser, B.; Prantl, L.; Schubert, T.E.; Zellner, J.; Fischbach-Teschl, C.; Spruss, T.; Seitz, A.K.; Tessmar, J.; Goepferich, A.; Blunk, T. In vivo development and long-term survival of engineered adipose tissue depend on in vitro precultivation strategy. *Tissue Eng. Part. A* **2008**, *14*, 275–284. [CrossRef]
103. Xu, J.; Chen, Y.; Yue, Y.; Sun, J.; Cui, L. Reconstruction of epidural fat with engineered adipose tissue from adipose derived stem cells and PLGA in the rabbit dorsal laminectomy model. *Biomaterials* **2012**, *33*, 6965–6973. [CrossRef]
104. Patrick, C.W., Jr.; Chauvin, P.B.; Hobley, J.; Reece, G.P. Preadipocyte seeded PLGA scaffolds for adipose tissue engineering. *Tissue Eng.* **1999**, *5*, 139–151. [CrossRef]
105. Fukushima, K. Poly(trimethylene carbonate)-based polymers engineered for biodegradable functional biomaterials. *Biomater. Sci.* **2016**, *4*, 9–24. [CrossRef]
106. Zhang, Z.; Kuijer, R.; Bulstra, S.K.; Grijpma, D.W.; Feijen, J. The in vivo and in vitro degradation behavior of poly(trimethylene carbonate). *Biomaterials* **2006**, *27*, 1741–1748. [CrossRef]
107. Jain, S.; Yassin, M.A.; Fuoco, T.; Liu, H.; Mohamed-Ahmed, S.; Mustafa, K.; Finne-Wistrand, A. Engineering 3D degradable, pliable scaffolds toward adipose tissue regeneration; optimized printability, simulations and surface modification. *J. Tissue Eng.* **2020**, *11*, 2041731420954316. [CrossRef]
108. Chiu, Y.C.; Cheng, M.H.; Engel, H.; Kao, S.W.; Larson, J.C.; Gupta, S.; Brey, E.M. The role of pore size on vascularization and tissue remodeling in PEG hydrogels. *Biomaterials* **2011**, *32*, 6045–6051. [CrossRef]
109. Lee, S.; Lee, H.S.; Chung, J.J.; Kim, S.H.; Park, J.W.; Lee, K.; Jung, Y. Enhanced Regeneration of Vascularized Adipose Tissue with Dual 3D-Printed Elastic Polymer/dECM Hydrogel Complex. *Int. J. Mol. Sci.* **2021**, *22*, 2886. [CrossRef]
110. Banyard, D.A.; Borad, V.; Amezcua, E.; Wirth, G.A.; Evans, G.R.; Widgerow, A.D. Preparation, Characterization, and Clinical Implications of Human Decellularized Adipose Tissue Extracellular Matrix (hDAM): A Comprehensive Review. *Aesthet. Surg. J.* **2016**, *36*, 349–357. [CrossRef]
111. Wang, L.; Johnson, J.A.; Zhang, Q.; Beahm, E.K. Combining decellularized human adipose tissue extracellular matrix and adipose-derived stem cells for adipose tissue engineering. *Acta Biomater.* **2013**, *9*, 8921–8931. [CrossRef] [PubMed]
112. Flynn, L.E. The use of decellularized adipose tissue to provide an inductive microenvironment for the adipogenic differentiation of human adipose-derived stem cells. *Biomaterials* **2010**, *31*, 4715–4724. [CrossRef]
113. Han, T.T.; Toutounji, S.; Amsden, B.G.; Flynn, L.E. Adipose-derived stromal cells mediate in vivo adipogenesis, angiogenesis and inflammation in decellularized adipose tissue bioscaffolds. *Biomaterials* **2015**, *72*, 125–137. [CrossRef]
114. Cheung, H.K.; Han, T.T.; Marecak, D.M.; Watkins, J.F.; Amsden, B.G.; Flynn, L.E. Composite hydrogel scaffolds incorporating decellularized adipose tissue for soft tissue engineering with adipose-derived stem cells. *Biomaterials* **2014**, *35*, 1914–1923. [CrossRef]
115. Choi, J.S.; Kim, B.S.; Kim, J.Y.; Kim, J.D.; Choi, Y.C.; Yang, H.J.; Park, K.; Lee, H.Y.; Cho, Y.W. Decellularized extracellular matrix derived from human adipose tissue as a potential scaffold for allograft tissue engineering. *J. Biomed. Mater. Res. A* **2011**, *97*, 292–299. [CrossRef]
116. Pati, F.; Ha, D.H.; Jang, J.; Han, H.H.; Rhie, J.W.; Cho, D.W. Biomimetic 3D tissue printing for soft tissue regeneration. *Biomaterials* **2015**, *62*, 164–175. [CrossRef]
117. Flynn, L.; Semple, J.L.; Woodhouse, K.A. Decellularized placental matrices for adipose tissue engineering. *J. Biomed. Mater. Res. A* **2006**, *79*, 359–369. [CrossRef]
118. Xu, M.; He, Y.; Li, Y.; Liu, K.; Zhang, Y.; Su, T.; Yao, Y.; Jin, X.; Zhang, X.; Lu, F. Combined Use of Autologous Sustained-Release Scaffold of Adipokines and Acellular Adipose Matrix to Construct Vascularized Adipose Tissue. *Plast. Reconstr. Surg.* **2024**, *153*, 348e–360e. [CrossRef]
119. Lee, E.Y.; Xia, Y.; Kim, W.S.; Kim, M.H.; Kim, T.H.; Kim, K.J.; Park, B.S.; Sung, J.H. Hypoxia-enhanced wound-healing function of adipose-derived stem cells: Increase in stem cell proliferation and up-regulation of VEGF and bFGF. *Wound Repair. Regen.* **2009**, *17*, 540–547. [CrossRef]
120. Irvin, J.; Danchik, C.; Rall, J.; Babcock, A.; Pine, M.; Barnaby, D.; Pathakamuri, J.; Kuebler, D. Bioactivity and composition of a preserved connective tissue matrix derived from human placental tissue. *J. Biomed. Mater. Res. B Appl. Biomater.* **2018**, *106*, 2731–2740. [CrossRef]

121. Magana, A.; Giovanni, R.; Essien, E.; Epel, B.; Kotecha, M.; Liu, S.; Mathew, M.T.; Hagarty, S.E.; Bijukumar, D. Amniotic growth factors enhanced human pre-adipocyte cell viability and differentiation under hypoxia. *J. Biomed. Mater. Res. B Appl. Biomater.* **2022**, *110*, 2146–2156. [CrossRef] [PubMed]
122. Ikegami, Y.; Mizumachi, H.; Yoshida, K.; Ijima, H. Heparin-conjugated collagen as a potent growth factor-localizing and stabilizing scaffold for regenerative medicine. *Regen. Ther.* **2020**, *15*, 236–242. [CrossRef]
123. Khanna, A.; Oropeza, B.P.; Huang, N.F. Engineering Spatiotemporal Control in Vascularized Tissues. *Bioengineering* **2022**, *9*, 555. [CrossRef]
124. Song, M.; Zhou, Y.; Liu, Y. VEGF heparinized-decellularized adipose tissue scaffolds enhance tissue engineering vascularization in vitro. *RSC Adv.* **2018**, *8*, 33614–33624. [CrossRef]
125. Murohara, T.; Shintani, S.; Kondo, K. Autologous adipose-derived regenerative cells for therapeutic angiogenesis. *Curr. Pharm. Des.* **2009**, *15*, 2784–2790. [CrossRef]
126. Todorova, D.; Simoncini, S.; Lacroix, R.; Sabatier, F.; Dignat-George, F. Extracellular Vesicles in Angiogenesis. *Circ. Res.* **2017**, *120*, 1658–1673. [CrossRef]
127. Ju, Y.; Hu, Y.; Yang, P.; Xie, X.; Fang, B. Extracellular vesicle-loaded hydrogels for tissue repair and regeneration. *Mater. Today Bio* **2023**, *18*, 100522. [CrossRef]
128. Mou, S.; Zhou, M.; Li, Y.; Wang, J.; Yuan, Q.; Xiao, P.; Sun, J.; Wang, Z. Extracellular Vesicles from Human Adipose-Derived Stem Cells for the Improvement of Angiogenesis and Fat-Grafting Application. *Plast. Reconstr. Surg.* **2019**, *144*, 869–880. [CrossRef]
129. Nie, F.; Ding, P.; Zhang, C.; Zhao, Z.; Bi, H. Extracellular vesicles derived from lipoaspirate fluid promote fat graft survival. *Adipocyte* **2021**, *10*, 293–309. [CrossRef]
130. Zhang, Y.; Liu, T. Adipose-derived stem cells exosome and its potential applications in autologous fat grafting. *J. Plast. Reconstr. Aesthet. Surg.* **2023**, *76*, 219–229. [CrossRef]
131. Yang, S.; Jiang, H.; Qian, M.; Ji, G.; Wei, Y.; He, J.; Tian, H.; Zhao, Q. MSC-derived sEV-loaded hyaluronan hydrogel promotes scarless skin healing by immunomodulation in a large skin wound model. *Biomed. Mater.* **2022**, *17*, 034104. [CrossRef] [PubMed]
132. Vyas, K.S.; Vasconez, H.C.; Morrison, S.; Mogni, B.; Linton, S.; Hockensmith, L.; Kabir, T.; Zielins, E.; Najor, A.; Bakri, K.; et al. Fat Graft Enrichment Strategies: A Systematic Review. *Plast. Reconstr. Surg.* **2020**, *145*, 827–841. [CrossRef]
133. Li, J.; Shi, X.; Chen, W. Influence of repeatedly injecting platelet-rich plasma on survival and quality of fat grafts in nude mice. *Zhongguo Xiu Fu Chong Jian Wai Ke Za Zhi* **2013**, *27*, 454–459.
134. Sasaki, G.H. The Safety and Efficacy of Cell-Assisted Fat Grafting to Traditional Fat Grafting in the Anterior Mid-Face: An Indirect Assessment by 3D Imaging. *Aesthetic Plast. Surg.* **2015**, *39*, 833–846. [CrossRef]
135. Gentile, P.; Di Pasquali, C.; Bocchini, I.; Floris, M.; Eleonora, T.; Fiaschetti, V.; Floris, R.; Cervelli, V. Breast reconstruction with autologous fat graft mixed with platelet-rich plasma. *Surg. Innov.* **2013**, *20*, 370–376. [CrossRef]
136. Lu, T.; Li, Y.; Chen, T. Techniques for fabrication and construction of three-dimensional scaffolds for tissue engineering. *Int. J. Nanomedicine* **2013**, *8*, 337–350. [CrossRef]
137. Shimizu, T. Cell sheet-based tissue engineering for fabricating 3-dimensional heart tissues. *Circ. J.* **2014**, *78*, 2594–2603. [CrossRef]
138. Villalona, G.A.; Udelsman, B.; Duncan, D.R.; McGillicuddy, E.; Sawh-Martinez, R.F.; Hibino, N.; Painter, C.; Mirensky, T.; Erickson, B.; Shinoka, T.; et al. Cell-seeding techniques in vascular tissue engineering. *Tissue Eng. Part. B Rev.* **2010**, *16*, 341–350. [CrossRef] [PubMed]
139. Nichol, J.W.; Khademhosseini, A. Modular Tissue Engineering: Engineering Biological Tissues from the Bottom Up. *Soft Matter* **2009**, *5*, 1312–1319. [CrossRef]
140. Brett, E.; Chung, N.; Leavitt, W.T.; Momeni, A.; Longaker, M.T.; Wan, D.C. A Review of Cell-Based Strategies for Soft Tissue Reconstruction. *Tissue Eng. Part. B Rev.* **2017**, *23*, 336–346. [CrossRef]
141. Zhang, Q.; Hubenak, J.; Iyyanki, T.; Alred, E.; Turza, K.C.; Davis, G.; Chang, E.I.; Branch-Brooks, C.D.; Beahm, E.K.; Butler, C.E. Engineering vascularized soft tissue flaps in an animal model using human adipose-derived stem cells and VEGF+PLGA/PEG microspheres on a collagen-chitosan scaffold with a flow-through vascular pedicle. *Biomaterials* **2015**, *73*, 198–213. [CrossRef] [PubMed]
142. Laschke, M.W.; Menger, M.D. Spheroids as vascularization units: From angiogenesis research to tissue engineering applications. *Biotechnol. Adv.* **2017**, *35*, 782–791. [CrossRef] [PubMed]
143. Di Stefano, A.B.; Urrata, V.; Trapani, M.; Moschella, F.; Cordova, A.; Toia, F. Systematic review on spheroids from adipose-derived stem cells: Spontaneous or artefact state? *J. Cell. Physiol.* **2022**, *237*, 4397–4411. [CrossRef]
144. Banerjee, D.; Singh, Y.P.; Datta, P.; Ozbolat, V.; O'Donnell, A.; Yeo, M.; Ozbolat, I.T. Strategies for 3D bioprinting of spheroids: A comprehensive review. *Biomaterials* **2022**, *291*, 121881. [CrossRef]
145. He, J.; Zhang, X.; Xia, X.; Han, M.; Li, F.; Li, C.; Li, Y.; Gao, D. Organoid technology for tissue engineering. *J. Mol. Cell Biol.* **2020**, *12*, 569–579. [CrossRef]

146. Mandl, M.; Viertler, H.P.; Hatzmann, F.M.; Brucker, C.; Grossmann, S.; Waldegger, P.; Rauchenwald, T.; Mattesich, M.; Zwierzina, M.; Pierer, G.; et al. An organoid model derived from human adipose stem/progenitor cells to study adipose tissue physiology. *Adipocyte* **2022**, *11*, 164–174. [CrossRef]
147. Strobel, H.A.; Gerton, T.; Hoying, J.B. Vascularized adipocyte organoid model using isolated human microvessel fragments. *Biofabrication* **2021**, *13*, 035022. [CrossRef]
148. Schmidt, V.J.; Hilgert, J.G.; Covi, J.M.; Leibig, N.; Wietbrock, J.O.; Arkudas, A.; Polykandriotis, E.; de Wit, C.; Horch, R.E.; Kneser, U. Flow increase is decisive to initiate angiogenesis in veins exposed to altered hemodynamics. *PLoS ONE* **2015**, *10*, e0117407. [CrossRef]
149. Leibig, N.; Wietbrock, J.O.; Bigdeli, A.K.; Horch, R.E.; Kremer, T.; Kneser, U.; Schmidt, V.J. Flow-Induced Axial Vascularization: The Arteriovenous Loop in Angiogenesis and Tissue Engineering. *Plast. Reconstr. Surg.* **2016**, *138*, 825–835. [CrossRef]
150. Fischer, K.S.; Henn, D.; Zhao, E.T.; Sivaraj, D.; Litmanovich, B.; Hahn, W.W.; Hostler, A.C.; Mojadidi, S.M.; Gonzalez, J.; Knochel, A.B.; et al. Elevated Shear Stress Modulates Heterogenous Cellular Subpopulations to Induce Vascular Remodeling. *Tissue Eng. Part. A* **2024**, *30*, 752–765. [CrossRef]
151. Schmidt, V.J.; Hilgert, J.G.; Covi, J.M.; Weis, C.; Wietbrock, J.O.; de Wit, C.; Horch, R.E.; Kneser, U. High flow conditions increase connexin43 expression in a rat arteriovenous and angioinductive loop model. *PLoS ONE* **2013**, *8*, e78782. [CrossRef]
152. Meyer, A.; Horch, R.E.; Schoengart, E.; Beier, J.P.; Taeger, C.D.; Arkudas, A.; Lang, W. Results of combined vascular reconstruction by means of AV loops and free flap transfer in patients with soft tissue defects. *J. Plast. Reconstr. Aesthet. Surg.* **2016**, *69*, 545–553. [CrossRef]
153. Debels, H.; Palmer, J.; Han, X.L.; Poon, C.; Abberton, K.; Morrison, W. In vivo tissue engineering of an adipose tissue flap using fat grafts and Adipogel. *J. Tissue Eng. Regen. Med.* **2020**, *14*, 633–644. [CrossRef]
154. Henn, D.; Chen, K.; Fischer, K.; Rauh, A.; Barrera, J.A.; Kim, Y.J.; Martin, R.A.; Hannig, M.; Niedoba, P.; Reddy, S.K.; et al. Tissue Engineering of Axially Vascularized Soft-Tissue Flaps with a Poly-(varepsilon-Caprolactone) Nanofiber-Hydrogel Composite. *Adv. Wound Care* **2020**, *9*, 365–377. [CrossRef]
155. Rnjak-Kovacina, J.; Gerrand, Y.W.; Wray, L.S.; Tan, B.; Joukhdar, H.; Kaplan, D.L.; Morrison, W.A.; Mitchell, G.M. Vascular Pedicle and Microchannels: Simple Methods Toward Effective In Vivo Vascularization of 3D Scaffolds. *Adv. Healthc. Mater.* **2019**, *8*, e1901106. [CrossRef]
156. Tanaka, Y.; Sung, K.C.; Fumimoto, M.; Tsutsumi, A.; Kondo, S.; Hinohara, Y.; Morrison, W.A. Prefabricated engineered skin flap using an arteriovenous vascular bundle as a vascular carrier in rabbits. *Plast. Reconstr. Surg.* **2006**, *117*, 1860–1875. [CrossRef]
157. Tanaka, Y.; Tamai, M.; Taguchi, N.; Niyazi, A.; Ueno, M.; Nagasao, T. Spontaneously generated large adipose flaps in vivo tissue engineering chambers. *J. Plast. Reconstr. Aesthet. Surg.* **2020**, *73*, 1889–1896. [CrossRef]
158. Lu, Z.; Yuan, Y.; Gao, J.; Lu, F. Adipose tissue extract promotes adipose tissue regeneration in an adipose tissue engineering chamber model. *Cell Tissue Res.* **2016**, *364*, 289–298. [CrossRef]
159. Horchler, S.N.; Hancock, P.C.; Sun, M.; Liu, A.T.; Massand, S.; El-Mallah, J.C.; Goldenberg, D.; Waldron, O.; Landmesser, M.E.; Agrawal, S.; et al. Vascular persistence following precision micropuncture. *Microcirculation* **2024**, *31*, e12835. [CrossRef]
160. Hancock, P.C.; Koduru, S.V.; Sun, M.; Ravnic, D.J. Induction of scaffold angiogenesis by recipient vasculature precision micropuncture. *Microvasc. Res.* **2021**, *134*, 104121. [CrossRef]

MDPI AG
Grosspeteranlage 5
4052 Basel
Switzerland
Tel.: +41 61 683 77 34

Biomolecules Editorial Office
E-mail: biomolecules@mdpi.com
www.mdpi.com/journal/biomolecules

www.ingramcontent.com/pod-product-compliance
Lightning Source LLC
Chambersburg PA
CBHW061618210926
51850CB00009B/2359
9783725867806